PLANT DISEASE CONTROL

Principles and Practice

Otis C. Maloy
Professor of Plant Pathology
Washington State University

JOHN WILEY & SONS, INC.
New York / Chichester / Brisbane / Toronto / Singapore

This text is printed on acid-free paper.

Library of Congress Cataloging in Publication Data:

Maloy, Otis C.
Plant disease control : principles and practice / Otis C. Maloy.
p. cm.
ISBN 0-471-57317-5
1. Phytopathogenic microorganisms—Control. 2. Plant diseases.
I. Title.
SB731.M364 1993
632'. 3—dc20 92-39600

Printed in the United States of America

10 9 8 7 6 5 4 3 2 1

To My Wife, Edna
Our Children, Michael and Margaret
And Our Grandchildren, Holly and Jared

PREFACE

Many books have been written that deal with plant disease control. Some are highly technical and concentrate on theoretical aspects of control. Others are simply listings of disease control recommendations, leaning heavily toward chemical control. Control measures are often an afterthought in a treatise on etiology, epidemiology, or symptomatology. I have used some of these books as reference material in a disease control course but have felt the need for a book that meshes theoretical principles with practical application. Such a book would be useful to students and practitioners of plants disease control in plant pathology, entomology, agronomy, horticulture, forestry, and related fields.

I first became interested in disease control in 1946 when employed in the *Ribes* eradication program of the Office of Blister Rust Control. After obtaining my education in plant pathology and brief experiences in several positions I was employed in 1963 by Washington State University as the only Extension Plant Pathologist for the entire state and necessarily covered all diseases of all crops. It did not take long to discover that my primary value to farmers was to help them control diseases in their crops. By 1978, addition of several more extension positions allowed me to take a teaching appointment in the Department of Plant Pathology where over the years I have taught several courses including one on plant disease control. The evolution of that course forms the foundation for this book.

I have tried to use a wide range of examples but am most comfortable with diseases I have seen or worked with so the text may appear to be heavy with examples from forest pathology and the Pacific Northwest. Nevertheless, a rust is a rust whether it is in Mozambique or in Montana. Perhaps it will challenge readers to adapt these examples to their situation. In the end I believe the examples are suited to the principle or strategy being discussed. Readers should note that, with few exceptions, examples are restricted to infectious diseases. Noninfectious diseases such as nutrient deficiencies, chemical and climatic injuries, genetic abnormalities, and the like are usually the concern of specialists in plant nutrition, air pollution, and other disciplines.

Included at the end of each chapter are several pertinent references to provide readers with leads to explore that subject further. It is not a complete bibliography and some may be offended by certain omissions. However, it is not always possible to identify the originator of every thought or concept and I believe that an error of omission is better than an error of commission. I also use only common names of diseases in the text unless referring specifically to pathogens. Scientific names and authorities for pathogens and host plants are listed separately in appendices.

Several colleagues graciously reviewed this book and made many helpful suggestions. G.W. Bruehl, R. Duran, and D.A. Inglis reviewed the entire manuscript. Various chapters were reviewed by R.E. Allan, R.J. Cook, W.J. Kaiser, M.G. Long, T.D. Murray, J.D. Rogers, G.S. Santo, C.B. Skotland, and R.E. Williams. G.E. Cook provided the fungicide label in Chapter 12. A.J. Gale assisted in obtaining photographs of orchard spraying. D.P. Richmond and R.M. Conway provided pictures of soil fumigation. J.M. Lawford and L.A. Winters helped process the manuscript.

I am grateful to John Wiley and Sons, Inc. for publishing this book and appreciate the help given me by P.C. Manor, D. Cisek, and S. Thedford of the Wiley staff.

Finally, I acknowledge that few of the concepts and ideas put forth in this book originated with me. They have been compiled, adapted, and synthesized from many sources and have evolved with my own personal view of plant disease control. I alone am responsible for the accuracy of the statements presented.

OTIS C. MALOY

Pullman, Washington
November 1992

CONTENTS

PLANT DISEASE CONTROL

1 Introduction

Control is the ultimate if not the primary objective of the plant pathologist. Control is the application of practices devised to reduce the damage or loss from plant disease. There are varying degrees of control, such as perfect (complete or absolute) control, partial control, and profitable or economic control.

Complete control is the total elimination of disease and is rarely achieved, even when extra effort is made to do so. Most control is partial and the extent of control depends on many factors arising from the intimate association and interaction of host plant, plant pathogen, and environment (= the disease triangle) that lead to disease development.

Profitable or economic control is the usual standard by which effective control is measured. It is the effort required to obtain the greatest reduction in disease loss at the lowest cost. This balance must tilt in favor of production because growers cannot afford to just "break even." If the gain in production just equals the additional cost of that gain then the producer might just as well have done nothing. Except for the potential increase in pathogen inoculum the end result would be the same.

The term "control" has been interpreted by some as being too absolute; "disease management" is considered a more realistic term. "Management" has some advantages over "control" but it also has limitations. For example, some argue that diseases are not managed, but rather that it is the various cultural practices that influence disease that are managed. That may be, but from a practical standpoint plant diseases can be "managed" just as easily as they can be "controlled."

Various dictionaries define "control" as the act of directing, restraining, regulating, and so forth and define "manage" as controlling, directing, and manipulating. Thus, it appears that these two terms are so similar as to be interchangeable. However, control has a stronger connotation. Some books, especially those with multiple authors or teams of authors, use the terms in different, often disjointed and contradictory ways.

A modification of this concept and an additional buzzword in recent years is the term "integrated pest management" (IPM). This is control using combinations of biological, cultural, chemical, and physical means to reduce or contain a pest population below an economic level. IPM has been widely used to control insect pests and has been "borrowed" by plant pathologists even though they have been practicing such integration of control measures for decades without calling it by this name.

Control of plant diseases is necessary because of the importance of plant diseases to man both economically and aesthetically. There are at least five major areas of damage from plant diseases.

1. *Loss of income to producers.* This is probably the most immediate loss and the easiest to demonstrate and quantify. The loss may be direct loss of production (reduced yield) and lower value (reduced quality), or indirect loss of markets. For example, some countries or states will not accept products that can harbor pathogens.

2. *Increased food costs to the consumer.* One of the hidden costs of plant diseases is the higher prices brought on by fewer products as a result of disease losses and the additional cost of production for disease control. These costs are ultimately passed on to the consumer.

3. *Restricted production of some crops.* Certain crops may not be grown profitably in some areas because of plant diseases. Coffee is no longer grown in Ceylon (now Sri Lanka) because of a coffee rust epidemic in the mid-1800s. In fact, coffee rust made tea drinkers out of the British. Pears cannot be profitably grown in the eastern United States without considerable cost and effort because of fire blight. In Washington state tomatoes are not a major commercial crop because of the devastation caused by the curly top virus. And in certain fields in the Walla Walla area of southcentral Washington the highly prized "Walla Walla Sweet" onion cannot be grown because of white rot.

4. *Loss of natural resources.* Natural populations of plants are also subject to extreme damage from plant diseases, especially where indigenous plants have evolved in the absence of a pathogen and then are exposed to a particularly virulent strain. Two examples of this phenomenon occurred in the United States in this century.

The American chestnut was a highly valued tree growing over much of the eastern third of the country. The tree was a valuable shade tree and provided several useful products. The nuts were highly prized and provided a major food source for wildlife such as deer and turkeys. The durable wood, used for construction timbers, lumber, panelling, shingles, and poles, was also fine enough for cabinets and other furniture. In addition, the bark was a major source of tannin for the expanding leather industry. In 1904 blight was discovered in the New York Zoological Park after being introduced into this country presumably from Asia. Within 50 years the chestnut was eliminated as an economic resource in America.

Whitebark pine is a small five-needle pine tree that grows at high elevations in the northwestern United States. It is highly susceptible to white pine blister rust, another introduced disease. The pathogen cannot spread directly from one pine to another. It requires an alternate host, *Ribes* species (currants and gooseberries), to complete its life cycle. The distance the fungus can spread from *Ribes* to pine is relatively short. *Ribes* eradication programs were carried out for many years in Glacier, Mount Rainier, Yellowstone, and Rocky Mountain National Parks to protect whitebark pine for scenic purposes despite reservations by some U.S. Department of Agriculture (USDA) officials from the Office of Blister Rust Control (BRC) about the cost effectiveness of the program. Eventually the entire *Ribes* eradication program in the west was discontinued and the fate of whitebark pine is not promising.

5. *Direct and indirect social and political costs.* Throughout modern history there have been many examples of changes in human progress resulting from plant disease.

Probably the best known of these is the great Irish potato famine resulting from the late blight epidemics of the 1840s. It is estimated that two and a half million Irish died or emigrated (many to the United States) in the period from 1845 to 1851. As recently as April 1981 the National Geographic magazine stated that the impact of this epidemic was still being felt.

Ergot, a disease of many cereal grains, especially rye, is associated with several historical calamities. Ergotism is the condition or disease in humans and other animals that results from consuming ergoty grain. Ergotism in humans is of two types: gangrenous, in which the vasoconstrictive effects of ergot restrict the blood supply causing dry gangrene of the extremities. This same vasocontricting property is used in treating human ailments such as migraine and postpartum bleeding. The second type of ergotism is convulsive, which may be caused by lysergic acid diethylamide (LSD) and other compounds in the ergot sclerotia. It causes twitching, convulsions, hallucinations, and other bizarre manifestations. These are thought to be the symptoms that resulted in the Salem witchcraft trials where several young women were condemned as witches and executed.

Ergot also played a role in the success or failure of conquering armies. In the ninth century France was invaded by Norsemen who met little opposition, presumably because the natives were debilitated by ergotism. And in the 1700s Peter the Great of Russia attempted to capture the Dardanelles and gain access to the Mediterranean Sea. His campaign failed because of the loss of men and horses from eating ergoty bread and grain.

Dutch elm disease has destroyed many valuable ornamental elms throughout Europe and the United States. In the United States considerable money and effort have gone toward slowing the progress of this disease, thereby prolonging the life of valuable trees. These efforts have involved two phases: the first is the enactment of various laws regulating the management (pruning, spraying, removal, etc.) of American and other elm hosts and the second is the development of a tree service industry that prunes, trims, removes, and treats elm to control the disease and the elm bark beetle that vectors the causal fungus.

The number of plant diseases affecting crops is staggering in the aggregate but it is rare that more than a few of the diseases afflicting a particular crop occur in the same place at the same time to the same extent. Usually a grower contends with only one or two diseases at a time. But the potential can be illustrated by the following table of the top 10 crops grown in Washington state. Table 1.1 compares the number of diseases reported in the 1960 USDA Index of Plant Diseases for each of these crops for the entire United States and for the Pacific Northwest (PNW). Even though the number of diseases occurring in the PNW is only a third of those reported in the United States, the number a grower has to consider is still sizable. But, fortunately, most are of minor occurrence and concern.

For a control program to be accepted it must have a sound biologic, economic, and ecological basis. Since a disease is the product of complex interactions between host and pathogen, the biology of each component must be understood to reduce the undesirable effects of such interactions. Of paramount importance is the correct

TABLE 1.1. Number of Diseases on 10 of Washington's Major Crops Reported from the Pacific Northwest Compared with the Number Reported in the United States.

	Diseases	
Crop	PNW	United States
Wheat	39	87
Alfalfa	21	66
Apples	60	193
Potatoes	59	145
Barley	27	46
Hops	9	20
Pears	35	90
Corn	12	91
Cherries	34	49
Asparagus	8	25
Total	304	812

Source: From USDA Agricultural Handbook 165.

identification of the pathogen since the procedures for controlling different major groups of pathogens, like fungi, bacteria, viruses, and nematodes, are often completely different and rarely cross-effective. The life cycle of a pathogen must be known so that weak links in the infection chain can be attacked. Certain characteristics of a host plant may be important in the management of a disease and should be recognized.

There must be some economic benefit to justify application of disease control measures. Economic justification varies considerably depending on relative and often vague values of the plant or crop. Costs in money and effort often cannot be justified to control a disease on a tree in an orchard or a forest because of the relatively small value of the individual in relation to the population, whereas the same tree may have considerable value as a shade or ornamental tree. The control program has to be justified on the value of the plant: in this case higher costs are justified for the ornamental and lower inputs for the tree in the forest or orchard.

The environment is important in a disease control program in two major ways. First the environment is the third component in the disease equation (= disease triangle) and ecological aspects of the disease are often key ingredients in effective disease control, for example, soil temperature, soil moisture, or relative humidity. Another major ecological consideration is that the control program, especially one involving pesticides or other chemicals, does not damage the environment. This has become of considerable political and social concern in recent years. Application of pesticides, field burning of crop residues, soil tillage, fertilization, irrigation, and other practices used to control or reduce losses from plant disease can be detrimental to the environment and thereby ecologically unsound.

For effective control of most plant diseases measures must be implemented early in the disease cycle and involves two steps.

1. *Accurate diagnosis of the problem.* Since the control measures are generally directed against specific pathogens and not against symptoms it is necessary to know the exact cause of the disease. This usually necessitates identifying the causal organism. Several examples encountered in Washington state illustrate this problem.

Cephalosporium stripe of wheat is caused by a soil-borne pathogen generally believed to enter the plant through broken roots and become established in the vascular system. It produces pronounced chlorotic streaks on the leaves. Stripe rust is a typical foliar rust that also produces pronounced chlorotic streaks on some wheat varieties. The similarity in names alone often causes misidentification of a disease and this is frequently compounded by the similarity of symptoms. But the control measures are completely different for these two unrelated fungi and the diseases they cause. What works for one is of little or no value for the other.

Similarity of symptoms can cause mistakes in identifying some virus diseases. An incident where leaves of young sweet cherry trees exhibited a mild chlorotic mottle that resembled symptoms of some sweet cherry viruses led to a preliminary diagnosis of Lambert mottle. The grower had applied a fungicide (zineb) believing that a foliar fungus was the cause and obtained a reduction of the mottling. This fungicide would have had no effect on a virus and the effect was discounted as a fluke. However, further study revealed that the problem was zinc-deficient soils and the zineb essentially provided a foliar feeding of zinc, which is a necessary micronutrient.

Another serious situation concerning misidentification of a disease occurred within recent years when a shipment of grass seed produced in Washington state was stopped by Canadian authorities for containing spores of the Karnal smut fungus, *Tilletia indica*. This fungus is present in Mexico but has not been reported from the United States. It does not infect grasses other than wheat and triticale but spores could be spread on grass seed or other plant parts. The United States, Canada, and other wheat-producing countries have quarantines against this pathogen. A great deal of money was at risk because of the importance of the grass seed industry to Washington state. Embargoes against a commodity can be serious for any industry. On further examination it was determined that the fungus was *Tilletia barclayana*, a smut with teliospores similar to those of *T. indica*. *T. barclayana* causes smut on rice and is endemic in California and some other rice-growing areas. The teliospores had come from the rice hulls used as extenders in grass seed to facilitate planting of these small seeds. When the true identity of the smut was confirmed the seed was allowed to proceed to its destination.

2. *Early application and proper timing of control procedures.* Despite postinfection application of some newer systemic fungicides, most disease control measures (both chemical and nonchemical) are preventive and must be applied *before* disease occurs. Once the pathogen has penetrated the protective surface of the plant and become established within the plant, disease control is essentially impossible. There are few

exceptions to this. The powdery mildews are one notable exception where the mass of the parasitic fungus remains outside the plant and derives nutrients from the host plant via haustoria within the epidermal cells. The bulk of the fungus thallus is exposed and vulnerable to fungicides applied after infection has occurred. This fungus is also subject to adverse environmental factors such as reduced relative humidity or even excess free moisture that disrupts fungal development and the infection process.

Timely application of control measures can stop pathogen and disease development early in the disease cycle. There may be only one cycle in diseases such as root rots, or this may be followed by secondary cycles such as occur in many foliar diseases.

Implementation of control measures usually must wait until the next cropping cycle after detection and identification of the exact cause of a problem. Because of the capricious nature of many plant diseases it is often difficult, if not impossible, to predict their occurrence. Sometimes a grower will know from experience whether a disease is likely to occur and the relative probability of that happening. The grower is then prepared to exercise whatever measures are needed to reduce chances of disease loss. However, disease control procedures are like insurance policies; many are applied that are not needed but it is better to have them in place and not need them than to need them and not have them. Unfortunately, this fact also increases the total production costs and should be considered a monetary loss to disease.

Of utmost consideration in controlling plant diseases is proper timing. This is especially true of fungicide (or other pesticide) applications for disease control. It is much more important "when" something is splashed on the plant than "what" (of the range of effective materials) is splashed on. The time period during which control is effective (sometimes called the plant–host timing window) can be highly variable over the range of diseases but most have a specific and vulnerable spot even within a broader effective period. Peach leaf curl is an example of a disease that can be effectively controlled by suitable fungicides while the trees are dormant. Once the leaf buds swell and leaf tissue is exposed infection will occur. Therefore, control measures must be applied while the buds are dormant.

The procedures employed in developing a plant disease control scheme involve a sequence at four distinct levels. The highest level is the principle or rule of action or conduct: exclusion, eradication, protection, or resistance. The principle describes how the control measure acts. Principles are the **WHAT** of disease control.

Plant pathologists define or interpret principles in various ways; therefore, a working definition must be selected. "Principle" is generally defined as (1) a rule of action or conduct, (2) a rule of science explaining how things act, or (3) the method of operation. The constant implication of action suggests that the principle explains what we are going to do. We are going to exclude or eradicate a pathogen or we will provide protection or resistance for a plant but the principle does not indicate how to do it. "Strategy" is (1) the planning and management of conducting a war (i.e., against disease), or (2) the act of devising or employing plans to accomplish a goal (i.e., disease control). Strategy is the general plan of action while tactics are the specific methods (i.e., technologies) used to combat plant diseases. "Technology" is the science of using specialized knowledge for practical purposes and "application" is the act of applying or using.

Once a principle has been selected, a strategy for conducting the battle must be developed. This is the plan to win the war against plant disease. Strategy is the **HOW** of disease control. After a strategy has been decided upon there must be a technology available to accomplish that plan. Technology is the **MEANS** by which disease is controlled. Finally the technology available must be applied in the field. Field application is the **USE** of disease control procedures. Table 1.2 illustrates the PRINCIPLE–STRATEGY–TECHNOLOGY–APPLICATION sequence as they apply in the control of several diseases.

When a disease is diagnosed a grower must obtain disease control or management information from some source. Application of specific control measures varies considerably according to area and situation. Such measures come in the form of recommendations usually based on experimental research or empirical experience from the region but in some cases derived from work done elsewhere. Moreover, disease control information comes from a variety of sources and in a variety of forms.

Books and technical journals generally do not present specific control measures but discuss control in general terms. But leaflets, bulletins, fact sheets, and other short publications often specifically deal with single diseases or groups of diseases on a single crop and usually include information about disease control.

Before stringent regulation of pesticides some books combined diagnosis and disease control by commodity group, such as diseases of fruit crops, ornamentals, vegetables, and field crops. Plant disease compendia discuss in detail diseases of specific plants, such as wheat, corn, potatoes, alfalfa, grapes, and so on. Although biological information is emphasized these usually include something on control. Because of pesticide laws and differences in cultural practices from region to region, state or regional disease control handbooks have evolved that deal specifically with diseases on crops in that region. Handbooks usually are updated periodically to keep abreast of changes in pesticide registration or availability, farming practices, new varieties, and changes in pathogen behavior such as new pathogenic races or pesticide resistance.

Oregon, Washington, and Idaho have common disease control recommendations set forth in the *Pacific Northwest Plant Disease Control Handbook*, which is revised annually. This originated as Oregon's *Plant Disease Control Handbook* but because of commonality of crops and diseases, Washington and Idaho joined with Oregon to develop a regional handbook. Even so, there are still instances where for some reason a disease control practice is accepted by one state but not the others and is so indicated in the handbook. Twelve northcentral states also produce regional publications such as *Diseases of Tree Fruits* and *Dry Bean Production Problems*.

Most states, through their Federal-State Cooperative Extension Service or Department of Agriculture, promulgate disease control recommendations by means of brief leaflets or news releases. These are generally very short, deal only with the specifics of control, and can be prepared, released, revised, or discontinued on short notice. These often result from a sudden and unexpected disease situation.

Use of the term "recommendation" has fallen into disuse in recent years because of legal issues where damages have been claimed for crop losses when a disease control recommendation resulted from improper or illegal use. As a hedge against the almost

TABLE 1.2. Examples of Selecting and Applying Principles, Strategies, and Technologies to Control Specific Diseases

Principle	Strategies	Technologies	Application	Disease Controlled
Exclusion	Prohibition	Quarantines	Embargo elms from Europe	Elm mottle virus
			Require import permit	Ascochyta blight of lentil
	Interception	Quarantines	Border inspections	Golden nematode
		Plant health testing	Examine seeds	Karnal bunt
	Elimination	Certification	Rogue diseased plants	Potato leaf roll
			Phytosanitary certificates	Bean halo blight
		Disinfection	Chemical seed treatment	*Verticillium* in alfalfa seed
			Hot water treatment	Nematodes in bulbs and corms
Eradication	Removal	Nursery and orchard inspections	Detect infected trees	Little cherry virus
		Roguing	Destroy infected plants	Blueberry mosaic
	Elimination	Destroy alternative hosts	Destroy weed hosts	Beet yellows
		Promote antagonists	Organic amendments	Potato scab
		Remove food base	Crop rotation	Take-all of wheat
	Destruction	Chemicals	Soil fumigation	Root-knot nematode
			Eradicants	Crown gall
		Fire	Destroy infected trees	Citrus canker
			Alter plant composition	Jarrah disease
			Destroy crop residues	Cephalosporium stripe of wheat
		Cultivation	Bury infected plants	Corn stalk rot
Protection	Prevent infection	Fungicide development and formulation	Spraying	Apple scab
			Dusting	Powdery mildew
	Escape infection	Modify environment	Pruning	Grape bunch rot
			Thinning	White mold
			Alter soil pH	Club root of crucifers
		Modify cultural practices	Alter seeding date	Barley yellow dwarf
			Alter seeding depth	Ergot
Resistance	Develop resistant plants	Selection	Selective breeding	Fusarium wilt of peas
		Hybridization	Cross-breeding	Curly top of tomato
		Irradiation	Induce mutations	Verticillium wilt of mint
	Cross-protection	Attenuated virulence	Induce resistance	Crown gall of fruit trees

dogmatic term recommedation, some publications use the softer terms of "guides" or "suggestions" even though the specifics of control remain the same.

Early history of plant disease control is not as complete as history of disease damage or of efforts to determine causes of disease. Early reports of plant disease control are vague and difficult to interpret in modern terms. Orlob (1973) states that the Greeks rarely mentioned disease control. According to Meyer (1959) there is a record of the use of a "fungicide" in India as early as 2000 B.C. but there is no indication of what the material was or what disease was controlled with it.

Most historians of plant pathology cite Homer's 1000 B.C. reference to "pest averting sulfur" but the pests were probably insects and not plant pathogens. There were early (B.C.) records of practices such as extra irrigation and crop rotation but these were not necessarily for control of pests.

Many early disease control efforts were associated with supernatural forces such as religion, magic, mythology, astrology, and the like or the application of noxious substances such as urine, dung, and residues. The Romans created the god Robigus (sometimes also the feminine form Robigo) and celebrated the festival of the Robigalia to appease the gods and ward off the effects of rust on cereal grains, especially wheat. According to historians Pliny recommended placing laurel branches in fields to force the rust to bypass fields of grain. Although the Aztecs of Central America had gods and goddesses who looked after the well-being of maize and other crops, there is no direct evidence to connect these dieties with smut that occurred on grain.

A number of palliative treatments were applied especially by the medically oriented practitioners of this early period. These included perfumed water, elephants milk, honey, meat broth, and various herbs. "Amurca," the name the Romans called the pulp remaining after olives are pressed to remove the oil, was sprinkled on plants to control blight, presumably powdery mildew. Pliny recommended soaking wheat seed in wine containing crushed cypress leaves. The alcohol and extractives from the cypress may have had some fungicidal action as might have the "Amurca" of the Romans. But mostly the suggested "remedies" had no logical basis.

Pliny included many cultural practices in his treatise on the control of tree diseases. These included loosening the soil,digging around the roots, making a drainage hole, applying manure, pruning away excessive growth, and draining sap. Other references recommend removal of moss and crooked, weak, fat, and malformed branches. Each of these has a modern counterpart and some measures apparently worked better than others, just as they do today.

Among the various noxious or unpleasant materials applied to control plant diseases were salt mixtures and pungent solutions to combat certain disorders of trees and grapevines, ashes and vinegar were sprinkled on grapes to prevent premature fruit drop, grain seed was steeped in oxgall before planting, goat and sheep dung was applied to injured grapevines, and grape branches that did not ripen were buried and drenched with urine. Frequent mention of manure and urine probably resulted from observations that these materials stimulated plant growth.

In 1629 John Parkinson in England wrote in a book for gardeners, "The canker is a shrewd disease when it happeneth to a tree; for it will eate the bark around and so kill the heart in a little space. It must be looked into before it hath runne too farre; most men do wholly cut away as much as is fretted with the canker and then dress it with vinegar, or cowes pisse, or sowes dung and urine, etc. until it will be destroyed and again with your salve before appointed." This description of a disease control procedure is almost modern in application if not in text. Essentially the writer describes the excision or cutting away of cankered tissue, treating the affected area with some compounds, and the possibility of repeating the treatment. It points to the urgency of early treatment and the compounds suggested have either acidic or alkaline properties that could be fungicidal.

Crop rotation, in the form of triennial rotation or the same crop every third year, apparently started in Europe around 1500 A.D. Also recommended about this time is to change seed every year, especially to use seed grown on land other than your own. This practice is still valid today, primarily to avoid increase of seed-transmitted pathogens.

No general guiding principles emerge from the maze of ancient plant health recipes except for the theory of opposites. If a tree was unthrifty apply manure, if it was overburdened drain the sap. It is difficult to know where or when modern plant disease control began but until the germ theory was established most disease control was based on trial and error, or accidental or coincidental observations.

Two chance observations led to widespread use of some disease control chemicals. The first occurred in the 1660s when a shipload of wheat sank off the British coast. Some of this seed was salvaged and planted. No smut developed on the crop. This accidental discovery gave rise to the general brining of seed grain in England.

Use of metallic fungicides to control plant diseases started in China about 1630 when seed treatment with arsenic was used to control soil-borne pests. The use of arsenic and mercuric chloride was mentioned as seed treatments for smut of cereals in 1755 in Germany but was criticized because of serious poisoning of persons engaged in agriculture. This is probably the first indictment of a fungicidal material detrimental to health or environment. Whether as a result of this concern or for other reasons, France, in 1786, banned all arsenic and copper compounds as seed treatments. Why copper was banned is uncertain. In 1808, marketing toxic substances such as arsenic and mercury was forbidden in Carlsruhe, Germany, because these materials could be replaced by nontoxic substances (presumably copper salts) for seed treatments.

The second accidental discovery of a valuable fungicide centers around the discovery of Bordeaux mixture, the active ingredient being copper hydroxide (one of a number of copper fungicides). Although sulfur in various forms, especially as elemental sulfur and lime sulfur, was used to control powdery mildews and certain other parasitic fungi as early as 1802 the era of modern fungicides is considered to have begun with development of Bordeaux mixture in 1882.

At least as early as 1800, Proust, a French chemist, studied the chemistry of a basic copper sulfate resulting from action of alkalies on bluestone (copper sulfate). This material was applied to grapevines by French farmers to give the grapes an unappetizing appearance and discourage theft. In 1882, Pierre Millardet, botany professor at Bordeaux University, observed that grapevines along roads and paths treated to deter pilfering were free of downy mildew. Millardet proceeded to formulate a standardized solution known as Bordeaux mixture. In the following years a range of Bordeaux substitutes or variants was developed. However, none of these gained the reputation of the original mixture, which remained the major, if not the only foliar fungicide until the 1900s.

Use of plants resistant to a disease is the ideal control. Undoubtedly early gardeners, farmers, naturalists, and others observed many instances of healthy plants while others around them were diseased. Whether this was true resistance or simply disease escape can never be known for certain but records and subsequent information demonstrate that some of these plants were disease resistant.

Charles Darwin (1868), in his book *The Variation of Animals and Plants Under Domestication*, which is essentially a literature review of the time, cites several plants that were less, or more, severely affected by disease than others of a different color. Darwin did not describe the diseases or used vague terms and we can only guess as to what they might be. Among those mentioned were purple-fruited plums being more affected than green- or yellow-fruited, and red sugarcane less affected by unnamed disease than white cane, green-fruited grapes suffering more vine disease (probably downy mildew) than red or black grapes, white verbenas being especially subject to mildew, and yellow-fleshed peaches more often affected than white-fleshed to yellows, whatever that might be. He also noted that the strawberry variety Cuthill's Black Prince was severely affected by mildew while other varieties were not affected. "Mildew" apparently was a commonly used term at the time and could refer to any of a number of diseases and conditions.

One other instance of color relationship to disease cited by Darwin was the report of M.J. Berkeley that white onions were more subject to attack by a fungus (*Colletotrichum circinans*) than colored onions. In the 1920s J. C. Walker, a University of Wisconsin plant pathologist, and his associates demonstrated that this difference in resistance was due to the presence of catechol and protocatechuic acid, fungistatic compounds, in colored onions.

Sharvelle (1979) reports that in the early 1800s in England certain wheat varieties were resistant to leaf rust. However, Chester (1946), in his extensive book on this disease, does not mention such an occurrence. Biffen, in 1905, began to cross plants with the expressed purpose of developing disease resistant varieties. In the succeeding years many disease resistant varieties have been produced. Some had the necessary agronomic or horticultural qualities to be accepted, some possessed a stable (durable) resistance that holds up today. Others did not and were not accepted, or the resistance was overcome by new pathogenic races. Breeding for disease resistance is a continuous effort.

REFERENCES

*Ainsworth, G.C., *Introduction to the History of Plant Pathology*. Cambridge University Press, New York, 1981.

Biffen, R.H., Mendel's laws of inheritance and wheat breeding. *J. of Agric. Sci., Cambridge* 1:4–48 (1905).

Chester, K.S., *The Nature and Prevention of the Cereal Rusts as Exemplified in the Leaf Rust of Wheat*. Chronica Botanica Co., Waltham, MA, 1946.

Darwin, C., *The Variation of Animals and Plants under Domestication*. 2 vols. Orang Judd, New York, 1868.

Meyer, K., *4500 Jahre Pflanzenschutz*. Verlag Eugen Ulmer, Stuttgart, 1959.

Orlob, G.B., Ancient and medieval plant pathology. *Pflanzenschutz Nachrichten* 26(2):65–294 (1973).

*Parris, G.K., *A Chronology of Plant Pathology*. Johnson and Sons, Starkville, MS, 1968.

Sharvelle, E.G., *Plant Disease Control*. AVI, Westport, CT, 1979.

*Whetzel, H. H., *An Outline of the History of Phytopathology*. W.B. Saunders, Philadelphia, 1918.

* Publications marked with an asterisk (*) are general references not cited in the text.

2 Principles and Concepts of Control

"Desperate diseases require desperate remedies"

The above adage attributed to Hippocrates applies to plant diseases as well as to human diseases. Growers often resort to extreme and sometimes outlandish remedies to reduce damage from plant diseases. Most of these remedies, however, have some factual basis of action, hence they follow some principle.

Various schemes have been suggested to illustrate disease control principles. Whetzel (1929) was one of the first to observe that all control measures would fit under one of four principles. These were the foundation of a course in plant disease control at Cornell University and Whetzel defined the "four fundamental principles of plant disease control" as follows:

> Exclusion—"preventing to a profitable degree the entrance and establishment of a pathogene [*sic*] in an uninfested area, as in a garden, field, region, state or country."
>
> Eradication—the "more or less complete elimination or destruction of a pathogene after it is established in a given area."
>
> Protection—"the interposition of some effective barrier between the susceptible parts of the plant and the inoculum of the pathogene."
>
> Immunization—"the development by natural or artificial means of an immune or highly resistant plant population in the area infested with the pathogene to be combatted."

Note that actions taken under the first two principles are directed at the causal agent (the pathogen) of disease, that is, the pathogen is kept out of an area where it is not present or it is eliminated from an area where it is established to a limited extent. Actions taken under the latter two principles are directed at the host plant (the suscept) without regard to, but assuming presence of the pathogen.

There are different usages of the terms principle and strategy. In at least one reference, exclusion, eradication, and resistance were considered strategies rather than principles. This source declared that a specific chapter would focus on principles, yet principles were never identified or defined. Another writer prepared a long list of principles of plant pathology including 18 "principles" of disease control. If we accept the definition that a principle is either a rule of action or a rule of science explaining action, many of these so-called principles are simply statements of fact or opinion.

There have been a number of other schemes outlining principles of plant disease control and each has its merits. Much as with taxonomists there are splitters and lumpers: some have expanded the number of principles and in some cases have elevated a strategy or technology to principle rank while others have lumped some strategies and technologies with others of a different category into a single principle, although it may not be identified as such.

Whetzel's system is preferred over others because of its simplicity, although some may argue that it is too simple. A beneficial change is to replace the principle term of "immunization" with "resistance." Immunization is defined as "the creation of immunity against a particular disease, especially treatment of an organism for the purpose of making it immune to subsequent attack by a particular pathogen." Immunization generally implies an antigen–antibody system that occurs in animals and the imparting of immunity by vaccines, antisera, and the like. No comparable system is known to occur in plants and, therefore, the term immunization is not appropriate for plants. Furthermore, plants are rarely immune to a pathogen or disease but are resistant in varying degrees. Therefore, resistance conveys the action involved and conveniently replaces immunization.

Gäumann (1950) was one of the first to depart from Whetzel's system of principles and instead of identifying principles outlined three general ways by which diseases are controlled:

1. Prophylaxis (i.e., protection from disease) against the infection process. This is accomplished primarily by chemical protectants and cultural methods and is comparable to Whetzel's protection.
2. Prophylaxis against development of disease. This is comparable to Whetzel's immunization.
3. Therapy. This involves treatment of the plant to eliminate the pathogen and is similar to Whetzel's eradication.

Sharvelle (1979) grouped the various methods for plant disease control into two major categories, immunization and prophylaxis, and outlined strategies for each in a dichotomous system (Figure 2.1). Sharvelle later revised this slightly and added a scheme depicting manipulation of the disease triangle to accomplish these control principles (Figure 2.2).

The National Academy of Science (NAS) (Anon. 1968) established a subcommittee on plant pathology in 1968 to outline general principles of plant disease control. Their system is similar to Whetzel's with some minor expansion. The principles were as follows:

1. Avoidance—avoiding conditions favorable to disease.
2. Exclusion—excluding inoculum.
3. Eradication—eradicating inoculum at its source.
4. Protection—rendering inoculum ineffective at site of infection.
5. Disease resistance—reducing the effectiveness of the causal factor in the host.
6. Therapy—curing infection in the diseased plant.

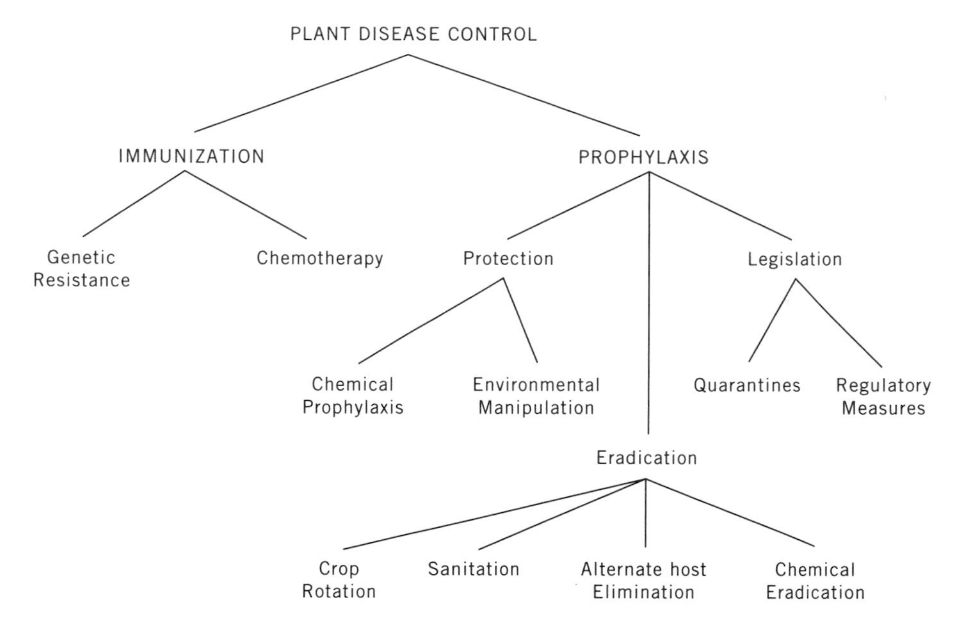

Figure 2.1. The immunization-prophylaxis system of classifying disease control principles. (From Sharvelle, *Plant Disease Control*, AVI, 1979. With permission.)

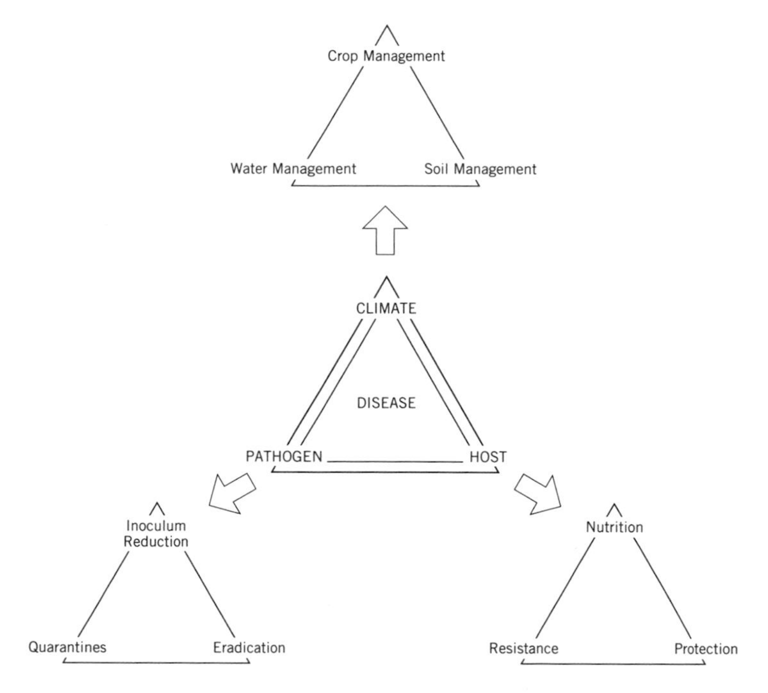

Figure 2.2. Manipulation of the plant disease triangle in disease control. (From Sharvelle, *Plant Disease Control*, AVI. 1979. With permission.)

The strategies involved for each are as follows:

1. Planting at times when, or in areas where, inoculum is ineffective, rare, or absent.
2. Preventing introduction of inoculum or establishment of a pathogen within an uninvaded area.
3. Reducing, inactivating, or destroying inoculum at the source either from a region or from an individual plant.
4. Interposing a toxicant or other effective barrier between susceptible host and a pathogen.
5. Employing resistant hosts, including altering physiological processes, structural nature, or habits of individual plants or populations to make them tolerate or resist infection.
6. Reducing severity of disease in an already infected plant.

Roberts (1978) used the same terms as did the NAS subcommittee, but split resistance into vertical resistance and horizontal resistance. It seems that if a principle is to be established on the basis of the mechanism of resistance others such as disease escape and disease tolerance also deserve principle rank.

Jones and Clifford (1983) do not refer to principles of disease control but rather to six "underlying objectives" or to the "method of application" of these objectives. These appear to be a mix of principles and strategies. Their second category, "methods of application," consists of technologic approaches rather than principles or strategies.

Underlying Objectives (= Principles)

Exclusion of the pathogen
Inoculum reduction
Host–plant protection
Pathogen eradication
Adjustment of environment
Increasing host–plant resistance

Methods of Application (= Technologies)

Regulatory
Cultural
Biological
Physical
Chemical
Integrated

Regardless of which scheme is used, a specific principle is selected (consciously or otherwise) at the beginning of a disease control program. This usually is based largely on (1) occurrence of the pathogen, i.e., is it established in the area and to what extent. If the pathogen is not established in an area, exclusion may be feasible; if present but to a limited degree, eradication is possible. (2) The nature of the crop, i.e., annual or perennial, irrigated or dryland. If the crop is annual, the use of resistant varieties is realistic; if perennial, cultural and chemical controls are the best means of control.

Having seen a variety of the schemes used to express principles of plant disease control, a look at some of the strategies developed to implement each of the four principles described by Whetzel is in order. A few specific examples follow. Additional examples of the technologies used for the application of these control measures will be given in subsequent chapters.

EXCLUSION

Strategies are developed that prevent pathogens, by whatever means and methods, from entering and establishing themselves in an area where they have previously been absent. This may involve one of three strategies.

Prohibition is commonly attempted through quarantines and embargoes. These are legal means of prohibiting or restricting importation of possible harborers of pathogens from infested areas. The United States prohibits importation of grapevines from outside the North American continent because of virus and other diseases.

Interception aims at intercepting pathogens by inspection of plant material at points where pathogens are likely to be brought into an uninvaded area, such as ports of entry, customs office along international boundaries, receiving points, or even farms and gardens of growers. Effectiveness of this method relies on natural barriers such as oceans, mountain ranges, and deserts across which pathogens cannot move by natural means. Golden nematode and citrus canker bacterium are two pathogens frequently intercepted at border inspection points.

Elimination involves some treatment (physical or chemical) either at point of origin or point of entry to ensure destruction or removal of pathogens. Disinfection and disinfestation of plants and soil and certification of seeds and plants are technologies commonly used. The South American leaf blight pathogen of rubber is excluded from Southeast Asia by requiring travelers that might be contaminated with spores of the fungus to shower thoroughly and their clothes be washed and heated to 75°C.

ERADICATION

Eradication involves removal, elimination, or destruction of a pathogen from an area (or individual plant) in which it is established. Like exclusion, its application is aimed directly at the pathogen. It is the logical step that follows unsuccessful exclusion and may be accomplished by one of three measures.

Removal is detection and effective disposal of diseased plants, parts of plants, or infested plant debris. An example is inspection of grafted nursery stock destined for shipment and removing those with evidence of crown gall.

Elimination utilizes indirect measures by which pathogens are rendered innocuous due to their inability to successfully perpetuate or propagate. This may be accomplished by removal of perennial weed suscepts (i.e., alternative hosts) and alternate hosts, crop rotation, encouragement of natural enemies (biological control), and destruction of sources of primary inoculum. Alternate hosts are essential for complete development of some pathogens, especially certain rusts; alternative hosts are simply other carriers of the pathogen. Burning weeds along drainage ditches eliminates many viruses and their vectors.

Destruction kills pathogens *in situ* through direct measures either on or in the suscept or in the enviroment. Such eradication is usually disinfestation or disinfection by toxic chemicals or heat. Burning infected citrus trees in nurseries and orchards destroys citrus canker bacteria.

PROTECTION

This principle places some protective barrier between suscepts and pathogens, the latter being generally present and in a position to bring about infection if conditions are favorable. Protective barriers involve three strategies.

Manipulation of environmental factors modifies certain factors necessary by themselves or in coordination with other factors for inoculation or invasion of the suscept. These usually involve utilizing or modifying cultural practices. Planting after soils are warm promotes plant vigor and protects against damping-off, seedling blights, and root rots.

Applications of protectants provide chemical barriers accomplished by spraying or dusting plants or plant parts with water-insoluble chemicals that become soluble or toxic in the presence of the host plant or active inoculum of the pathogen. Fungicides such as fixed coppers form a chemical barrier over the surface of grape leaves to prevent infection by the downy mildew fungus. Small quantities of copper are released from the chemical film to kill the fungus spores.

Barriers such as screenhouses, plastic covers, space (distance), and other protective barriers can prevent pathogens from contacting host plants, thereby preventing disease. Windbreaks can prevent movement of fungal spores such as those of southern corn leaf blight and protect fields on the lee side of such barriers.

RESISTANCE

Resistance is a principle of disease control in which physiological processes, structure, or habits of individual plants or of plant populations are temporarily, or permanently, changed in such ways that plants escape, become tolerant, or no longer respond, or only mildly so, to infecting activities of plant pathogens. Whetzel proposed six strategies for disease resistance in plants.

Selection is the simplest approach and involves detection of tolerant, resistant, or immune individuals within a plant population, and propagation and multiplication of such plants. Fusarium resistant cabbage and curly top tolerant sugarbeets are examples of disease resistance obtained by selection.

Hybridization consists of crossing susceptible but otherwise desirable individuals with less desirable but tolerant, resistant, or immune individuals with the expectation of combining desirable characteristics from one parent with disease resistance of the other. Fusarium wilt resistance in watermelon and strawbreaker foot rot resistance in wheat were obtained by crossing different kinds of plants.

Grafting involves grafting disease resistant root, stem, or top onto an economically valuable (desirable) but susceptible plant. The use of disease resistant rootstocks is common in tree fruits and ornamental plants. Some apple varieties grafted onto certain rootstocks are more resistant to collar rot.

Nutrition involves fertilizing plants to bring about temporary, nonheritable freedom from disease. Nitrogen, phosphorus, and potassium are the nutrients most commonly applied to increase plant vigor, harden, and mature tissues, thereby rendering plants more tolerant or resistant to diseases. Phosphorus applications to wheat make the plants more resistant to take-all and other diseases.

Medication (chemotherapy) introduces chemical substances toxic to pathogens into plant tissues for prevention of infection or therapy of disease. This method has long been tried but has had little or limited success. Infusion of tetracycline antibiotics into plants infected by mycoplasma-like organisms (causes of diseases such as pear decline and x-disease of stone fruits) is probably the most effective use of this method so far.

Vaccination involves the introduction of vaccines, serums, and antitoxins similar to those applied in human and animal medicine but it is an inappropriate term for plants. Because plants lack both an antibody system and a circulatory system this method has little application to plant disease control. The use of cross-protection with plant viruses is probably the closest thing to "vaccination" in plant pathology. One use of this strategy is the initial inoculation of tomato seedlings with a mild strain of tomato mosaic virus to protect plants against a severe strain of the same virus.

A seventh strategy should be included, and that is to induce mutations by physical or chemical means. This technique has resulted in mint plants resistant to Verticillium wilt. Genetic engineering and plant transformation also should be considered in this general area.

Although we can assign each specific control measure to a given principle most plant diseases are actually controlled (or managed) by combinations of measures, rarely by a single practice. They are usually combinations of (1) chemical, (2) cultural, and (3) resistance measures. These measures are not listed in order of preference but rather in the order in which they can be applied in a timely fashion once the need for treatment is apparent. If a grower needs to control a disease in an established crop, chemical control is most immediately available, assuming, of course, that an effective chemical control (technology) has been developed for that disease.

Cultural practices usually cannot be applied until the appropriate time in the growing cycle of a crop. For example, seeding, cultivation, irrigation, fertilization, and other practices must be carried out at the proper time or there could be more damage created than that caused by the unattended disease. In turn, a resistant variety cannot be used until the next cropping cycle. With annual crops such as wheat the delay until a resistant cultivar can be used is a year or less, whereas with a short-lived perennial crop such as alfalfa or strawberries the conversion to a new variety may be several years away. And it is even longer with long-lived perennial crops such as tree fruits, blueberries, or grapes.

The initial decision in the disease control process identifies which principle will be utilized. This decision is based on distribution and occurrence of pathogens. If a pathogen is not present in a region but has the potential of being introduced, then exclusion is a feasible action. If, however, a pathogen has been introduced but is restricted in the extent of its distribution and otherwise not well established, eradication is possible.

When a pathogen is widely distributed, firmly entrenched, and capable of maintaining its position, exclusion and eradication are not options. The choice is between protection and resistance. The latter is preferred for several reasons. Unfortunately, a good stable source of resistance combined with requisite agronomic or horticultural traits is not usually available. Then protection becomes the only option.

Now a strategy must be developed to accomplish the selected principle. There are at least two general ways to protect a plant from a pathogen that is present in an area. These are chemical barriers and environmental or cultural barriers. If there is no effective chemical protectant available or no system for applying it, only the establishment of some environmental or cultural barrier remains. If disease severity is determined by the stage of plant growth when infection occurs and infection takes place only in a certain temperature range, altering seeding date may prevent these two events (i.e., susceptible growth stage and favorable temperature) from coinciding, thus protecting the plant.

Since the strategy for combatting this disease has settled on modifying the seeding date, seeding techniques must be developed to change the seeding date without adversely affecting crop production. In the case of wheat that is normally fall seeded (a winter wheat variety), spring seeding will not result in a crop unless the plants are vernalized. So suitable spring wheats must be planted or facultative varieties be developed. Or, if the seeding date is delayed in fall, the technology must allow seedlings to develop to a stage that they can withstand winter damage.

Finally, a grower must eventually apply this new seeding technology to the field, perhaps adapting it further to meet specific local or area conditions. A technology must be accepted by growers or it will not be applied, no matter how effective it controls the disease in question.

REFERENCES

Anon., *Plant Disease Development and Control*. National Academy of Sciences, Washington, D.C., 1968.

*Clifford, B. C. and E. Lester (eds.), *Control of Plant Diseases: Costs and Benefits*. Blackwell, Palo Alto, CA, 1988.

*Fry, W. E., *Principles of Plant Disease Management*. Academic Press, New York, 1982.

Gäumann, E., *Principles of Plant Infection*. (trans. by W.B. Brierly). Lockwood, London, 1950.

Jones, D. G., and B. C. Clifford, *Cereal Diseases: Their Pathology and Control*. John Wiley, New York, 1983.

*Heitefuss, R., *Crop and Plant Protection: The Practical Foundations*. John Wiley, New York, 1989.

*Martin, H. H., *The Scientific Principles of Crop Protection.* Edward Arnold, London, 1959.

Roberts, D. A., *Fundamentals of Plant-Pest Control.* W.H. Freeman, San Francisco, 1978.

Sharvelle, E. G., *Plant Disease Control.* AVI, Westport, CT, 1979.

*Sill, W. H., Jr., *The Plant Protection Discipline.* John Wiley, New York, 1978.

*Thomason, I. P. and E. P. Caswell, *Principles of nematode control.* In Brown, R. H., and B. R. Kerry (eds.), *Principles and Practice of Nematode Control in Crops.* Academic Press, New York, 1987. Pp. 87–130.

Whetzel, H. H., The terminology of plant pathology. *Proc. Int. Cong. Plant Sci.*, Ithaca, N.Y. 1926:1204–1215 (1929).

*Yarwood, C. E., Some principles of plant pathology. II. *Phytopathology* 63:1324–1325 (1973).

* Publications marked with as asterisk (*) are general references not cited in the text.

3 Plant Disease Losses

"It is a poor plant pathologist who cannot inflate a minor problem into a major epidemic." That quote by a former plant pathology colleague reflects the frequent comments by farmers that if they believed the various experts who appraise crop losses they would lose one- third to disease, one-third to insects, and one-third to weeds, leaving nothing to harvest. Such cynicism reflects the vagueness and exaggeration of many disease loss estimates. Knowledge of losses due to diseases is essential to planning control programs.

The possibility of negative yields from exaggerated losses is one argument against the use of the term "loss". This view maintains that a farmer cannot lose what he does not have and compares the less-than-expected yield to the "shortfall" of tax revenue that is less than the expected amount. It is a valid point since using this concept overcomes the tendency to derive negative yields from cumulative losses. Nevertheless, the term loss is firmly entrenched and probably will continue to be used.

Estimates of disease losses involve many considerations, some biological, some economic, and some political, but two approaches are generally used. The first is disease appraisal (or assessment), which estimates the amount of disease in a crop. This may be reflected by percent of plants affected, leaf area diseased, reduction in growth or yield, or any other measure of effects of a disease on a plant or plant population. Disease appraisal quantifies what *has* happened. The second approach is disease forecasting, which predicts what *might* happen. Forecasting is discussed in Chapter 5.

As early as 1918, G. R. Lyman, the pathologist in charge of the Bureau of Plant Industry Plant Disease Survey, asked: "How can we expect practical men to be properly impressed with the importance of our work and to vote large sums of money for its support when in place of facts we have only vague guesses to give them and when we do not take the trouble to make careful estimates?" He went on to state: "Determination of loss is a difficult and complicated matter, but I believe that we should seriously attempt it. We should develop quantitative methods and make careful counts in restricted areas." This statement was made in a paper read to the American Phytopathological Society and went on to outline some of the ways the newly recognized Plant Disease Survey could provide needed quantitative information.

Ten years later, E. P. Meinecke (1928), a forest pathologist discussing impact of diseases to the forest industry, commented that "apart from its purely scientific interest, accurate determination of the loss caused by a given disease offers the only safeguard in a rational policy of control."

It was not until 1950 that an extensive analysis of plant disease appraisal was made by K. S. Chester. He recognized three aspects: (1) standarized, accurate appraisal of disease losses, (2) translation of disease intensity into crop loss in terms of production units, and (3) interpretation of this loss on human welfare. However, he considered only the first two aspects, believing the third to be the domain of economists and sociologists. Nonetheless, social impact of plant diseases should be considered here, albeit briefly. Concern about the social aspects of agriculture are again in vogue as reflected by programs like LISA (Low Input Sustainable Agriculture) and others.

In 1949, pathologists and other plant specialists in various states identified some effects of plant disease losses on crop industries and farm life since the 1900s. Many instances were recorded of fields and areas being abandoned for certain crops because of unacceptable, even catastrophic, disease losses. In Alabama, Austrian winter peas had been grown as a green manure to improve the soil but were abandoned because of Ascochyta blight. Apricot and prune orchards killed by Armillaria root rot in California were replanted to walnuts. In one California county loganberries had been grown for 30 years but were wiped out by viral diseases. Sugar beet factories shut down in Idaho and other western states as a result of curly top of sugar beets. In Texas Phymatotrichum root rot eliminated cotton as a crop in some areas, attempts to establish a pear industry were abandoned because of fire blight, and one rose nursery was closed because of crown gall. Oregon was, and is, a major grass seed-producing state, especially of perennial ryegrass seed but blind seed disease resulted in no income because the seed could not be sold. Infected seed fields were often sold to unsuspecting buyers who could not sell subsequent crops. Ergot in ryegrass pastures caused sickness and occasional death of sheep.

Farm life was also impaired by crop disease losses. Among the many negative results were reduced income, education neglected or not possible, farmsteads deteriorated, standard of living and health declined, inability to repay loans, bankruptcy, etc. Many farmers took outside jobs to support the farm, and sometimes the wife and children had to work. Some farmers changed from crops to cattle, not always successfully.

A number of extensive reports have been published that relate crop losses from many causes. Most of these comprehensive reports were made more than 25 years ago. Nothing approximating their depth has been attempted since, despite the need for more current loss estimates. However, these surveys provide some comparative estimates. Hepting and Jemison (1955) evaluated the effect of diseases and other causes to the forest industry and considered two general areas of damage: growth reduction and mortality. Most of the loss, 4.3 billion cubic feet, was from growth reduction and 0.8 billion cubic feet from mortality. Of the total loss about 45% was from disease (compared with 16% from insects and 15% from fire). Thirty-three percent of the 45% was from heart rots (about 73% of the total impact). Table 3.1 shows the estimated losses from selected diseases.

The high loss from heart rots probably reflects the relatively high proportion of old growth timber remaining in 1952, but as this resource is depleted the balance of diseases will shift: fewer heart rot losses, more from other diseases like root rots, dwarf mistletoes, and rusts.

TABLE 3.1. Losses from Diseases in Commercial Forest Land in the United States and Coastal Alaska in 1952

Disease	Total Losses (%)
Douglas fir root rot	2.3
Heart rots	72.9
White pine blister rust	3.2
Dwarfmistletoe	2.9
Fusiform rust	1.4
Needle diseases	0.5
Birch disease	2.5
Miscellaneous diseases	14.3
	100.0%

Source: From Hepting and Jemison (1955). In *USFS Timber Resource Review.*

Another source of loss is the cost of control programs directed at some forest tree diseases. Table 3.2 gives some of these costs for 1952. It has been estimated that the white pine blister rust control program cost about 150 million dollars over the 50 years it was conducted, with a peak allocation of 6 million dollars in 1947.

Similar but less detailed estimates were presented in a U.S. Forest Service report in 1988 (Anon., 1988), Only three diseases were considered in the report: root diseases in general, dwarf mistletoes, and fusiform rust. Only the acres affected for each disease category and suppression costs for dwarf mistletoes were given.

Cramer in 1967 published a comprehensive analysis of losses from diseases, insects, and other pests throughout the world. Despite the time lapse since this report,

TABLE 3.2. Expenditures for Direct Control of Forest Diseases in 1952.

Disease	Total Losses (%)	Total Cost ($)
White pine blister rust	3.2	3,603,700
Oak wilt	0.2	127,600
Brown spot of longleaf pine	0.3	51,800
Dwarfmistletoe	2.9	19,000
Fusiform rust of southern pines	1.4	9,500
Larch canker	?	6,400
Miscellaneous diseases	11.9	39,200
	19.9%	$3,857,000

Source: From Hepting and Jemison (1955). In *USFS Timber Resource Review.*

it was still being cited in 1991. Nothing comparable has replaced it. Based on USDA publications Cramer estimated losses in the United States from plant diseases to be 0.3 billion dollars in 1937: 2.9 billion in 1954 and 3.25 billion in 1965. He pointed out the difficulties of estimating indirect losses such as those caused by the limitation or restriction against certain crops because of quarantines. Cramer gives proportional losses due to specific plant diseases for many crops as illustrated in Table 3.3 for wheat.

Cramer also published tables giving specific annual losses for various crops in tons, dollars, and as a percentage of potential value. For example, wheat losses worldwide due to disease were over 33 million tons with a value of more than 2 billion dollars and equal to 9.1% of potential value. A 1987 estimate of global wheat losses from plant diseases amounting to 12.5% of total production was obtained by adjusting Cramer's data (James, et al., 1991). If these estimates are accurate and comparable it means that despite many improvements in disease control, losses have actually increased. This same 1987 estimate listed losses for both potatoes and sugar crops (sugar beets and sugarcane) due to plant diseases as 33%, which seems high. Table 3.4 gives the worldwide percentage losses due to plant disease cited by Cramer for several major crops and the corresponding 1987 percentage losses calculated from the estimates derived by James.

These estimates are surveys of plant pathologists and others. While they are useful these estimates are usually so broad that they have little value in localized situations except to show what might happen.

In 1953 the USDA estimated the average annual loss from plant disease at 3 billion dollars but this is probably not a precise value. In the 10-year period 1951-1960 the USDA estimated annual losses due to disease at 14% for wheat and 21% for beans. Total annual loss for all crops was estimated at 4.5 billion dollars.

Some diseases are catastrophic in some years. In 1930 over 23% of the wheat crop in the United States was lost to stem rust. In 1889 there were 6.1 million peach trees in production in Delaware but 20 years later, in 1909, only one million were in

TABLE 3.3 Losses from Various Wheat Diseases in the United States

Disease	Annual Loss (%)
Stem rust	28.6
Leaf rust	17.8
Fusarium diseases	7.1
Septoria leaf blights	7.1
Wheat streak mosaic	7.1
Loose smut	5.7
Strawbreaker foot rot	4.3
Other	22.3
	100.0

Source: Compiled from Cramer. *Pflanzenschutz Nachrichten* 20:1–524, 1967.

TABLE 3.4. Estimated World Yield Losses for Major Crops in 1967 and Adjusted Losses for 1987

	Estimated Loss to Disease (%)	
Crop	1967	1987
Wheat	9.1	12.5
Rice	8.9	17.0
Corn (maize)	9.4	14.9
Potatoes	21.8	32.8
Sugar crops	16.5	33.4
Vegetables	10.1	15.4
Fruits	16.4	22.9

Source: Compiled from 1967 data from Cramer and 1987 values calculated from data modified by James, et al. (1991) from Cramer.

production because of peach yellows. Sugarcane mosaic in Louisiana resulted in a drop in average yearly production from 400,000 to 50,000 tons. In the 1970 southern corn leaf blight epidemic an estimated one billion dollars was lost.

Worldwide, plant disease losses have taken a toll in lives as well as in dollars. We have already mentioned the Irish potato famine. The Great Bengal rice famine in India in 1942, where yield losses of up to 91% occurred in some rice varieties, resulted in 2 million people dying from starvation. Similar famines have occurred throughout history partly as a result of plant disease.

Losses take many forms and can be grouped as direct and indirect losses. Direct primary losses may be summarized as five types: (1) quantity, (2) quality, (3) cost of control, (4) added harvesting and processing costs, and (5) reduced value of replacement crops.

Quantitative losses are the first to be recognized and may involve the total mass of product (e.g., tons of sugar beets per acre) or yield of usable product (e.g., percent of sugar content per ton of beets). These losses are usually indicated as reductions in production of crop as tons or bushels per acre, boxes per tree, bags per acre, etc.

Losses in quality are often more difficult to ascribe and are usually based on some industry standard that may be quantified, such as specific gravity, sugar and protein content, color, size, shape, presence of blemishes, or other defects. Many of the qualities that constitute this type of loss are based largely on consumer preference and perceptions. For example, potatoes with a few shallow lesions of common or powdery scab are just as flavorful and nutritious and yield just as well as blemish-free tubers, but are downgraded because of appearance. The same is true of apples with one or two apple scab lesions. But because of appearance these products may be culled and either discarded or used for lower value by-products.

Costs of control measures are additional losses to disease since they are extraordinary measures not routine in production of a crop. Whether they are needed or not, disease control costs must be considered a loss, or expense, caused by the potential of disease. These costs can vary considerably. For example, the cost of wheat seed treatment with the fungicides hexachlorobenzene (HCB) or mercury for smut control was about 15 cents a bushel but increased to about a dollar a bushel when these materials were no longer available and growers had to use a newer, more costly fungicide. This is also true of application costs. If a wheat grower sprays for rust the cost of application alone is about 6 dollars an acre, so if the treatment must be repeated, 6 dollars plus cost of material is an additional loss to disease.

Many diseases also result in additional costs in harvesting or processing of the crop. Strawbreaker footrot of wheat causes not only a reduction in grain yield and kernel weight but also lodging requiring combine headers to be lowered and speed reduced making effecient recovery of grain more costly. Smaller sized fruit as a result of disease increases picking time per box or bin and thereby increases harvest costs.

Perishable produce such as fruit and vegetables often has to be sorted several times during storage to eliminate diseased products. Pinpoint scab and bull's eye rot of apples are diseases that begin in the orchard but do not develop until the fruit has been in storage for some time. Fruit must be examined and sorted periodically to maintain high quality, increasing the cost of the product, a cost chargeable to disease. Wood products such as lumber, posts, and framing timbers are often cut from trees containing incipient decay. This decay may continue to develop in the wood after it is placed in service. Additional treatment such as kiln drying or chemical preservatives may be required and increase cost of the product. This is a cost that would not be necessary except for disease.

Added costs occur as a result of seedling disease in cotton. If a field is not replanted and the stand is poor losses to disease result from wasted fertilizer, additional costs of weed control, weak plants more subject to other diseases and insects, and more difficult harvesting. If the field is replanted extra costs result from additional seed, wages, use of equipment, insect control,late planting, late harvest, and increased boll rots, as well as loss of weed control chemicals and soil moisture to the first seeding.

Losses from disease also occur when less valuable crops are planted to replace those eliminated by disease. In the Palouse region of eastern Washington two crops prevail: cereals, of which fall-seeded wheat is the more valuable crop, but also including barley, spring wheat, and oats, and edible legumes, primarily peas and lentils. The latter are often grown in rotation with wheat but because of smaller markets these crops are not competitive with wheat. The other cereals are usually lower in yield and price than winter wheat and are not as valuable. However, occasionally a disease such as barley yellow dwarf, strawbreaker foot rot, or Cephalosporium stripe will force a grower to grow one of these lower value crops in place of winter wheat and essentially suffer a loss because of the threat of disease.

Losses are relative to the value of a crop. Loss of a tree in a forest where it has value for timber is less than the loss of the same tree in a park where it has value as an ornamental. Generally the more intensive the culture the greater the loss. In some cases the same crop plant may have considerably different values depending on its end use. Cabbage may go to three different markets. Heads of mature cabbage may bring only a few cents a pound at the market but the same head carried over for seed production may have a potential value of several dollars. Cabbage seedlings sold as transplants for the home garden may cost 50 cents or more each. Thus, the estimated loss of any one will depend on its potential value.

Sugar beets used for livestock fodder have a lower value per ton than for sugar production. Potatoes sold on the fresh market bring a higher price than those sold by the ton for processing and these, in turn, are worth more than tubers used for starch production. Tubers sold for seed potatoes may bring the highest price of all but the market is much smaller and the cost of production higher. Similarly apples held in controlled atmosphere (CA) storage for later sale in a better market are higher valued than those sold immediately at a road side stand. And both bring a better price than those sold for juice which is still better than the culls sold for livestock feed.

Losses may occur at various levels in the commercial chain. The first and most obvious level is at the production level but sometimes greater losses occur after the product has been harvested and is in storage. Soft fruits such as cherries, peaches, and apricots are usually held briefly between harvest and marketing because of the high incidence of fruit decays. Apples and pears may be stored for several months before marketing. Apples, especially, are stored for long periods under CA conditions.

Additional losses occur after postharvest storage. The product goes to the wholesaler for a relatively short time, then to the retailer for varying periods, and finally to the consumer where it may be held too long to be fit for consumption. These are still considered disease losses even though they are caused by nonparasitic agents or weakly parasitic organisms.

Losses may be indirect as illustrated by the effect of coffee rust on loss of income from foreign trade and reduction of labor needs. Dependence on coffee varies considerably from country to country but some south and central American countries are highly dependent on coffee sales for a major part of their foreign exchange (Table 3.5).

TABLE 3.5. Foreign Exchange from Coffee Sales (1968)

Country	Percent of all foreign exchange
Colombia	67.7
El Salvador	42.7
Brazil	41.2
Haiti	38.9
Guatemala	33.0
Costa Rica	31.4
Ecuador	17.9
Nicaragua	14.0
Honduras	12.6
Dominican Republic	11.9
Mexico	6.3

Source: Reproduced with permission, from Schieber (1972), Annual Review of Phytopathology Vol. 10, © 1972 by Annual Reviews Inc.

In 1968, before coffee rust was known to be established in the western hemisphere, several writers suggested the economic damage to countries from Panama to Mexico would be about 22 million U.S. dollars for each 5% rust damage increment. Thus, 20% rust damage would reduce income by 88 million dollars. There would also be less need for labor to pick and process the reduced crop. This reduction in hand-labor was estimated to be 7.75 million man-days for each 5% increment of rust damage so that the same 20% rust damage would reduce the need for labor by 31 million man-days. Similar reductions can occur in other labor intensive crops as a result of any calamity such as disease, insects, and weather.

In the early days of lumbering, loggers did not fell trees if they were severely decayed because payment was based on sound wood scale of the tree. If time was consumed cutting decayed trees for no pay loggers suffered a loss of income directly attributable to disease.

Chester used the term "appraisal" when referring to estimates of disease losses; more recent writers have used the term "assessment." The difference in word choice is probably one of preference rather than meaning since both terms refer to an evaluation. Herein the terms interchangeably coincide with cited material.

Chester describes several features considered desirable for estimating losses. Briefly these are as follows:

Disease appraisal measures should measure disease intensity and translate into crop losses. This is not always possible because of several complications in correlating disease and yield losses. These complications include the following:

1. Compensation often results when diseases eliminate plants randomly throughout a planting. Surviving plants can use nutrients, water, and space freed by missing (dead) plants to produce higher than expected yields. On fruit trees where some blossoms or young fruit have been killed by a disease such as brown rot, surviving fruit may be larger, thereby compensating in part for the damage.
2. Weather may be optimum for crop production. Cool temperatures and abundant moisture are probably the two environmental factors that most favor fungal and bacterial diseases. They often promote excellent plant growth and negate some disease damage.
3. Other limiting factors such as weeds, nutrients, moisture, or light deficiencies may overshadow disease losses making it difficult to accurately establish amount of loss.
4. Stage of infection may have a direct bearing on losses. Several wheat diseases illustrate this point. The earlier infection by barley yellow dwarf virus occurs the greater the yield reduction. A 10% infection of plants in the seedling stage may result in greater loss than 100% infection at the flag leaf stage.

Maximum losses from strawbreaker foot rot occur when infection takes place after the seedling stage and before late tillering. Very early infection tends to be less severe because young plants (seedlings) are more resistant than older plants. Late infections may not give sufficient time for the fungus to penetrate the several layers of leaf sheaths to colonize stems where severe plant damage occurs. A similar phenomenon can occur with snow mold of wheat. And late infection by the stripe rust fungus may result in little damage in wheat varieties with adult plant resistance.

5. Some stages in the disease cycle may have a greater impact than others. In mummyberry disease of blueberries the initial inoculum is from ascospores. One ascospore can infect and blight an entire shoot, killing all blossoms on it. Later, conidia infect individual blossoms resulting in mummified berries. In this disease one ascospore infection can destroy many blossoms while one conidium eliminates only one berry. Since ascospores do the most damage control measures should be directed at them.

Disease appraisal methods should be comprehensive. They should include and distinguish losses from all major diseases and other causes. In many instances the most visible or dramatic disease is not causing the major damage. For example, powdery mildew may be very apparent on wheat, especially on old leaves early in the season, yet have no appreciable effect on yield. However, some minor diseases such as Pythium root rot of wheat, root-lesion nematodes, and some leaf spots may produce no apparent symptoms except for a slight stunting yet reduce yields measurably in the aggregate.

Disease appraisal should have a practical degree of accuracy. It is important to recognize the difference between accuracy, which is the closeness of an estimate to the true value, and precision, which is reproducibility (repeatability) of an estimate. An estimate might be repeated year after year to give similar results yet be far from the actual loss, and likewise, an estimate may be close to the mark (i.e., accurate) one time but cannot be repeated often enough to be reliable.

Disease appraisal should be comparable from one worker, location, or season to another. This is often difficult to accomplish because of differences in methods, complicating factors such as other diseases, cultural practices, and climates. Efforts have been made to standardize disease appraisal as much as possible. For example, the Food and Agricultural Organization of the United Nations (FAO) produced an assessment manual that permits considerable standardization throughout the world.

Disease appraisal should be objective. While objectivity is always desirable it is difficult to eliminate personal and professional bias or influence from evaluations, especially where they often are based on subjective or empirical criteria.

Disease appraisal should embrace all forms of disease losses. Some types of losses have been described above, such as quantity, quality, and cost of control, and should be considered when arriving at disease loss values. Inclusion or exclusion of one or more of these can be used to tilt results of disease appraisal and is a common reason for differences in appraisals by different workers.

Disease-loss appraisals may involve several steps, approaches, or techniques. The relationship of disease to crop loss should be determined under conditions that eliminate or minimize yield reductions from other causes such as fertilizer, moisture, soil type, topography, temperature, and similar variables. Such field experiments characterize the relationship between a disease and yield reduction (i.e., losses).

Field experiments can utilize natural infections and measure yield differences from plants with varying amounts of disease. Or yields recorded on plants infected at different stages of development from seed germination to harvest. Yields of healthy plants can be compared with those from inoculated plants. Effects of viral diseases on potato yields can be measured by planting virus-free tubers beside tubers infected with viruses such as leafroll. These experiments generally characterize the nature and extent of yield reductions.

A disease must be studied throughout the season to determine the total range and magnitude of losses. Growth of healthy and diseased plant populations must be monitored throughout the season to determine if losses are sporadic or cumulative, or if there is some yield compensation in surviving unaffected plants.

Aids must be developed to identify different growth stages of crops that often categorize disease development or control application. For cereals the Feekes scale divides the development of the plant into 11 stages (Table 3.6A). This scale was revised to a broad decimal system (Zadoks, et al., 1974) in which the principal growth stages are given a one-digit code and secondary stages given a two-digit code (Table 3.6B).

Chemical sprays are often applied to deciduous fruit trees based on bud development. Charts are available that describe and illustrate these stages. For example, dormant, bud swell, bud break, green tip, pre-pink, pink, popcorn, king bloom, and full bloom describe the degree of development to fruit formation. These stages can be coded: 0 for dormant to 7 for king bloom or full bloom.

Fungicides must by applied to slash pine at the correct stage of female flower (strobilus) development for best control of southern cone rust. Female strobilus stages range from twig bud to stage 3, the pollen receptive stage, to stage 4, where the strobilus has become a conelet or juvenile cone.

Disease appraisal methods are needed to measure disease and ultimately estimate losses. Appraisal involves both disease incidence and disease severity. Disease incidence is the occurrence of disease in a plant population. It is the number of plant units affected and may be expressed as percent of plants, leaves, fruit, or some other unit.

Disease severity refers to the area of plant tissue affected as a percentage of total area. FAO uses the term "intensity" to cover both incidence and severity.

There may be three different relationships of plant damage by disease:

1. *Incidence–intensity* is applied mostly to foliar disease as percent of leaf surface affected (as by rust, powdery mildew, leaf spots, etc.) and is the relationship most commonly described in disease assessment keys discussed below.
2. *Disease–damage* reflects injury by defoliation (percent of leaves dropped), progressive debilitation (25% reduction in shoot growth, etc.), mortality of some or all of plant parts (10% of plants killed).
3. *Disease–quality* where the loss may be some biochemical measure, such as reduction in protein or sugar content, or change of appearance (less color, poor shape, smaller size, etc.) or flavor (too sour, too acid, bland, mealy, etc.).

Disease assessment keys are developed to facilitate disease appraisal. One of the earliest was the Cobb scale for wheat rust developed in 1890–1894. This rated rust intensity on leaves on a scale of 1 to 5 and represented percentage of leaf surface covered by rust from 1 to 50%. This scale was modified in 1922 by the USDA and expanded slightly to six categories representing leaf infections of 5 to 100%.

TABLE 3.6. Cereal Growth Stages Expressed by the Feekes Scale (A) and the Revised Scale of Zadoks (B)[a]

	A. Feekes Scale			
	General Stage			
	Tillering	Jointing	Heading	Ripening
Numerical range	1 – 5	6 – 10	10.1 – 10.5	11

B. Zadoks Scale		
Stage	One-digit code	Two-digit code
Germination	0	
Seedling growth	1	
First leaf through coleoptile		10
First leaf unfolded		11
Tillering	2	
Jointing (stem elongation)	3	
Flag leaf visible		37
Booting	4	
Boot swollen		45
Inflorescence emergence	5	
Anthesis	6	
Milk development	7	
Dough development	8	
Ripening	9	
Caryopsis hard		91
Seed dormant		95
Seed not dormant		97

[a] These scales have been condensed for this table.

Illustrated keys primarily for cereal and forage crops as aids in standardizing appraisal of diseases have been produced in Canada. Similar keys are available from various regions to help appraise local diseases. For example, Minnesota has published a leaflet illustrating bean rust severity ratings.

Appraising severity of soilborne diseases creates a special problem. It is more difficult to quantify root rots than to count rust pustules or apple scab lesions. A variety of systems have been devised to establish a relationship between disease and damage in these examples. One of those frequently used is a root rot index (RRI) in which individual plants are rated into one of several categories carrying an assigned value. Such values are summed, and an average derived to give an index value. For Fusarium root rot of beans affected plants are rated in one of six classes from 0 to 5, 0 being healthy plants (clean roots, no stunting), and 5 being dead plants. Each class carries a respective numerical value of 0, 20, 40, 60, 80, or 100. The number of plants in each class is multiplied by the corresponding value, and the total is summed and divided by the total number of plants to give the root rot index.

A more current approach to quantifying disease losses uses a variety of mathematical formulas usually based on incidence of disease at some stage of plant growth times a constant to obtain yield losses. A number of formulas are used in the United Kingdom to assess losses of cereals in disease surveys. To estimate powdery mildew losses in winter wheat the formula $y = 2.0Xi$ is used where y = percent loss in grain yield, Xi is the percent of disease on leaf 3 at growth stage 58 of Zadoks' scale, and 2.0 is a constant. A more straightforward formula for losses caused by *Septoria* is $y = Xiii$ where $Xiii$ is the percent of disease on the flag leaf at growth stage 75. While these formulas provide standardization and allow workers in one area to compare losses with losses from another region or year, they will probably be used more in research and intensive surveys than in commercial farming operations.

A variety of methods used to assess damage from plant diseases consist of direct visual assessment, remote sensing, and a range of indirect methods.

Visual examination is the oldest and simplest way to assess the amount of disease in a crop but is frequently inaccurate. Horsfall and Barratt (1945) observed that when less than 50% of an area is diseased the eye assesses the disease but when it is more than 50% the eye assesses the healthy area. Chester later noted that when disease is less than 50% there is a tendency to overestimate and when disease is over 50% the tendency is to underestimate.

The Horsfall–Barratt scale illustrates that the eye is most accurate when disease incidence is very low (0–12%) or very high (87–100%). Relative accuracy at estimating level of plant disease is poorest when disease incidence ranges from 25 to 75%. The Horsfall–Barratt scale stresses that the eye sees on a logarithmic scale, rather than a linear or arithmetic (equal increments) scale. Thus, a logarithmic scale is more valid because most pathogens multiply at a logarithmic rate while time advances at a linear rate.

Extent of disease loss is directly related to value or importance of a crop and disease loss percentages should be adjusted to reflect that value. McCallan (1946) devised an *index of importance* for a given disease loss derived from the log of disease loss in percent times the log value of the crop in units of $100,000. For example, apple scab causing an average of 6.8% loss in a $95,000,000 apple crop (950 units) would have an index of importance of 2.5 [0.83 (the log of 6.8) times 2.98 (the log of 950)]. A crop having only a total value of $2,000,000 (20 units) would require an average of 85% loss from disease to have an equivalent index [1.93 (the log of 85) × 1.3 (the log of 20) = 2.5]. Nonetheless, McCallan recognized that his system has two major limitations: (1) loss figures are usually only estimates and therefore of uncertain accuracy and reliability, and (2) the method does not distinguish between endemic and epidemic diseases.

Remote sensing uses photography in some form to assess losses. It utilizes film sensitive to infrared wavelengths reflected from plants stressed for some reason. For example, potato plants wilting from insufficient water, excess fertilizer, or Verticillium wilt will show up in the same way on infrared film. The photographs show the extent and outline of areas in which plants are affected. A ground check is usually necessary to determine the cause of a problem, and extent of damage.

Remote sensing is a useful monitoring program, but it cannot by itself reveal plant disease losses. However, remote sensing data can be incorporated into computer programs that can yield disease loss estimates.

Indirect methods include a number of techniques developed to better appraise the damage caused by diseases. Very often these are techniques originally used for some other purpose but have been adapted for detecting and measuring disease losses.

Spore counts are often used to forecast occurrence of diseases such as apple scab, early blight of potato, and onion leaf blight but cumulative spore counts of some cereal rust fungi are related to the amount of disease. However, as a rule, counting rust pustules or rust lesions is a more reliable measure of the amount of disease than are spore counts.

Accurate sampling is an essential part of establishing disease losses. The fewer samples needed to obtain an accurate and reliable estimate the more efficient the procedure. The number of samples required when disease is very high or very low will probably be lower than when disease is in a mid-range but in most surveys a fixed number of samples is taken. This may not be efficient and where possible the number of samples taken should not be excessive.

Sequential sampling is one way to allow adequacy of the number of samples to be determined as sampling proceeds. This technique has long been used to estimate insect populations and other types of surveys but its potential for plant diseases has not been fully employed.

Fungicides are used to prevent disease development and thus allow comparisons of sprayed and unsprayed plots. Rust-free plots are routinely maintained in Washington state by using fungicides thus allowing a comparison of disease-free yields with those in unsprayed plots. In 1980, comparisons were made of yields from paired sprayed and unsprayed plots of 16 winter wheat varieties at two locations. Average yield for all varieties was 25% less in unsprayed plots than in sprayed plots. There was, however, great variation in yields among varieties with varying levels of genetic resistance.

Soil fumigation is used to measure reduction of growth and yield from soilborne diseases. However, increased yields in fumigated plots may in part reflect the growth response commonly seen following fumigation, even where no disease is present. This growth response may result from elimination of minor pathogens that otherwise go unnoticed or from release of nutrients by dead soil microorganisms.

Isogenic lines are plants genetically identical in all respects except for a single variable such as susceptibility (or resistance) to disease. Isogenic lines can be used to measure effects of a particular disease on yield by eliminating variance from genetically controlled factors. Varieties that differ in resistance can be used in the absence of isogenic lines even though the amount of inherent variability will be greater.

Diseases may be simulated in a number of ways. For example, missing plants may simulate plants eliminated by disease or foliage may be removed mechanically to simulate defoliation by disease. These simulated effects may not exactly mimic the intended disease and are not always truly comparable.

Comparisons can be made between yields of disease-free planting stock with those infected by seedborne or tuberborne pathogens, but care must by taken that only the pathogen under study is present and that the disease is not transmitted to disease-free plants thereby compromising the comparison. Comparisons of yields from plants growing in infested soil with those in noninfested soil can also produce disease loss estimates as long as conditions are favorable for disease to develop. This technique is most feasible in contained areas where danger of contaminating soil with an unwanted pathogen is minimal.

REFERENCES

Anon., Forest Health Through Silviculture and Integrated Pest Management: A Strategic Plan. *USDA Forest Service* 576–488 (1988).

*Bissonnette, H. L., and H. G. Johnson, Pinto Bean Rust Control with Fungicides. *Univ. Minnesota Ag. Ext. Plant Pathology Fact Sheet* No. 20, 1972.

Chester, K. S., Plant disease losses: Their appraisal and interpretation. *Plant Dis. Reporter Suppl.* 193:196–347 (1950).

*Chiarrapa, L., *Crop Loss Assessment Methods.* Supplement 3. FAO/ Commonwealth Agric. Bur., Farnham Royal (UK), 1971.

*Cook, R. J., Editorial: Use of the term "crop loss." *Plant Dis.* 69:95 (1985).

*Cook, R. J., and J. E. King, Loss caused by cereal diseases and the economics of fungicidal control. In Wood, R. K. S., and G. J. Jellis (eds.), *Plant Diseases: Infection, Damage and Loss.* Blackwell, London, 1984. Pp. 237–245.

Cramer, H. H., Plant protection and world crop production. *Pflanzenschutz Nachrichten* 20:1–524 (1967).

Hepting, G. H., and G. M. Jemison, Forest protection against destructive agencies. In *Timber Resource Review.* USDA Forest Service, Chap. 4, Sect. A, 1955. Pp 1–69.

Horsfall, J. G., and R. W. Barratt, An improved grading system for measuring plant disease. Phytopathology (Abst.) 35:655 (1945).

*James, W. C., An illustrated series of assessment keys for plant diseases: Their preparation and usage. *Can. Plant Dis. Surv.* 51:39–65 (1971).

*James, W. C., Assessment of plant diseases and losses. *Annu. Rev. Phytopathol.* 12:27–48 (1974).

James, W. C., P. S. Teng, and F. W. Nutter, Estimated losses of crops from plant pathogens. In Pimentel, D. (ed.) *CRC Handbook of Pest Management in Agriculture*, Vol. 1, 2nd ed. CRC Press, Boca Raton, FL, 1991. Pp 15–53.

*Large, E. C., Measuring plant disease. *Annu. Rev. Phytopathol.* 4:9–28 (1966).

*Line, R. F., Losses caused by rust. *Am. Phytopathol. Soc. Fungicide-Nematicide Tests*, 36–44 (1980-1988).

Lyman, G. R., The relation of phytopathologists to plant disease survey work. *Phytopathology* 8:219–228 (1918).

*Maloy, O. C., and W. H. Burkholder, Some effects of crop rotation on the Fusarium root rot of bean. *Phytopathology* 49:583–587 (1959).

*Maloy, O. C., and F. R. Matthews, Southern cone rust: distribution and control. *Plant Dis. Rept.* 44:36–39 (1960).

McCallan, S. E. A., Outstanding diseases of agricultural crops and uses of fungicides in the United States. *Contr. Boyce Thompson Inst.* 14:105–115 (1946).

Meinecke, E. P., The evaluation of loss from killing diseases in the young forest. *J. For.* 26:283–298 (1928).

Schieber, E., Economic impact of coffee rust in Latin America. *Annu. Rev. Phytopathol.* 10:491–510 (1972).

Zadoks, J. C., et al., A decimal code for the growth stages of wheat. *Weed Res.* 14:415–421 (1974).

* Publications marked with as asterisk (*) are general references not cited in the text.

4 Disease Development

Understanding how disease develops is important in developing control measures. The study of plant disease development is called "epidemiology", although this word is derived from epidemic, which refers to outbreaks of disease in humans. Some plant pathologists, especially Whetzel, argue that "epiphytotic" should be used for disease in plant populations, but this term has not been wholly accepted and most readers understand the context in which epidemic is used for major outbreaks of plant diseases.

Diseases may be *endemic* and occur more or less generally in a population in the same area, year after year. An *epidemic* occurs when a disease develops sporadically in relatively high and damaging proportions. Diseases such as Fusarium root rot of beans caused by soilborne pathogens are usually endemic but can become epidemic under some conditions. Epidemic diseases are more common in rusts, powdery mildews, and other diseases in which the causal agent is airborne. Although there are many plant diseases caused by noninfectious agents (e.g., nutrient deficiencies, toxic chemicals, genetic malformations, and unfavorable climate) these are not included in this book because their diagnosis and treatment usually involve disciplines other than plant pathology.

Disease development takes place at two levels. Development in the individual plant follows a cycle that is basically the same for all infectious diseases regardless of whether the disease is endemic or epidemic. Progressive disease development in a plant population can lead to an epidemic. Understanding disease development at both levels is important in effective disease control.

For any disease to develop, and particularly for those diseases caused by some infectious agent (a fungus, bacterium, virus, nematode, or seed-producing parasite) at least three components (= ingredients or factors) must be present. These are (1) a susceptible host plant, (2) a causal agent (the pathogen), and (3) a favorable environment. The interaction of these components of disease has been conceptualized or illustrated in several ways. The most prevalent has been the disease triangle, but other symbols with fixed components (three-legged stool, tripod, pyramid, etc.) have been used. These representations are unsatisfactory because they imply inflexibility of the three components, yet each is extremely flexible and variable. Two highly variable biological systems (host plant and pathogen) intimately interact in response to a highly variable physical system (the environment). This interaction can be illustrated by three overlapping circles representing the three components with the triangular area in the center representing disease. Each circle represents one of the three components (Figure 4.1A). The circles are free to move in any direction. Only when they overlap does disease develop. If all

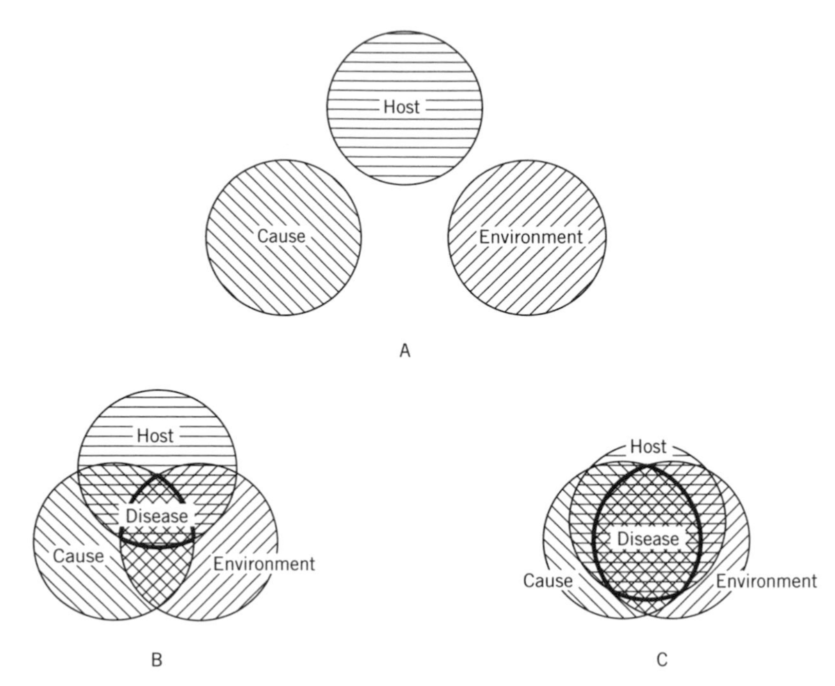

Figure 4.1. Interaction of host, causal agent, and environment in disease development. (A) No interaction = no disease; (B) slight interaction = mild disease; (C) much interaction = severe disease.

conditions are ideal, the circles will be almost one on top of the other (Figure 4.1C). This would represent a very severe disease situation as sometimes occurs in epidemics. If conditions are average or moderate, only small portions of each circle would overlap (Figure 4.lB) and disease would be limited. If one or more of the components is not present under otherwise favorable circumstances, the circles do not overlap (Figure 4.1A) and no disease occurs.

This example has two advantages over fixed or rigid symbols. First it allows for varying extent of disease. Second it shows that the three components can interact in pairs without involving the third so that even though disease does not develop there may be an ecological influence on either host plant or causal agent. This concept illustrates that a disease is not a tangible object but is a process resulting from the interaction of a pathogen on a host plant. It is equally important to recognize that this interaction must take place under favorable conditions for disease to occur. The basic principles of disease control are also conveyed since eliminating or negating any one of the three components prevents disease development. Using resistant varieties or nonsusceptible species, eliminating causal agents with chemicals or other protective barriers, and modifying the environments to be less conducive to disease development are practical means of disease control.

Each of the three main components of disease consists of many variables. The host plant varies in its heritable susceptibility to disease. It also serves as a source of inoculum that can infect other plants . Host age, vigor, and stage of growth may influence disease development. The host, by shading the soil or removing nutrients, may modify the environment. Similarly, the host, by acting as a genetic screen, can modify a pathogen and give rise to new races and strains. The population or density of plants may determine whether a disease remains minor or explodes to epidemic proportions. Even the rootstocks of grafted plants may influence disease development in the top.

The causal agent, if a living organism, may vary in pathogenicity, which is its ability to cause disease, or in its virulence, which is its relative aggressiveness as a pathogen. The length of time a pathogen survives away from its host and its ability to reproduce are important to its survival and subsequent development of disease.

Host plants vary in their genetic makeup and so can causal agents. Pathogenic variation is common in some pathogenic fungi, as in rusts and smuts. The quantity or mass that a pathogen produces may be a critical factor in its ability to cause disease. The Armillaria root rot fungus must build up on dead substrates such as roots and stumps before it can invade living roots. Sometimes associations of certain organisms produce different or more serious disease than can be produced by either alone.

The environment consists of many components but temperature and moisture are probably the two most consistently influential factors.

All of these components can relate to disease control, some more than others. Very often the means of removing these influences are the tactics of disease control.

INFLUENCES OF THE HOST PLANT

Hosts influence disease development in several ways. Probably the most important is the relative susceptibility of a host to disease: how readily it becomes infected, how extensively it is colonized by a pathogen, and how rapidly and severely symptoms and signs of disease develop. Some plants are infected and die very soon after onset of disease while others tolerate pathogens and diseases without apparent damage.

An infected host also becomes a source of inoculum for successive waves of infection. This is especially damaging in polycyclic diseases that increase exponentially in a cropping cycle as a result of large numbers of infective propagules produced on relatively few plants.

Age of the host may directly influence severity of a disease. Wheat plants may be susceptible to stripe rust as seedlings but develop resistance as they mature. In strawbreaker foot rot, wheat seedlings are more resistant than older plants. Severity of choke disease in grasses increases as plants age, whereas apples become increasingly resistant to apple scab infection as they mature.

Vigor may also influence disease severity. Some pathogens such as obligate parasites like rust and powdery mildew fungi or some nonobligate parasites like the fire blight bacterium or apple scab fungus occur more regularly on young vigorous growth than on older, senescent, or low vigor tissues. Some organisms such as wood decay fungi invade only physiologically inactive tissues. Organisms such as the gray mold fungus or soft rot bacteria that are usually saprophytes can attack only tissues weakened from some cause. Obligate parasites (rusts, etc.) are organisms that require live host tissue to complete their development. Facultative saprophytes (e.g., apple scab fungus) spend most of their active life in live plant tissue but can develop to a limited extent in dead tissue. Facultative parasites (e.g., soft rot bacteria) usually live on dead plant material but under some circumstances can invade living tissue.

Stage of growth at which a plant is subject to disease can also vary widely. Using wheat smuts as examples we see that common smut and flag smut fungi infect soon after seed germination (i.e., seedling infection). The dwarf smut fungus infects after the plant has tillered but before heading (shoot infection). The loose smut fungus infects only during flowering (embryo infection).

The host can modify a pathogen by serving as a genetic screen allowing those strains of the pathogen that can infect, colonize, and reproduce in that plant to survive and multiply. The host thus serves as a source of pathogenic (physiologic) races of pathogens.

Plants can, individually or collectively, modify the environment by shading soil, thereby lowering surface and soil temperatures, reducing air movement, and hence raising relative humidity. Plants also remove nutrients and moisture from soil, which may in turn alter its ability to resist or tolerate a pathogen or disease.

Composition of plant populations may inluence disease development. In nature most plant populations are mixtures of plants where diseases rarely reach epidemic proportions but maintain more or less low and constant levels. Agriculture, however, frequently involves monoculture where the same crop plant is grown over a wide area. This practice has resulted in some severe epidemics. If the particular plant is highly susceptible, or often even if it is not, there are no gaps in the host plant continuum to break or slow the progress of a disease. Plant density may further compound effects of monoculture or even species mixtures. Widely spaced (less dense) plants allow more aeration and light. This not only results in warmer, drier microclimates around plant foliage that are less favorable for pathogen development, but also enhances growth and vigor, resulting in good plant health.

Root–scion combinations may have an effect on development of certain diseases. In the apple industry it is common practice to graft or bud onto rootstocks that regulate size of mature trees to produce dwarf, semidwarf, or standard trees. Certain of these rootstocks tolerate Phytophthora root rots much more than others.

INFLUENCE OF THE PATHOGEN

Causal agents have certain intrinsic properties that influence their success as pathogens and thereby increase disease severity. The first is pathogenicity or the ability to cause disease. Although this is virtually requisite for plant pathogens there are a few true saprophytes that live only on dead or moribund tissues. Perhaps best known of these are the wood decay fungi that are sometimes called perthophytes rather than parasites or saprophytes because they live on dead portions of living plants. Pathogenicity is a qualitative trait; either an organism is pathogenic or it is not.

Along with pathogenicity a causal agent is endowed with virulence, its relative ability to infect a plant. Some pathogens are highly virulent and cause severe disease in a host; others are mildly virulent (i.e., weak parasites) and cause only mild disease.

Longevity of a pathogen, both in its active state or in survival units, is a major factor in its ability to cause disease. Organisms such as root rot and wood decay fungi live for many years in an active relationship with their hosts. Others such as cyst nematodes and fungi that form sclerotia can survive for many years completely inactive. Most plant pathogenic bacteria have short life spans but rely on rapid reproductive rates to survive. A few, like the potato ring rot bacterium, can survive for several years in a desiccated state. The time that infective units can survive varies greatly with conditions but may be only a few minutes for basidiospores of some rust fungi or 15–20 years for sclerotia of the onion white rot fungus, 28 years in dry storage for the wheat gall nematode, and 10 for golden nematode cysts.

The reproductive capacity of a pathogen has a direct bearing on the rate at which disease epidemics develop. A fungus such as a rust that produces spores exponentially and where one spore can initiate an infection that will produce a thousand spores increases at a much faster rate than one such as the onion white rot fungus that produces no spores. The first fungus initiates a polycyclic ("compound interest") disease and the latter initiates a monocyclic ("simple interest") disease.

Some pathogenic fungi produce high numbers of propagules. A uredial pustule of the leaf rust fungus, for example, produces about 2000 spores a day for 2 weeks. It has been projected that 100 trillion urediospores are produced from an acre of heavily infected wheat in 2 weeks. A single apothecium of the white mold fungus can produce 30 million ascospores and a single white mold infected bean plant can give rise to many sclerotia, each producing one to several apothecia. The Indian paint fungus, which causes a decay in true firs and hemlock, produces about 25,000 spores per square millimeter of sporophore surface per week at the peak of sporulation. A single common smut ball contains 8-12 million spores and spore trapping studies have shown that 2 million spores may fall on a square foot of soil one-quarter of a mile from the source. A single gall of cedar-apple rust 1.75 inches in diameter will produce about 7.5 billion basidiospores. The cherry leaf spot fungus produces as many as 200 to 500 ascocarps per square inch of leaf surface, each has hundreds of asci, each of which contains 8 ascospores. A single perithecium of the apple scab fungus can have as many as 240 (usually 50–100) asci yielding a total of 400 to 2000 ascospores per perithecium. And there are many perithecia in each diseased leaf.

Bacteria also reach very high numbers. If a bacterial cell divides every 30 minutes, a single bacterium can produce about 281.5 trillion cells in 24 hours (2 to the 48th power). A milliliter of apple nectar can hold as many as 800 million fire blight bacteria. Although the numbers are huge very few of the individual propagules will cause infection.

Host range, the number of different plants that a certain pathogen can colonize, greatly affects its ability to survive. This in turn influences its management. The wider the host range of a pathogen, the more difficult it is to eliminate, or even greatly reduce its population. Some pathogens such as *Botrytis cinerea*, *Sclerotinia sclerotiorum*, *Phymatotrichum omnivorum*, *Pseudomonas solanacearum*, *Rhizoctonia solani*, *Verticillium dahliae*, *Pratylenchus penetrans*, and *Erwinia carotovora*, to name a few, can attack hundreds of hosts. At the other extreme many obligate parasites like the chrysanthemum rust fungus attack only one single host.

It is important to distinguish between infection units and dispersal units because they may or may not be the same structure. Many spores such as rust urediospores or conidia of many fungi are both dispersal units and infection units. A zoosporangium is usually a dispersal unit that produces swarm spores (zoospores) that are the infection units but sometimes sporangia germinate directly forming a germ tube that functions as an infection unit. Many sclerotia such as those produced by *Sclerotinia sclerotiorum* and *Typhula* species are survival units and in turn produce spores that are both dispersal and infection units. Sclerotia of fungi such as *Rhizoctonia* and *Sclerotium* germinate directly to produce infection hyphae and thus serve as both dispersal and infection units. Nematode cysts and eggs are dispersal units while larvae are the infection units.

Variation is a common if not a standard characteristic of plant pathogens. Pathogens vary in many ways, such as color, growth rate, and reproductive capacity, but with respect to disease control, variation in pathogenic races is of primary concern. Proliferation of pathogenic races is particularly common in rust and smut fungi but occurs in most pathogens, especially where a sexual stage provides the means for genetic recombination.

A second type of variation becoming increasingly important is resistance to pesticides. Development of resistance to fungicides and ways to minimize its effects are discussed in Chapter 13.

A certain quantity, or "critical mass," of a pathogen sometimes is required for infection. This is particularly apparent with fungi such as *Armillaria mellea* where the fungus requires a food base on which to build before it can attack healthy roots. The same applies to some bacteria that require a certain population of cells to initiate disease.

Mobility of pathogens has a great effect on the distances they can spread. Most pathogens are passively dispersed and their movement depends on wind, water, insects, or other dispersal agents, including man. Some are actively motile and can direct their dispersal to some extent, although the distance moved may be relatively short. For example, nematodes can migrate through soil for short distances but long distance spread is via infested soil, water, or plant parts. Dwarf mistletoe seeds are actively projected to new infection sites and may be shot horizontally for distances of 100 ft (30 m), but dispersal is random and not directional. Zoospores of some phycomycetes are motile and may be attracted by root exudates, but the distance moved is not more than a few centimeters. Fungal spores are sometimes classed as dry spores, which are generally carried by air currents, and wet spores, which are usually spread by splashing or running water. The former are usually spread long distances whereas the latter spread only short distances from their point of origin. Viruses require some means of dispersal and this is often via insects with piercing-sucking mouthparts, primarily aphids and leafhoppers. But throughout history the greatest disperser of pathogens over long distances is man by the movement of infected or infested plants, plant parts, seeds, or soil.

In some diseases the cause is an association of organisms rather than a single pathogen, even though one organism may predominate. Such associations mostly occur in soilborne diseases where the environment favors microbial interaction. Verticillium wilt of peppermint is often more severe when parasitic plant nematodes such as root-lesion or root-knot nematodes are present. Their effect is indirect and affect physiology of the host. A number of root-rot complexes such as black root rot of tobacco and strawberry involve root fungi and nematodes.

INFLUENCE OF THE ENVIRONMENT

Temperature and moisture are undoubtedly the most important of the many environmental factors that influence disease development, and are often interrelated. These are generally the two parameters used in disease forecasting systems.

Most plant pathogens are active at temperatures ranging from just above freezing to about 45°C which is the upper limit for most living organisms. The thermal death point for many fungi is in the range of 40–60°C, but only a few specialized fungi can survive at the higher temperatures. The majority of plant disease organisms do best at 5 to 30°C. A few such as the various snow mold fungi (*Typhula* spp., *Fusarium nivale*, *Herpotrichia nigra*, etc.) do best at temperatures just above freezing under an insulating cover of snow. Pathogens such as *Pseudomonas solanacearum* and *Sclerotium rolfsii* have temperature optima of about 35°C and occur largely in warm climates. *Rhizina undulata*, an ascomycete that causes a root rot of conifer seedlings has ascospores that are activated by temperatures of 35 to 45°C. This fungus can cause serious disease losses in areas where logging debris has burned at high temperatures.

Moisture may be available in a variety of forms and influences many diseases. The most frequent form is free water on plant surfaces. A film of free water is often the critical limiting factor for spore germination, bacterial invasion, and zoospore formation, motility, and infection. The length of time free moisture is on a plant surface is an important component of many disease forecasting systems. Moisture may also be present as vapor and is generally expressed as relative humidity. Some pathogens, particularly powdery mildews, are especially responsive to high relative humidity but are damaged by free moisture. Soil moisture may influence disease development. Diseases such as take-all of cereals or Cephalosporium stripe of wheat are more severe where soil moisture is plentiful. This is also true of root diseases caused by phycomycetes because motile zoospores need moisture. Other diseases such as dryland root rot of cereals are favored by dry soils.

There are many other environmental factors that influence one or more diseases to some extent. Some such as elevation, aspect, or exposure may alter some other factor such as temperature or moisture and thus have only an indirect effect.

Oxygen is required by all plant pathogenic organisms except for a few anaerobic bacteria of uncertain pathogenic status such as those associated with Clostridium rot in potato and slime flux of deciduous trees. Oxygen is a limiting factor only in the soil and in aquatic environments. Most soilborne organisms such as root rot and decay fungi are most active in the upper layers of soil, generally the upper 12 inches, where oxygen is readily available. This is easily seen by damping-off of seedlings or the decay of fence posts or poles that rot off at or just below the soil surface.

Air movement (ventilation) is important, primarily in its effect on moisture in the air space around plant parts. Increasing air movement by pruning and thinning, plant spacing and orientation, or greenhouse construction is an important technique for controlling foliar diseases.

Soil pH has a strong influence on many soilborne diseases. Common scab of potatoes and club root of crucifers are two classical examples of diseases that have been controlled by modifying soil pH. Common scab is most severe in alkaline soils and practices such as burning debris on a garden spot or adding wood ashes to soil intensify this disease. Applications of sulfur have been made to soil to lower soil pH and reduce damage from scab. Club root is favored by acid soils and finely ground lime is mixed with soil to raise pH and minimize losses.

Soil fertility influences plant disease in several ways. First are nutrient deficiencies or toxicities. Deficiencies of nitrogen, iron, zinc, sulfur, and other essential elements may result in various noninfectious diseases of many crops. Under extreme conditions as in very acid soils, elements such as aluminum and manganese can be toxic to certain crops. The second is an indirect effect on susceptibility to disease. Potassium and phosphorus are elements that increase resistance to disease by hastening tissue maturation and hardening cell walls. Third, some soil nutrients can alter soil pH and have an indirect effect on disease severity. For example, nitrate forms of nitrogen raise soil pH and increase severity of diseases such as take-all of wheat whereas the ammonium form lowers soil pH and reduces severity. The two forms of nitrogen mentioned have the opposite effect on Cephalosporium stripe of wheat, which is more severe at low pH values.

Light influences pathogen activity such as spore or seed production and dispersal, activity of insect vectors, and symptom expression. Light intensity can be modified by pruning and thinning to open the plant canopy, wider spacing of plants, or establishing shade of some sort. Light, in turn, may alter temperature and moisture levels.

Insects and mites may be direct causes of plant diseases, for example, the toxic reactions induced by feeding of aphids, leafhoppers, and some true bugs, or the galls produced by a number of insects and mites. These arthropods also can weaken plants and make them more vulnerable to minor pathogens. But the most important role of insects and mites is as dispersal agents for pathogens, be they fungi, bacteria, viruses, or even nematodes. Insects have their greatest influence on diseases as vectors of viruses. Frequently the primary means of controlling some plant diseases is to control the insect vectors, a responsibility generally left to entomologists.

Higher animals, including mammals and birds, may affect plant disease development by injuring plants, thereby creating infection courts through which pathogens enter plants, or weakening them to the extent that they become more vulnerable. These animals also serve as important long-range disseminators of plant pathogens.

Cultivation and other farming operations alter the environment by eliminating weeds that harbor pathogens, loosening soil to increase aeration and oxygen content, improving drainage, and reducing compaction. Cultivation, on the other hand, can injure plant roots, creating infection courts for root pathogens, it can cause dirting around the bases of plants and create moisture pockets in which stem fungi are active, and travel of heavy farm machinery can increase soil compaction. Cultivation also can bury infested plant residues and decrease the amount of available inoculum.

THE DISEASE CYCLE

The disease cycle has active and inactive phases. In the active phase there are three stages, inoculation, incubation, and infection, leading to disease. The disease cycle parallels the life cycle of a pathogen and with some obligate parasites, such as rust, smut, and powdery mildew fungi, and dwarf mistletoes, it is impossible to separate the two cycles.

Although plant pathologists have long recognized the importance of the disease cycle, Gäumann was one of the first to critically analyze the many steps involved in the infection process. He used the term "infection chain" to include the sequence of events—infection, colonization, sporulation, dispersal, and so on—that leads to establishment of disease. This infection chain consists of many components or links (Table 4.1). These are the major components of a disease cycle that apply to all pathogens. Also given are some equivalent terms for fungal pathogens. The major terms apply to any parasitic disease, but the equivalent terms will vary depending on whether the pathogen is a fungus, bacterium, virus, or nematode.

There is a wide variety of such components when one considers the range of plant pathogenic organisms, and the disease cycle becomes more complex when these specific components are inserted. Control measures are directed at weak links in the infection chain, that is, vulnerable points in the disease cycle. Figure 4.2 illustrates where different control principles or strategies are applied at various points in a disease cycle.

There may be only one primary disease cycle in a given cropping sequence (year, crop, generation, or other growing period). Diseases with a single cycle are called monocyclic or "simple interest." They increase arithmetically (i.e., linearly).

TABLE 4.1. Major Components of the Disease Cycle and Equivalent Terms for a Fungal Pathogen

Components of the Disease Cycle	Terms for Fungal Pathogens
Initial inoculum	Sporulation
Quantity	Sporophore production
Survival	Spore production
Dispersal	Spore maturation
Inoculation	Dissemination
Duration	Spore liberation
Distance	Spore dispersal
	Spore deposition
Infection process	Germination
Virulence	Penetration
Duration	Colonization
Pathogen reproduction	
Rate (frequency)	
Quantity	
Secondary cycles	

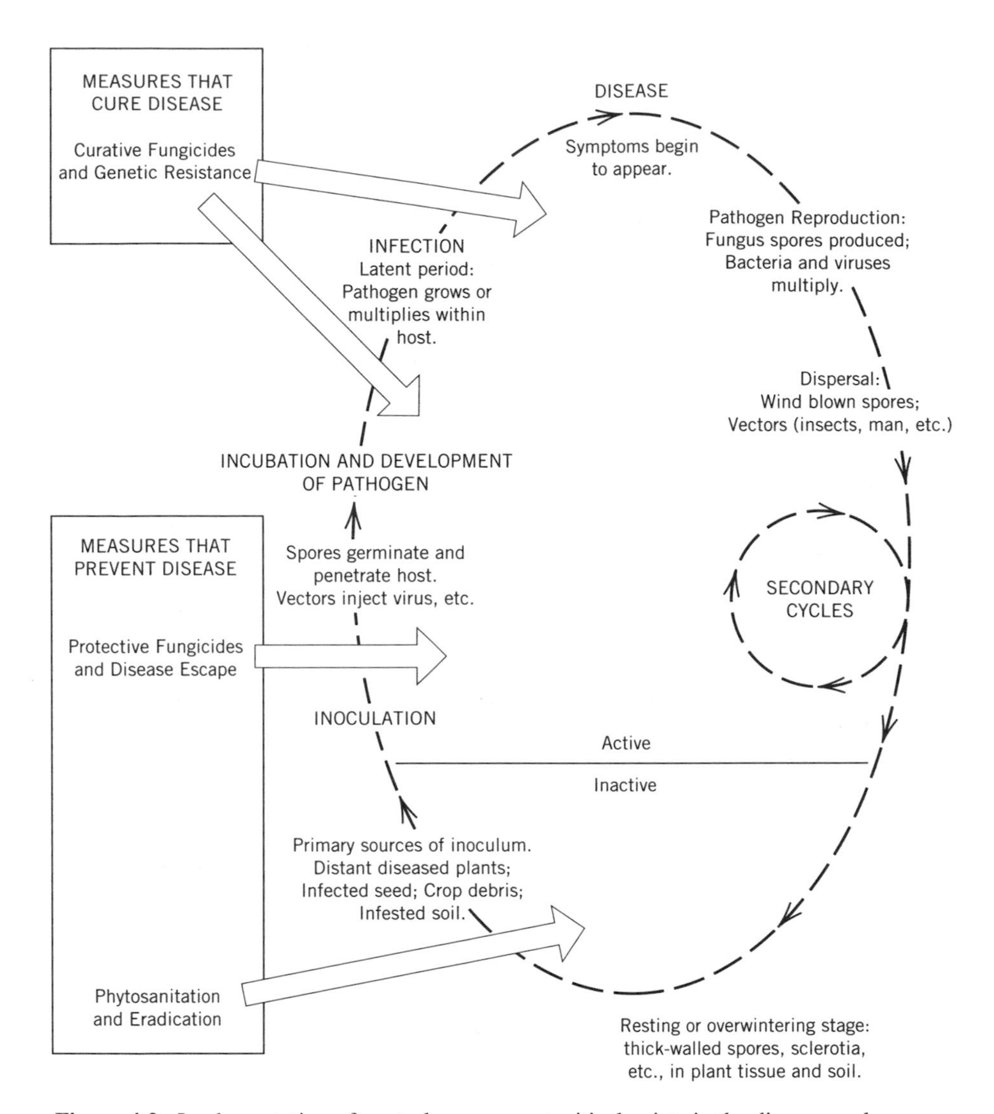

Figure 4.2. Implementation of control measures at critical points in the disease cycle.

Conversely, there may be one or more secondary cycles, each involving the same three stages of activity as the primary cycle during the cropping sequence. These are polycyclic or "compound interest" diseases. They increase exponentially at an accelerating rate demonstrated by a sharp upward curving line of progress.

The theory and mathematics of plant disease epidemiology was brought to prominence by Van der Plank in several books starting in 1963 and more recently in 1975. There are now many books on this subject, a few of which are listed at the end of this chapter and readers are referred to these for a detailed explanation of this complex topic. Below is given only a very simplified concept of plant disease progress curves.

The slope of the curve, that is, the progress of disease, is a function of the amount of initial inoculum (y) and the rate of infection (r) (r = the number of plants infected per unit of time). Control aims at reducing either y (by sorting and elimination, quarantines, eradicants, crop rotation, residue management, etc.) or r (by chemical protectants, cultural practices, resistance, etc.).

We can use two exagerated examples to illustrate the development of disease from initial inoculum (y) and rate (r) over time (in any unit such as weeks, months, years). The first graph (Figure 4.3A) shows the progress of disease (or accumulation of capital)

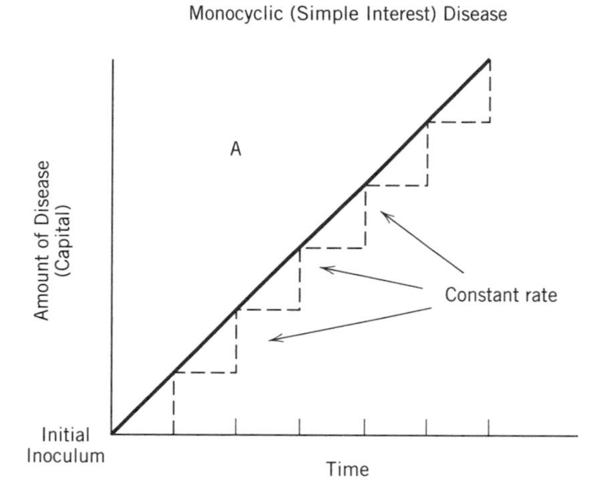

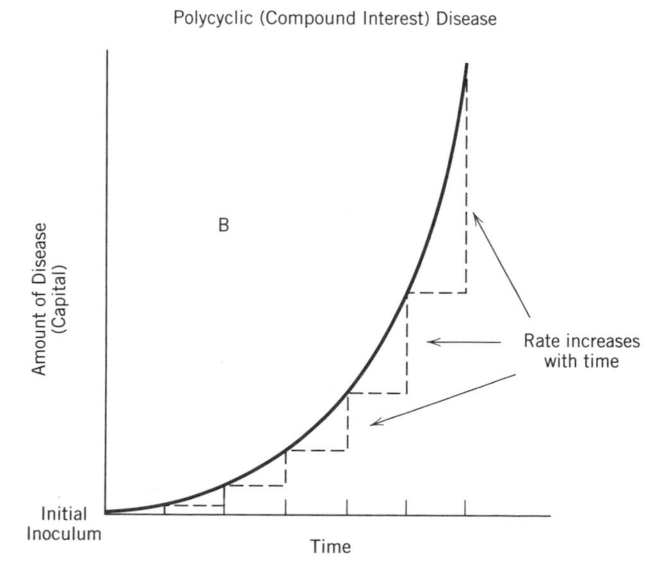

Figure 4.3. Disease progress curves expressed as accumulation of capital. (A) Monocyclic disease increase. (B) Polycyclic disease increase.

for a monocyclic disease where the accumulated increments (i.e., rate of infection or proportion of diseased plants) remains the same throughout the course of the disease cycle. The second graph (Figure 4.3B) illustrates the progress of a polycyclic disease where the accumulated increments increase with time as the rate of plants infected per unit of time increases.

As an example of disease development assume that a root rot pathogen is present at a low level (e.g., initial inoculum level is capable of infecting 5% of the plants) and an additional 5% of the plants are infected per unit of time (each infection cycle) to give a 5% rate of infection (slope a in Figure 4.4A). If initial inoculum is higher (e.g., 20%) but the rate of infection remains the same (5%) the slope of the line will be the same but of

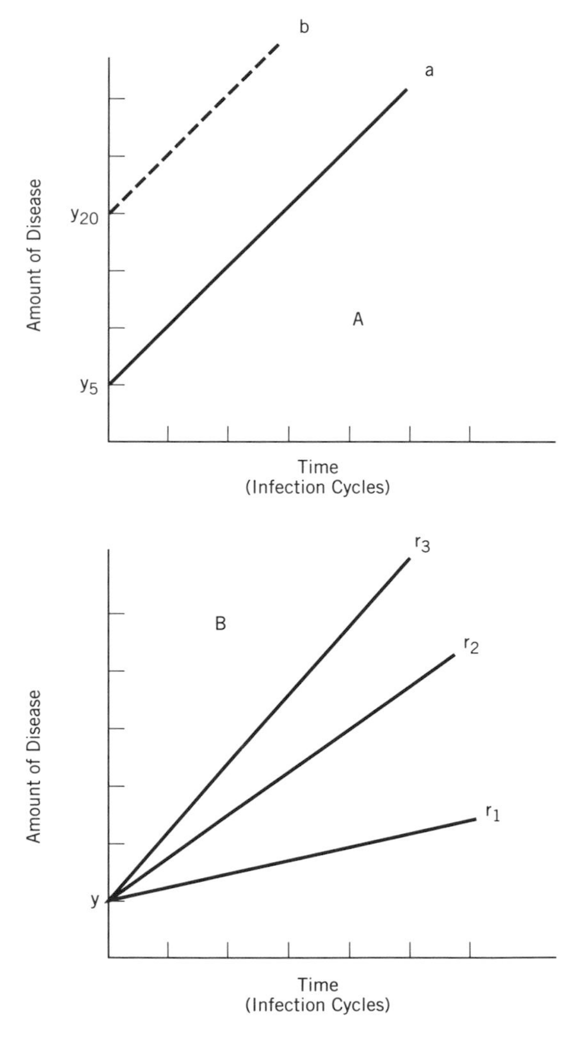

Figure 4.4. Disease progress (logarithmic) when (A) initial inoculum varies but rate of infection is the same; (B) initial inoculum is the same but rate of infection varies.

a higher magnitude (slope b). If the rate of infection changes then the slope of the line changes. For example, if initial inoculum is 5% but infection rates are 1, 5 and 10%, the resulting slopes are different as illustrated by *r*-l, *r*-2, and *r*-3 in Figure 4.4B. In considering these examples it is assumed that inoculum is evenly distributed, conditions are uniformly favorable for infection, and infection progresses at a uniform rate. Actually none of these is likely to occur.

The configeration (shape) of disease progress curves is also influenced by the population in which the disease is measured. If it is a finite population, such as the number of plants in a field, then the disease progress curve can extend only as far as uninfected plants remain. Progress stops when 100% of the plants are infected. However, if the population measured is infinite, such as tree leaves during the growing season where new leaves continue to be produced, the disease can continue until the environment (especially weather) halts disease development and plant growth. There are so many influencing factors in disease development that cannot be measured with any degree of accuracy that disease progress curves should be used only as broad general indicators rather than absolute predictors. Disease forecasting systems basically project disease progress and the above limitations render many of these systems too uncertain to have practical value.

Figure 4.5 illustrates disease progress curves for both monocyclic (A) and polycyclic (B) diseases in relation to stages of the infection process. The main difference

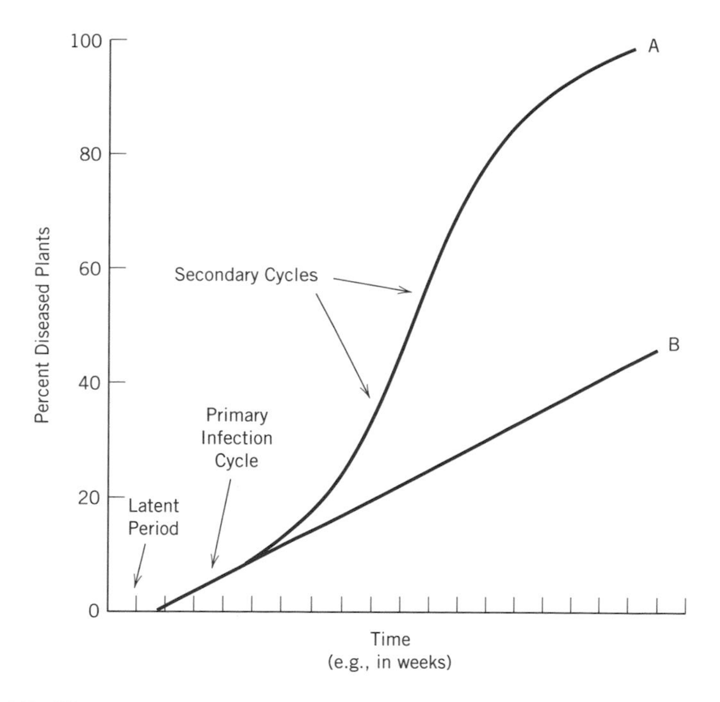

Figure 4.5. Disease progress curves. (A) Rapid disease development due to secondary infection cycles (e.g., a foliar disease); (B) slower, steady development with no secondary cycles (e.g., a soilborne disease). (From Hill and Waller, *Pests and Diseases of Tropical Crops*, Longman. 1982. With permission.)

between the two is the occurrence of secondary cycles. A polycyclic disease starting at a low level of inoculum but doubling in each disease cycle (100% increase each time) will develop rapidly. Polycyclic diseases are typically caused by pathogens with high rates of reproduction and short infection cycles as in rusts, powdery mildews, and many leaf spot and blight fungi.

Examples of Epidemic Development and Major Factors Involved

Suscepts and virulent pathogens are often ubiquitous so that some environmental factor limits disease development. Often this factor is temperature, interacting with moisture. The following diseases demonstrate the influence of temperature on disease epidemics.

In the Pacific Northwest stripe, leaf, and stem rust fungi attack wheat. Urediospores are the repeating stage and are the only spores important in epidemics. The three fungi have different temperature optima for urediospore germination, and hence infection. These optima are 7°C for stripe rust, 19°C for leaf rust, and 20°C for stem rust. Stripe rust can begin activity much earlier in the spring (often being active throughout the winter) than leaf or stem rusts and gets a decided lead. Leaf rust cannot develop until late spring or early summer. Stem rust gets an even later start since, in addition to having higher temperature requirements for urediospore germination, this fungus overwinters in the telial stage and must infect its alternate host, common barberry, and produce aeciospores before it can infect wheat. Figure 4.6 shows development of the three rusts over a cropping year in the Pacific Northwest. Notice that the scale of rust intensity is logrithmic and each increment represents a 10-fold increase.

Stem rust is the major rust on wheat through the central North American wheat growing area where wheat is grown in a continuous belt from northern Mexico into southern Canada. The fungus can overwinter in the southern part of this belt in the

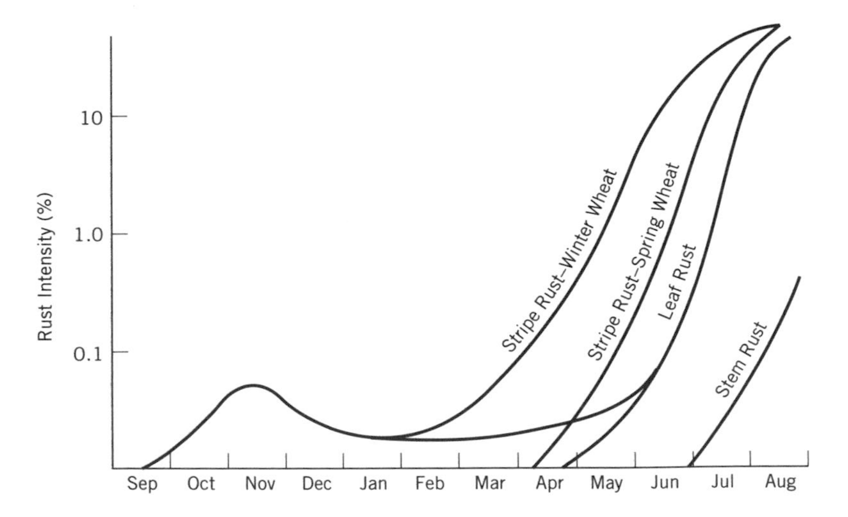

Figure 4.6. Rust development on wheat in the northwestern United States. (With permission of R. Line.)

uredial stage so that primary inoculum is readily available. Overwintering can occur north to the 34th parallel (northern Texas and southern Oklahoma) and urediospores almost always survive in trace amounts to the 30th parallel (southern third of Texas). From this overwintering base the rust moves northward in waves until by the end of June it has reached the Canadian border.

The brown rot fungus of stone fruits can infect most stages of the tree from blossoms and expanding shoots to mature fruit. When moisture is adequate conidia can germinate within 2 hours at 26°C causing high levels of infection in a relatively short time.

Apple scab infection is also a function of temperature but is correlated with free moisture on plant surfaces. Ascospores can germinate in 9 hours at 16–24°C if moisture is present but at higher or lower temperatures longer periods of moisture are required for ascospores to germinate. This relationship is the basis for an apple scab forecasting system discussed in Chapter 5.

Disease Development in Relation to Levels of Plant Organization

Diseases affect plants at various levels of plant organization (Figure 4.7A). Pathogens invade and colonize cells and tissues while altering lower order systems such as plant

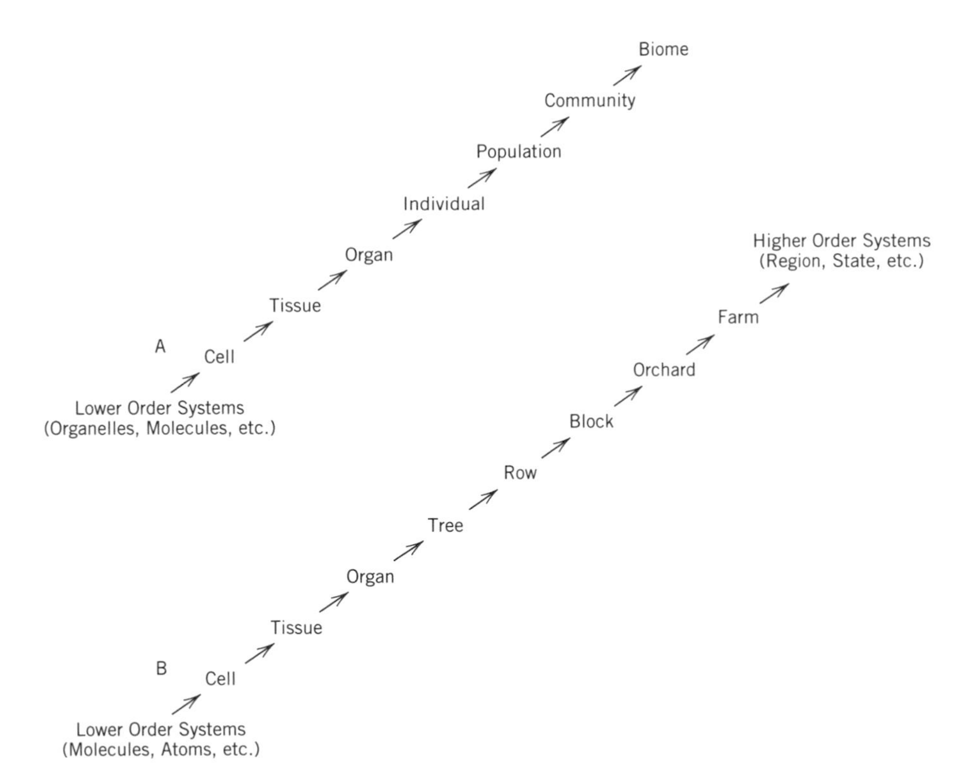

Figure 4.7. Levels of organization in a plant hierarchy (A) and parallel levels in an agricultural system (B).

cytoplasm and other cellular constituents. Diseases develop and are apparent (i.e., recognized as disease) on organs and individual plants but they become economially important (except for ornamentals) mainly in plant populations (fields, orchards, groves, transplant beds, nurseries, etc.). Control is generally directed at plant populations (only occasionally at individual plants) although control measures may have an indirect effect on higher orders such as communities or biomes. Levels of organization can also be expressed in terms of agricultural units (Figure 4.7B). This is where practical disease control occurs.

REFERENCES

*Campbell, C. L., and L. V. Madden, *Introduction to Plant Disease Epidemiology*. John Wiley, New York, 1990.

Hill, D. S., and J. M. Waller, *Pests and Diseases of Tropical Crops*, Vol. 1, *Principles and Methods of Control*. Longman, London, 1982.

*Kranz, J. (ed.), *Epidemics of Plant Diseases*, Springer-Verlag, New York, 1974.

*Leonard, K.J., and W.E. Fry (eds.), *Plant Disease Epidemiology*, Vol. 1, Macmillan, New York, 1986.

*Palti, J., and J. Kranz (eds.), *Comparative Epidemiology—A Tool for Better Disease Management*. Centre for Agr. Publ. and Doc., Wageningen, 1980.

Van der Plank, J. E., *Principles of Plant Infection*. Academic Press, New York, 1975.

*Zadoks, J. C., and R. D. Schein, *Epidemiology and Plant Disease Management*, Oxford University Press, New York, 1979.

*Publications marked with an asterisk (*) are general references not cited in the text.

5 Disease Forecasting

Disease forecasting is the ability to predict when a disease is likely to develop in important amounts before it actually occurs. Such a prediction days or even months before a disease occurs allows growers to develop strategies and take action in a timely and efficient manner. It helps in making management decisions as to what varieties to plant, when to seed, fertilization, irrigation, rotation, and other cultural practices. A negative prediction, moreover, can reduce or eliminate application of unnecessary pesticides. This is economically and environmentally advantageous, especially in this era of concerns over pesticides in the environment.

Disease forecasting is generally based on three criteria.

1. *Available inoculum.* Demonstration or assumption of viable and pathogenic inoculum is the first requirement for disease development in a population of susceptible plants.
2. *Requisite environmental conditions.* Usually these include temperature and/or moisture that favor disease development. Often temperature and moisture are interrelated. For example, low temperatures result in dew formation, which provides free moisture required by some pathogens for spore formation and germination.
3. *Stages of plant growth.* Some plants are susceptible for brief periods of time and others throughout their life cycle. Seem and Russo (1984) have illustrated the interrelationship of pathogen development and host development as it affects decisionmaking by growers (see Figure 5.1). They propose that the period of time the host is susceptible and the overlapping period when the pathogen can cause damage is the timeframe or window during which control options are effective. They call this timeframe the "plant-host timing window."

Four diseases demonstrate the extreme areas that the "plant-host timing window" can enclose. Brown rot of peaches has a very large window, loose smut of barley has a very small one. Those of cone rust of slash pine and fire blight of pear are intermediate.

The brown rot fungus can produce spores (inoculum) throughout the entire season and the host, in this case peach, is susceptible from bud break (blossom and shoot infection) until after harvest (fruit rot) so that the periods of host susceptibility and injurious action by the pathogen are very long. This requires decision making and action over a long period.

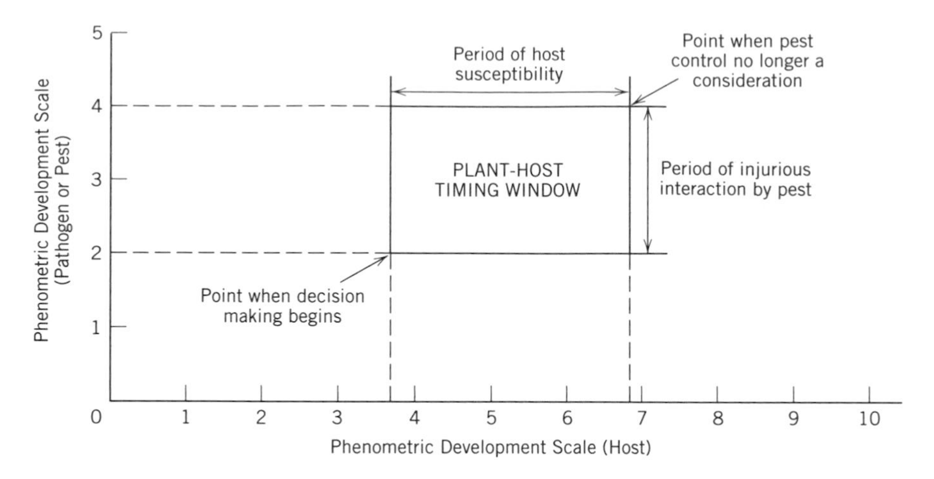

Figure 5.1. Management timing window in plant–pest interactions. (Adapted from: Seem, R. C., and J. M. Russo, *Plant Dis.* 68:656–660)

Basidiospore inoculum of cone rust of slash pine is produced over several months, but female strobili on individual trees are susceptible for only a few days. Strobilus development (stages 1 through 4) takes about 15 days and strobili are most susceptible in stages 1–3 so the infective period is probably no longer than 10 days. Fire blight of pears reverses the length of these phenological occurrences. The pathogen is most active, and infection occurs, when the mean daily temperature is between 60 and 80°F. In some climates this may be only a few weeks. Infection is primarily through the blossoms (specifically through the nectaries) and because of the tendency of pears to produce second or late bloom the period of host susceptibity may extend over several months.

Barley loose smut inoculum is produced by smutted plants at flowering so that inoculum is available for a very short time. Likewise, infection is through open blossoms, only during anthesis; individual plants are thus susceptible for only a short period.

Disease forecasting apparently started about 1923 in Europe correlating rainfall in May with incidence of apple and pear scab. Some 20 years later (1944) W. D. Mills at Cornell University prepared a graph relating temperature and hours of continuous wetting of foliage with levels of apple scab development. This later became known as the Mill's table or scale and is still used today to predict outbreaks of scab.

The earliest organized forecasting system began, not unexpectedly, with the historic disease, late blight of potato. Workers in Holland evolved four rules (the "Dutch rules") on which to base the appearance of late blight. These rules were (1) dew during at least 4 hours at night, that is, night temperatures below the dew point for 4 hours, (2) a minimum temperature of 10°C or above, (3) a mean cloudiness on the next day of 0.8 (80%) or more, and (4) measurable rainfall of at least 0.1 mm during the next 24 hours. When all four conditions prevailed, control measures were recommended.

Modern adaptation of these predictive systems for apple scab and late blight is computerized so that weather parameters are measured and processed electronically to identify infection periods. One such system for predicting potato late blight serving the eastern United States was called BLITECAST. Another is a microcomputer Apple Scab Predictor developed in the north central United States.

Waggoner and Horsfall were among the first to use computers to analyze disease development. They determined factors involved in development of *Alternaria solani*, the cause of early blight of potato and tomato, to model a disease epidemic and called their system EPIDEM. They did not propose it for disease forecasting but, nonetheless, the system contains all the elements of a disease forecasting system.

Development of a disease forecasting system requires knowledge and understanding of several phenomena and the availability of some technology. The primary requirements are as follows:

1. To understand environmental and host influences on the pathogen and disease development. In some cases a few dominant factors (such as leaf wetness and temperature for apple scab) control disease development.
2. Technology to detect, monitor, and analyze pathogen and disease development.
3. Knowledge of pathogen and disease dynamics. Factors influencing inoculum production, dissemination, germination, and other components of the infection chain, infection of, and reproduction in the host, and symptom development, must be understood.

Since the presence of inoculum is a key requisite for disease development all disease forecasts are based to some extent on the assessment of either primary or secondary inoculum. The methods used can be grouped into one of three categories on the basis of the stage in the disease cycle that is evaluated. These categories are (1) source and amount of inoculum, (2) survival of inoculum, and (3) development of inoculum.

An effective forecast combines two or more of these methods. For example, in some areas spore traps are used to detect the initial availability of apple scab inoculum (ascospores). This establishes the starting point of the plant-host timing window. Then, periods of wetness and temperature are used to detect infection periods and determine how long control measures are needed. Japan has developed a very elaborate forecasting system for rice diseases, particularly for blast, which is considered the main rice disease because of its wide distribution and destructiveness under favorable conditions. This system is justified because of the large quantity of chemicals otherwise used for disease control, and the high support price maintained for rice. The purpose of the system is to determine as precisely as possible the optimum time to apply fungicides. Forecasts for blast are based on weather, especially dew formation and wind velocity during spore dispersal. Blast incidence is correlated with the number of spores collected in traps. A formula incorporating dew period and airborne spores is used to develop 5-day forecasts.

One of the major benefits of accurate disease forecasts is the reduction in number of fungicidal sprays needed to control diseases. Control of downy mildew in hopyards in Europe illustrates this. When fungicides are applied on a calendar basis an average of 14 sprays is applied during the year. When based on disease predictions only 8 sprays are needed. Resistant varieties require only four sprays a year. The relative savings in costs are 43% using disease forecasting and 72% combining resistant varieties with disease forecasting.

SOURCE AND AMOUNT OF INOCULUM

Several diseases are predicted by determining presence and quantity of spores, nematodes, viruses, or other inoculum. This is done directly using spore traps, soil sieving to quantify nematodes or fungal sclerotia, plate counts on selective media for bacteria and fungi, or indexing plant material for viruses with indicator plants or serological tests. *Verticillium dahliae* can be detected directly in soil by dispersing a small amount of soil on nonnutrient media such as water agar containing low concentrations of ethanol. Such media restrict the amount of mycelial growth but provide sufficient carbon to allow production of microsclerotia, thus giving a direct measure of the inoculum potential in the soil.

Spore traps are most effective for spores that are readily identified such as *Alternaria*, *Fusarium*, *Helminthosporium*, *Venturia*, some smut spores, and urediospores of rusts. It is less effective for spores of fungi such as *Botrytis* that are less distinctive. Spore traps have been used in Colorado, Idaho, and to a limited extent in Washington to detect the distinct and identifiable spores of the potato early blight fungus. An initial step in predicting leaf rust of wheat in Oklahoma has been to examine plants in late winter and early spring to determine if a supply of inoculum is present. If so, and weather conditions remain favorable through spring, forecasts of rust can be very reliable.

Bioassays are indirect methods for determining presence of inoculum and are most commonly used to detect soilborne pathogens. Before planting fields to potatoes, bioassays are used to determine levels of root-knot nematodes in soil. The technique involves growing indicator plants (tomatoes) in aliquots of field soil in greenhouses and after suitable incubation periods examining the roots for galls. The technique works specifically for *Meloidogyne* species that produce conspicuous galls on roots.

The 1990 discovery of potato virus Y necrotic strain (PVY-N) on Prince Edward Island, Canada, resulted in an intense monitoring program using the highly susceptible tobacco varieties, Samsun and Delgold planted in suspect fields. Absence of symptoms provided evidence that some fields were free of the virus.

Apple powdery mildew can be assayed by cutting samples of dormant shoots in late winter or early spring (before budbreak) and forcing them in greenhouses. The relative amount of mildewed shoots indicates the potential for disease in an orchard and the grower can prepare to spray to prevent development of secondary inoculum (conidia).

Pea root rot, especially Fusarium root rot, is evaluated by growing peas in field soil in greenhouses and indexing the plants after a suitable time. Indexing is also used to measure the amount of inoculum carried in vegetatively propagated plants like potatoes. The Washington State University Potato Seed Lot Trials are used to demonstrate amounts of various pathogens like the leaf roll virus and ring rot bacterium in seed potato stocks. However, since these lots comprise the potato seed already planted by growers, the information is of little use for that season but can indicate sources of low (and high) infection.

A variety of forecasting systems has been tested for bacterial leaf blight of rice in Japan. An indirect bioassay system to determine bacteriophage populations is the most advanced and usually quite accurate. It is based on samples of irrigation water added to cultures of sensitive bacteria on an agar medium. Plaque counts, where the bacteria have been dissolved by the phage, reflect populations of the bacteriophage in the water and are related to presence of host bacteria. Low plaque counts represent low pathogen populations.

SURVIVAL OF INOCULUM

Some pathogens are pressed to survive unfavorable weather, especially low temperatures. Stewart's wilt of corn is caused by the bacterium *Xanthomonas stewartii*, which overwinters in the gut of the corn flea beetle. In 1934, N. E. Stevens devised a disease forecasting system based on winter temperatures. Mild winters, where average temperatures for December, January, and February are above 32°F, allow beetles to survive in large numbers. Stevens used sums of mean temperatures for the winter months to derive a "temperature index." Wilt was usually absent when this index was below 90 (a cold winter) and present when the index was above 100 (a mild winter).

Powdery mildew of apples is one of the few powdery mildews known to overwinter as mycelium in dormant buds. Infected buds are more sensitive to low winter temperatures than noninfected buds and are killed at temperatures below –20°C.

Stripe rust overwinters in the Pacific Northwest in the uredial stage on live wheat tissue. If winter temperatures are low enough much green tissue is killed, which also destroys the rust fungus. The fungus requires mild (cool) temperatures during early spring for urediospore development and subsequent plant infection. A forecasting system has been developed based on knowledge that mild winters and springs favor stripe rust. The predictive model (forecasting system) uses a base temperature of 7°C, the optimum temperature for urediospore germination and infection. Degree days are calculated by subtracting 7 from the minimum temperature for each day. Accumulate negative degree days for the coldest period (December 1 through January 31) and positive degree days for the spring period (April 1 through June 30). If negative degree days total less than 500, or if positive degree days total less than 440, there is a high probability of severe rust. Simply stated, the milder the winter and spring the greater the chance of rust.

DEVELOPMENT OF INOCULUM

Forecasting systems are based largely on certain weather conditions that favor development of inoculum and/or infection of the host.

Late blight of potato was one of the first diseases for which a forecasting system was developed and it continues to receive considerable attention in that warning systems are used both in the United States and Europe. Several methods have been used and generally the more precise the index used, the more reliable the forecast.

Three systems have been used to forecast late blight. The moving graph method calculates blight-favorable days, periods when the 5-day mean temperature is less than 26°C, and the 10-day total rainfall is 1.2 inches or more. Initial blight is forecast after 10 consecutive blight-favorable days (Figure 5.2). Notice that the susceptible stage of growth extends over the entire production cycle (from planting to harvest) necessitating spray applications as needed through the growing season.

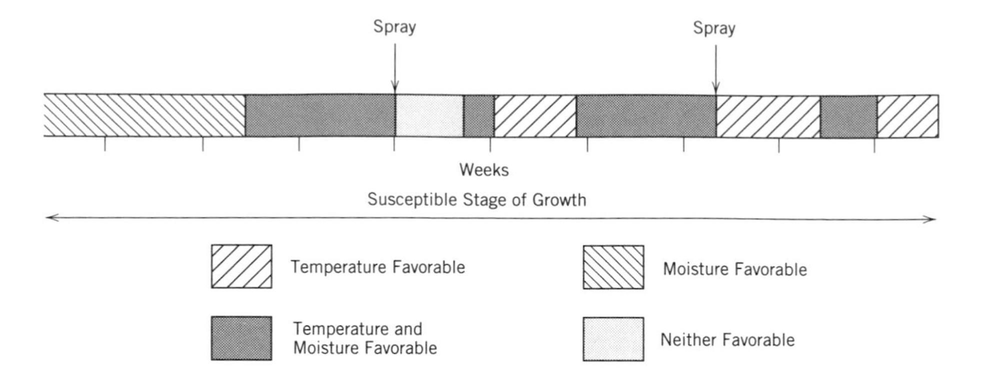

Figure 5.2. A moving graph combining temperature and moisture to determine when conditions are favorable for a disease such as potato late blight and sprays required.

The 90% relative humidity method predicts severe blight when the relative humidity is 90% or more and ranging from 10 hours when the mean temperature is 7°C to more than 25 hours at 27°C.

The 70% relative humidity method is based on periods when relative humidity is 70% or more and temperature is 10°C or higher for 48 or more consecutive hours.

Of these three methods, the first is the most reliable and requires least equipment. The third method is the least accurate.

Apple scab is another disease that has received much attention, primarily to tell growers when to apply protective sprays. Apple scab infection is predicted on the basis that spore discharge and germination and subsequent infection are related to rain (continuous wetness) and temperature. The Mill's scale (Figure 5.6a) is sometimes expressed as tables (Figure 5.6b and e) or in semicircular form (Figure 5.6c and d) to simplify its use. This forecast is based on hours of continuous leaf wetness on a leaf surface required for spore germination and infection at different temperatures. For example, if the average daily temperature is 50°F, leaves must be continuously wet for 14 hours before infection can occur. If the average daily temperature is 65°F, then only 9 hours of continuous wetness is required for infection. The hours indicated are for ascospore (primary) inoculum. Conidia (secondary inoculum) require only about two-thirds this amount of time, that is, 9.5 and 6.5 hours, respectively.

Most forecasting systems are valid only when initial inoculum is available and it is not always possible to predict with certainty when inoculum comes in contact with an infection court. Workers in Canada and the eastern United States have developed a model to predict discharge of apple scab ascospores. The model is based on accumulated degree days (base 0°C) after the silver tip stage of bud development. Table 5.1 relates number of days after silver tip to accumulated degree days. If the number of actual degree days exceeds that of the table for the corresponding elapsed days, ascospore discharge is likely. This establishes the starting point of the plant-host timing window. Once spore discharge is established infection periods (hours of continuous wetness at a given temperature) determine application of fungicides (see Figure 5.3).

TABLE 5.1. Cumulative Degree Days from Silver Tip to First Ascospore Discharge

Number of Days from Silver Tip	Accumulated Degree Days
2	8
4	30
6	52
8	73
10	95
12	116
14	138
16	170

Source: Adapted from Proctor, J. T. A., et al., Highlights of Agricultural Research in Ontario Vol. 6, No. 3. Sept. 1983, with permission.

Sometimes determining onset of ascospore production is more direct than using bud development. Infected leaves from the previous season are placed in an orchard and microscope slides suspended over the leaves. At intervals slides are removed and examined for the characteristic two-celled ascospores. When spores are found the forecasting system is activated.

Continuous monitoring of weather is essential for success of this system and microcomputers have been designed to record and analyze weather data, triggering a

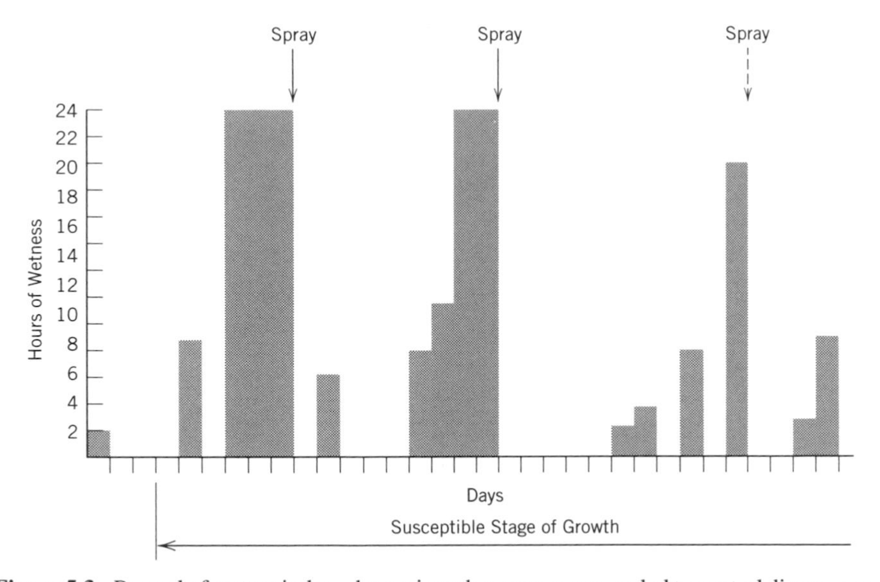

Figure 5.3. Record of wet periods to determine when sprays are needed to control diseases such as apple scab.

signal to alert growers when infection periods occur. One of these is the Apple Scab Predictor™ developed in north central United States where apples are grown extensively.

Control of fire blight of pear and apple depends on protecting blossoms, which are primary infection courts, with bactericidal sprays and timing is critical. The bacteria multiply in the nectaries at temperatures between 60 and 80°F. A prediction system developed in Illinois is based on accumulation of degree days from the latest prebloom freeze (32°F) to early bloom. At least 30 degree days with maximum temperatures between 70 and 80°F during early bloom favor blossom infection. In addition there should be adequate rainfall preceding bloom to provide normal growth, and occasional light rains with a relative humidity of 70% or more during early bloom. Degree days are the accumulated degrees above a base of 65°F. If the maximum temperature the day after a freeze was 68°F, there would be 3 degree days, if the next day was 70°F, there would be 8 degree days (3 plus 5), if the third day was 75°F, there would be 18 degree days (3 plus 5 plus 10), and so on. When the total reaches 30 and the tree is still in bloom, a bactericidal spray is applied. However, if the temperature drops to 32°F all accumulated degree days are eliminated and the count starts over.

A fire blight prediction system for Washington state is much simplified and is based on the 60°F minimum temperature required for infection. Sprays are advised whenever the mean daily temperature (maximum + minimum/2) is over 60°F *and blossoms are present* (see Figure 5.4). Sprays are also advised when alfalfa or other legumes in the vicinity are cut for hay because of the influx of leafhopppers and other sucking insects to the fruit trees where they cause punctures that serve as entry points for the bacteria.

The California system is not a prediction system but essentially a calendar spray where sprays are applied during bloom when the mean temperature reaches 60°F and repeated at 5–7 day intervals until the end of bloom.

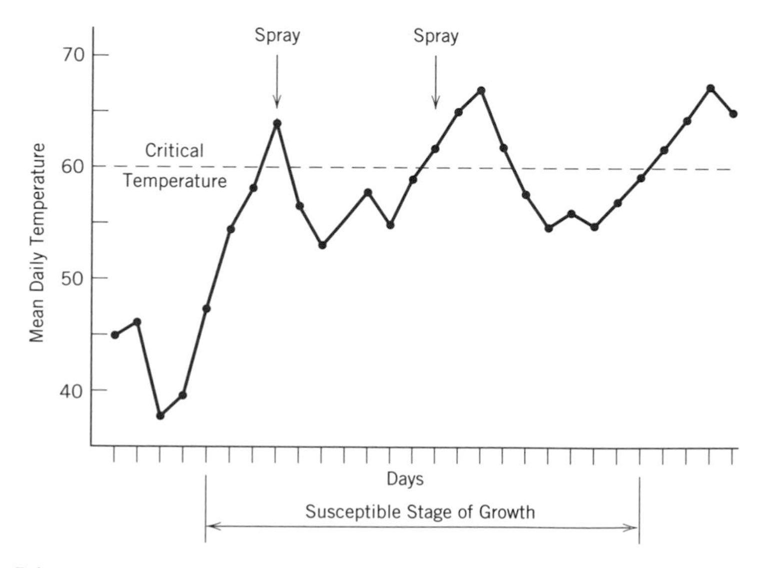

Figure 5.4. Charting mean daily temperatures to determine when sprays are needed to control fire blight of pears. Infection occurs only at certain temperatures mainly through blossoms.

Positive degree day accumulation has been used to predict early blight of potato in Colorado. In this model negative degree days were not included but the base was 7.2°C (45°F) similar to that for stripe rust. Appearance of first lesions and sporulation of *A. solani*, which provides secondary inoculum, is related to plant maturity and plant maturity is a function of degree days. This predictive model uses a damage threshold of 361 degree days Celcius (DDC) or 650 degree days Fahrenheit (DDF). When this total is reached growers have 7-10 days to apply sprays.

Downy mildew of hops in Washington state overwinters in infected crowns and initial sporangial inoculum is produced on the early shoots (basal spikes). A forecasting system used to establish the need for fungicidal sprays is based on number of infected spikes, night temperatures above 5°C, and relative humidity above 70%. The first factor indicates potential inoculum; the latter two favor sporulation.

Disease forecasting is not feasible for all diseases and to be useful a forecasting system must satisfy at least four conditions.

1. The disease should be widespread. Cost of developing and implementing a forecasting system cannot be justified for a disease that occurs only in narrow ecological niches or is erratic.
2. The disease must be consistently damaging yet be sporadic. Systems for diseases of minor economic importance are not cost effective. And if a disease consistently appears year after year it is predictable and no forecasting system is needed.
3. Effective management technology (i.e., control measures) must be available. It is of no avail to predict disease if control measures are lacking.
4. Rapid communication of forecasts to growers (crop producers, managers, etc.) is essential. The purpose of disease forecasting is to allow growers to respond in a timely manner. If growers cannot get a forecast until after a disease develops it will be of little benefit. Most forecasting systems are designed to transmit results immediately to growers by telephone, radio, or other devices.

The Plant Disease Survey was begun in 1916 to collect information on plant diseases in the United States and to make such information immediately available to persons concerned with disease control. Originally detection and reporting of new and threatening diseases were emphasized. By the 1940s a warning service was initiated to report outbreaks of potato late blight and other diseases so that control chemicals in short supply during World War II could reach places where needed. By 1948, disease warnings also were issued for tomato late blight, tobacco blue mold, and cucurbit downy mildew. Weekly weather reports of temperature and rainfall were the major parameters used to prepare these forecasts. Development of the disease warning system (Figure 5.5) suggested that disease forecasting was feasible and efficient on a regional, state, or local basis.

Disease forecasting is contraindicated if the disease is endemic, that is, regular in occurrence, and spraying is done routinely, or growers are unable to apply corrective measures for some reason. In some cases forecast may be too late. Peach leaf curl infection occurs soon after budswell and detection is not possible until leaves start to

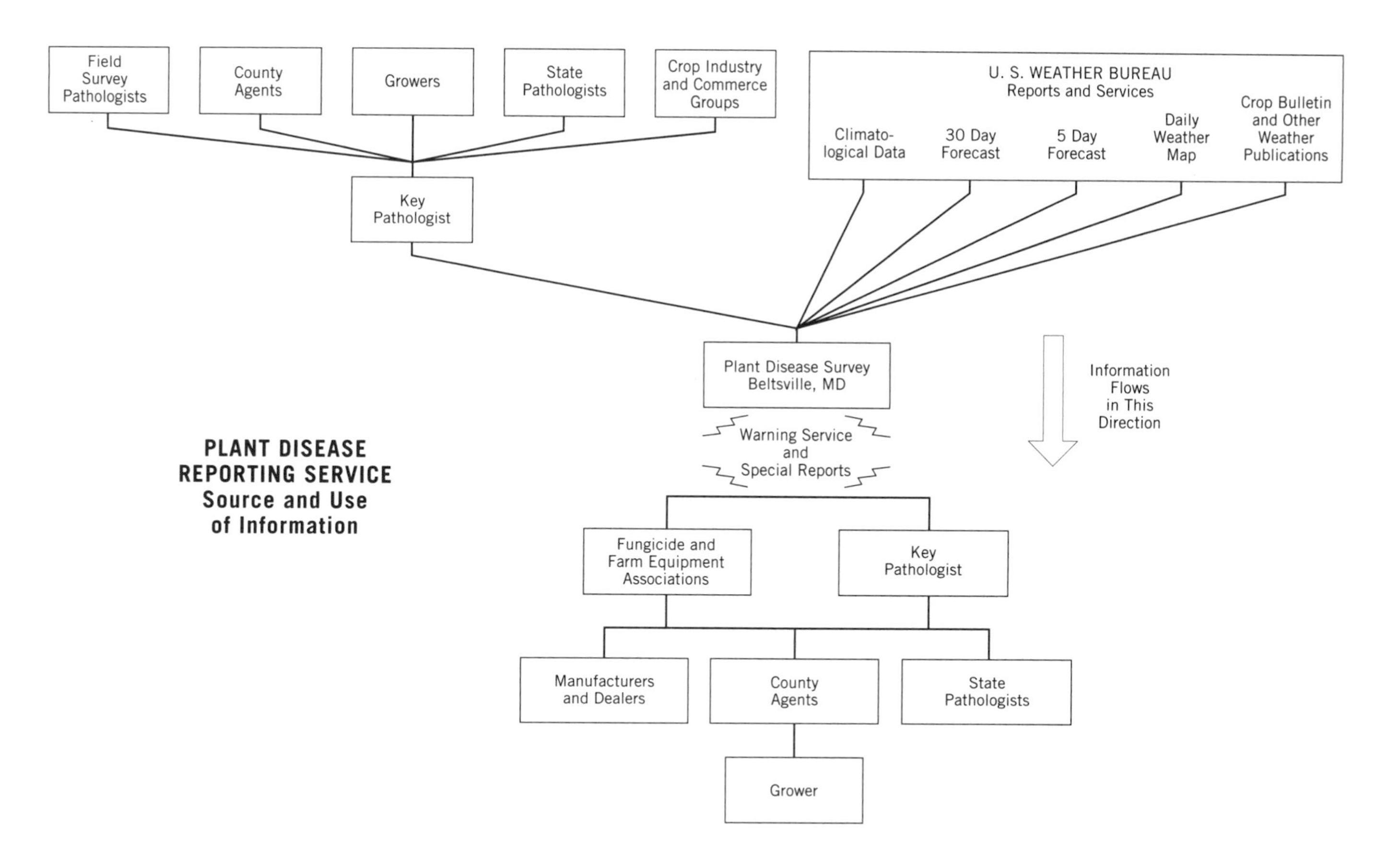

Figure 5.5. Organization of the Plant Disease Survey. [From Miller, *Plant Disease Reporter Supplement* 195:471-482 (1950).]

develop. By the time leaf curl can be predicted spray applications are useless. Or a grower may have too many acres to cover in a reasonable time and forecasts are of limited value. Similarly, if growers rely on contract applicators the demand may exceed availability of such services. Forecasting narrows the time span for effective action and contract sprayers may be too busy to service all crops that need to be sprayed.

Some disease forecasting systems, while technically successful, have not been accepted by growers. BLITECAST is an example. This system offered the possibility of reducing the number of sprays, resulting in savings of $50 to $100 per acre. However, growers were not willing to risk a crop worth $5000 per acre for that saving. Many growers view routine (i.e., calendar) spraying as insurance—better to have it and not need it than the other way around.

DISEASE FORECASTING SYSTEMS AND DECISION AIDS

Weather Instruments

Since disease forecasting usually is based on weather conditions, particularly temperature and moisture, weather instruments of various types are primary tools used to collect data pertinent to forecasts. Instruments include thermometers, rain gauges, hygrometers, wetness meters, wind gauges, and the like, many of which record data for prolonged periods.

Computerized Monitors and Warning Devices

Time for application of control measures can be limited when faced with potential outbreaks of disease. Rapid collection, analysis, and integration of weather parameters, such as hours of leaf wetness and temperature (e.g., in apple scab development), are essential to most forecasting systems. The development of computer programs speeds and automates this process.

In addition to weather-based systems some forecasts are derived from forecasts of sporulation rather than infection. Sporulation detection or indexing has been used for primary inoculum of apple scab, overwintering uredial inoculum of cereal rusts, primary inoculum of hop downy mildew, and numbers of airborne spores of the potato early blight fungus. Spore collection devices of various kinds, from grease-coated slides to complex automated, mechanical spore collectors have been used. For these to be useful the pathogen sought must produce a readily identifiable spore.

Threshold levels similar to those used in monitoring insect populations have not been as useful for disease monitoring or forecasting since the latent period between inoculation/infection and disease development permits many diseases to expand beyond the point of feasible control. In such cases monitoring is more likely to tell what *has* happened rather than what *might* happen!

Critical disease level has been used to time initial fungicide application for controlling Botrytis leaf blight on onion. This involves scouting onion fields and counting lesions. The first spray is applied when an average of 1.0 lesion per leaf occurs in a field. A similar approach is used to determine when to apply fungicides to control

a) Approximate Hours of Wetting Necessary for Primary Apple Scab Leaf Infection in an Orchard Containing an Abundance of Inoculum

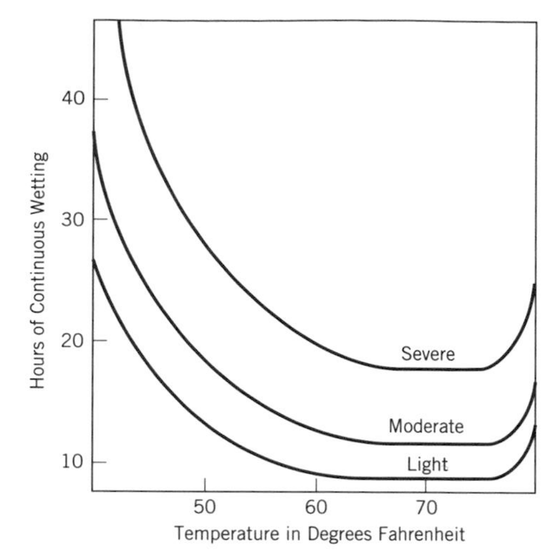

b) Time Needed at Various Temperatures for Winter Spores of Apple Scab to Cause Infection*

Average Temperature	Hours to Infect Leaves: Light	Moderate	Heavy	Days Until Lesions Appear
Degrees F.	Hours			Days
33–41	Over Two Days			
42	30	40	60	–
43	25	34	51	–
44	22	30	45	–
45	20	27	41	–
46	19	25	38	–
47	17	23	35	–
48	15	20	30	17
49	14	20	30	17
50	14	19	29	16
51	13	18	27	16
52	12	18	26	15
53	12	17	25	15
54	11	16	24	14
55	11	16	24	14
56	11	15	22	13
57	10	14	22	13
58	10	14	21	12
59	10	13	21	12
60	9	13	20	11
61	9	13	20	10
62	9	12	19	10
63	9	12	18	9
64	9	12	18	9
65	9	12	18	9
66	9	12	18	8
67	9	12	18	8
68	9	12	18	8
69	9	12	18	8
70	9	12	18	8
71	9	12	18	8
72	9	12	18	8
73	9	12	18	8
74	9	12	18	8
75	9	12	18	8
76	9	12	19	9
77	11	14	21	9
78	13	17	26	10

* Under constantly wet conditions. Summer spores germinate in two-thirds of the time required for winter spores.

c) Graph Showing the Number of Hours Leaves Must Remain Wet for a Primary Scab Infection to Occur

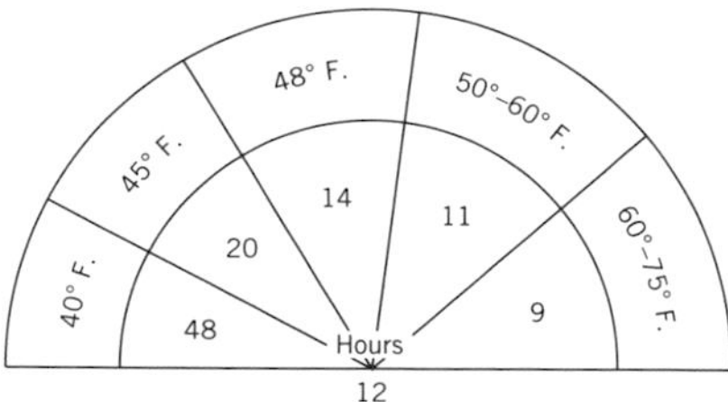

d) Graph Showing the Number of Days Required for Visible Scab Lesions to Appear After an Infection Has Taken Place

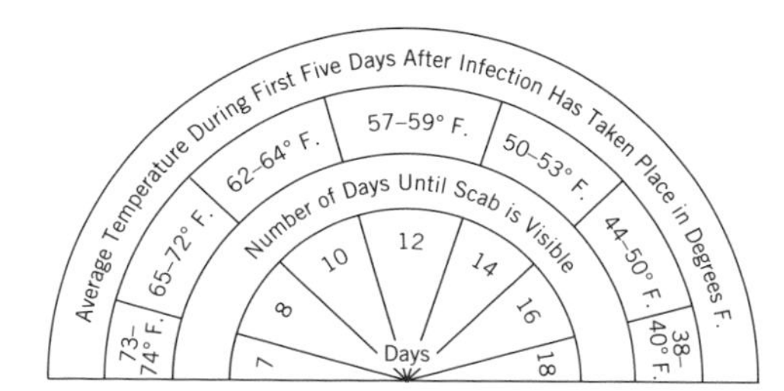

e) Approximate Hours of Wetting Required for Fruit Infection[1]

Mean Temperature (°F.)	Weeks After Full Bloom: 5	10	15	20	25
50	26	37	45.5	52.5	59.0
60	16.7	23.8	29.3	33.8	38.0
70	12.3	17.5	21.6	24.9	27.9

[1] This is the hours of wetting required for 2% infection of apple fruit.

Figure 5.6. The original Mill's scale and various modifications in use.

strawbreaker foot rot of wheat. In Washington state wheat growers monitor fields after tillering (Feekes growth stage 3) but before jointing (Feekes growth stage 6) and examine at least 50 tillers. Fungicidal sprays are warranted if 10% or more of the tillers have visible eyespot lesions under the outermost leaf sheath. This assumes that there are twice as many latent infections as observed infections for a total of 30% and agrees with recommendations in the United Kingdom where 25% infection at this growth stage is considered the critical disease level for spraying. In Australia foliar sprays to control stripe rust of wheat are recommended when 1% of the leaf area is infected with rust. This is equivalent to 35–40 affected leaves per 100 examined. However, other factors such as the variety grown and time of season must also be considered.

Disease thresholds could be used to determine control needs if (1) the technology is available to detect, quantify, and control the disease and (2) growers are able and willing to assess diseases before applying fungicides. Experience with some disease prediction systems (e.g., BLITECAST) suggests growers are not willing to take the risk or expend the effort.

DISEASE PREDICTION MODELS

Several computer programs for simulating, and thereby forecasting, plant diseases have been described and often are given acronyms derived from their function:

EPIDEM is a computerized simulator of potato early blight based on the environmental factors—temperature, moisture, light, leaf density, wind, etc.—that influence many, or all, of the links in the infection chain.

EPIMAY, a program simulating epidemics of southern leaf blight of corn, is as complicated in its derivation as is EPIDEM from the same authors.

EPIPRE is a more complete but also complex disease and pest management system for winter wheat based on a range of data including soil type, cultivation, seeding date, fertilization, and expected yield from individual fields plus weather data from a series of stations. The name is derived from EPIdemic and PREdiction (and/or PREvention).

BOTCAST is a system developed for timing fungicide applications to control Botrytis leaf blight of onions and is based on periods of moisture on leaves that favor inoculum production (i.e., sporulation) and infection.

BLITECAST has received the most attention and has gone through several forms. The latest combines two earlier systems based on rainfall patterns and temperature correlated with disease severity values. Severity value is the probability of infection, estimated from the number of hours of 90% or higher relative humidities at various temperature ranges. The accumulated severity values and rainfall (low or high) produce four recommendations in ascending order of urgency: (1) no spray, (2) late blight warning or alert, (3) spray every 7 days (=moderate spray), or (4) spray every 5 days (=heavy spray).

BLITECAST was simplified to make growers, rather than research, extension, or other government personnel, more responsible for the final decision making.

Simplification eliminated the need to collect some data in the original BLITECAST program, but requires a brief, nonmathematical analysis of weather data by growers.

Decision aids are methods and devices that allow growers and others to make decisions based on forecasting criteria available. These aids usually translate some numerical value(s) such as degrees of temperature and/or hours of wetness or high relative humidity into some action category, such as spray, no spray, delay action, or other option. The aids may include equations that relate the critical component (e.g., hours of relative humidity over 90%) to likelihood of infection. They may be nomograms of various sorts, charts, graphs, tables, or scales (Figure 5.6), that permit the location of two (or three) intersecting lines to establish a point for decision making. Nomograms such as Mill's table are fixed in time in that when the conditions for a given period (for example 12 hours) indicate a disease likelihood, action (i.e., spray) is taken.

Many nomograms are converted to moving graphs of one type or another (see Figures 5.2–5.4) in which pertinent data such as rainfall or temperature are plotted daily and when a favorable period has accumulated fungicidal sprays are applied or other suitable action taken. Moving graphs inherently contain negating features that cancel favorable periods when some nullifying condition such as dry weather or low temperature occurs.

Decision aids derived mathematically in computer programs are best understood when expressed as graphs and decision rules. This was reflected in the simplified version of BLITECAST. The aid might be as simple as a "rule of thumb" such as "spray after every rain during the bloom period." Such rules usually result in excessive spraying.

Schedules of management practices can play a major role in the decision process. Because of cost, spray applications frequently are more acceptable to a grower if the spray can be applied as part of some other application. When possible wheat growers in the Pacific Northwest, for example, are inclined to apply fungicides for strawbreaker control with routine applications of herbicides. This is satisfactory if (1) the fungicide can be applied both legally and effectively as a tank mix and (2) timing of the fungicide application is not jeopardized. Irrigation is a major cultural operation that can greatly influence disease development. Growers can reduce losses by withholding an irrigation cycle or even changing the form of irrigation (for example, from overhead sprinkling to rill or drip irrigation). Also, seeding dates can be used to either eliminate or enhance fungicidal application.

In practice decisions usually are based on a combination of the above aids, seldom on one alone.

REFERENCES

*Brent, K. J., and R. K. Atkin (eds.), *Rational Pesticide Use*. Cambridge University Press, New York, 1987.

*Bourke, P. M. A., Use of weather information in the prediction of plant disease epiphytotics. *Annu. Rev. Phytopathol*. 8:345–370(1970).

*Coakley, S. M., R. F. Line, and L. R. McDaniel, Predicting stripe rust severity on winter wheat using an improved method for analysing meteorological information. *Phytopathology* 78:543–550 (1988).

*Colhoun, J., Effects of environmental factors on plant disease. *Annu. Rev. Phytopathol.* 11:343–364 (1973).

*Krause, R. A., and L. B. Massie, Predictive systems: Modern approaches to disease control. *Annu. Rev. Phytopathol.* 13:31–47 (1975).

*Miller, P. R., The place of the Plant Disease Survey in plant pathological investigations. *Plant Disease Reporter Suppl.* 195:471–482 (1950).

*Miller, P. R., and M. O'Brien, Prediction of plant disease epidemics. *Annu. Rev. Microbiol.* 11:77–110 (1957).

*Scott, P. R., and A. Bainbridge (eds.), *Plant Disease Epidemiology.* Blackwell, London, 1978.

*Seem, R. C., Disease incidence and severity relationships. *Annu. Rev. Phytopathol.* 22:133–150 (1984).

Seem , R. C., and J. M. Russo, Simple decision aids for practical control of pests. *Plant Dis.* 68:656–660 (1984).

*Sutton, J. D., T. J. Gillespie, and P. D. Hildebrand, Monitoring weather factors in relation to plant disease. *Plant Dis.* 68:78–84 (1984).

*Young, H. C., Jr., J. M. Prescott, and E. E. Saari, Role of disease monitoring in preventing epidemics. *Annu. Rev. Phytopathol.* 16:263–285 (1978).

*Zadoks, J. C., A quarter century of disease warning, 1958–1983. *Plant Dis.* 68:352–355 (1984).

* Publications marked with an asterisk (*) are general references not cited in the text.

6 Legal Bases of Exclusion

The principle of exclusion was applied to prevent spread of communicable diseases among humans long before the germ theory was proposed or accepted. In biblical times visitors were held at city gates so that they could be checked for leprosy. In medieval times visitors could be excluded from cities because of fear of bubonic plague and the afflicted were locked up, that is, quarantined. Ships were held in harbors for a period of 40 days to detect signs of latent smallpox, cholera, or other disease. "Quarantine" is derived from the Italian word quarantina that in turn derives from the Latin quadraginta which means 40. The significance of 40 days is uncertain but may have had a religious or other special connotation. When contagious diseases became known, such as smallpox, quarantines were imposed, not necessarily for 40 days but long enough to establish the presence or absence of disease.

Today quarantine refers to laws restricting movement of objects (including plants, animals, and humans) that might harbor undesirable organisms (insects, pathogens, weeds, etc.). A quarantine may involve specific times, areas, and commodities.

The explorations of Marco Polo, Columbus, Magellan, and others opened the world to the commerce, exploitation, and colonization that followed. By the 1800s people traveled more widely than before and in this period mechanical improvements revolutionized agriculture. New technology created a demand for new crops and man began to introduce exotic plants from other areas to satisfy this demand. Along with the introduced plant materials came insect pests, diseases, and weeds of various kinds.

Many diseases have been introduced into the United States from various parts of the world and some have been introduced to Europe from the United States. Table 6.1 illustrates a few such introductions.

Some diseases were introduced so early that date of introduction, as well as their source, cannot be determined. Several writers have listed diseases introduced into the United States. Those introduced in the 1800s include potato blackleg, tomato leaf spot, black rot of crucifers, white pine blister rust, downy mildew of melons, asparagus rust, sorghum head smut, celery late blight, chrysanthemum rust, onion downy mildew, hollyhock rust, grape anthracnose, gooseberry leaf spot, cherry leaf spot, olive knot, peach leaf curl, rice smut, wheat stripe rust, and corn brown spot.

Diseases introduced in the early 1900s include citrus canker, potato powdery scab, potato wart, chestnut blight, poplar canker, sugar beet cyst nematode, alfalfa rust, bean anthracnose, alfalfa leaf spot, and blackleg of crucifers.

TABLE 6.1. Some Plant Diseases Introduced into the United States or Europe

Disease	Introduced from
Downy mildew of grape	U.S. to Europe
Powdery mildew of grape	U.S. to Europe
Powdery mildew of gooseberry	U.S. to Europe
Chestnut blight	Asia to U.S.
White pine blister rust	Europe to U.S.
Citrus canker	Japan to U.S.
Onion smut	Mideast to U.S.
Dutch elm disease	Europe to U.S.
Black leg of crucifers	Europe to U.S.
Flag smut of wheat	Australia to U.S.
European canker of apple	Europe to U.S.

Diseases introduced in the mid-1900s include flag smut of grasses and wheat, Dutch elm disease, bacterial ring rot of potato, strawberry red stele, Sigatoka disease of banana, and golden nematode.

Viral diseases are conspicuously absent from the above lists. This probably reflects the limited knowledge of viral diseases during the times indicated and the general difficulty in diagnosing them.

By the mid-1800s Europe had imported many plants and introduced pests were creating many problems. The first quarantine was passed in Germany in 1873. It was intended to prevent introduction of the Colorado potato beetle by prohibiting importation of potatoes from the new world. By 1890 virtually every developed country except the United States had quarantine laws.

The first plant quarantine legislation in the United States was passed in 1905 against insects but did not include plant pathogens. The catastrophic results of introducing gypsy moth, chestnut blight, and white pine blister rust resulted in the introduction in 1909 and passage in 1912 of the Plant Quarantine Act. This law has since been revised many times. The Plant Quarantine Act has 15 sections dealing with various aspects of the Act such as authority (Secretary of Agriculture), certification (in country of origin), public hearings, violations, and penalties.

The Federal regulations and revisions concerning quarantines are published in the Code of Federal Regulations (CFR), which has 50 titles dealing with all aspects of the federal government of the United States. Title 7 concerns the Department of Agriculture and Chapter 3 of Title 7 deals with plant quarantines. The agency responsible for administering plant quarantines originally was called the Federal Horticultural Board. This became the Bureau of Plant Industry (BPI), which became the Bureau of Entomology and Plant Quarantine (BEPQ), which in turn became the Animal and Plant Health Inspection Service (APHIS), the present name of the agency.

Chapter 3 of Title 7 contains all of the quarantines in force at a given time. Quarantines are revised, replaced, and lifted as conditions change. There are two types of federal quarantines, foreign and domestic. Foreign quarantines restrict importation of commodities into the United States. Domestic quarantines restrict the interstate movement of commodities as well as establish eradication areas. Quarantine number 1, promulgated September 12, 1912, was a foreign quarantine prohibiting importation of five-needle (i.e., white) pine seedlings from Europe and Asia into the United States. This quarantine was promulgated to prevent introduction of *Cronartium ribicola*, the white pine blister rust fungus, thereby protecting the valuable white pine and sugar pine stands of the United States. The quarantine was revised May 21, 1913 as Quarantine number 7, and amended to prohibit importation of currants and gooseberries (*Ribes* species), the alternate hosts of *Cronartium ribicola*. The revision added Canada and Newfoundland as quarantined countries.

Quarantine number 26, promulgated April 21, 1917, was a domestic quarantine that prohibited or restricted the movement of five-needle pines and *Ribes* species west of the Mississippi River, including Minnesota, Iowa, Missouri, Arkansas, and Louisiana. White pine blister rust had not yet been found in the western United States (although it probably was introduced into the northwest before 1910) and the Great Plains of central United States appeared to be a natural barrier to spread of the disease. However, blister rust was discovered at Vancouver, British Columbia in the summer of 1921 and later the same year in northwestern Washington state. Quarantine 26 also authorized eradication of the European black currant, which was widely planted in home gardens and known to be one of the most susceptible of the alternate hosts of *Cronartium ribicola*. Quarantine 26 was amended March 1, 1922 as Quarantine number 54, and revised March 2, 1923 to include the state of Washington as a quarantined area. These were succeeded August 27, 1926 by Quarantine 63, which restricted interstate movement of five-needle pines and *Ribes* species throughout the United States. None of the white pine blister rust quarantines is now in force but the importation and interstate movement of *Ribes* species is now restricted by Quarantine number 37, which requires postentry holding of nursery stock for 1 or 2 years.

Similar changes frequently occur in other quarantines as circumstances change. For example, Quarantine 19, a foreign quarantine, was enacted January 1, 1915, after citrus canker had devastated citrus groves and nurseries in Florida. This quarantine restricted importation of citrus nursery stocks but exempted fruit and seeds. This oversight was corrected August 1, 1917, by Quarantine 28, which prohibits importation of citrus fruit and peel. However, 28 did not prohibit importation of the Unshu orange (*Citrus reticulata* var. *unshu*), also known as Satsuma orange or Mandarin, grown in Japan and imported under permit to most northern United States. Prohibited areas included citrus growing and adjacent states. The Unshu orange is not affected by citrus canker because of a specialized stomatal structure that prevents entry of the bacteria, although fruit can carry the pathogen on their surface. Several stringent conditions are required for importation into the United States including being grown and packed in isolated, canker-free export areas established by the Japanese Plant Protection Service. The export area must be surrounded by a 400-m (437 yard)-wide buffer zone in which no susceptible citrus are grown. The oranges are inspected by Japanese and United States plant pathologists in groves prior to and during harvest, and in packing houses. This includes bacteriophage testing of washings from fruit to indicate the presence of causal bacteria. The oranges must by surface sterilized before packing and each shipment must have a certificate from the Japanese Plant Protection Service declaring that the fruit is apparently free of citrus canker bacteria. Importation is allowed only through specified ports of entry and is prohibited through ports in quarantined states. The oranges are subject to a final examination at ports of arrival by USDA inspectors.

The citrus canker quarantine was strenghthened further September 15, 1947, by domestic Quarantine number 75 restricting movement of citrus fruit from Hawaii, then a territory of the United States. For some reason Quarantine 28 was revised in the 1960s to limit importation and movement of Unshu oranges to the Pacific Northwest states of Alaska, Washington, Oregon, Idaho, and Montana, as well as Hawaii but by 1988 further revision quarantined only citrus growing and adjacent states and again expanded the area to which Unshu oranges could be imported.

Many quarantines are lifted once it becomes apparent that a pathogen is established in an area and the quarantine is no longer effective. However, quarantines may be left in place to prevent more virulent forms of the pathogen from entering a region. Quarantine 39, a foreign quarantine, originally was directed at the exclusion of flag smut and take-all of wheat. After it was determined that take-all was widely established in the United States this disease was removed from the quarantine. Even though flag smut occurs in localized areas the quarantine remains in force to prevent introduction of new races of the flag smut fungus. The quarantine has been amended to include Karnal smut, a wheat disease recently found in Mexico but not yet established in the United States.

Golden nematode is one of the most serious pests of potatoes and potato-growing areas throughout the world diligently guard against its introduction. Importation of potatoes has been prohibited from practically all of Europe since 1912 but despite stringent quarantines golden nematode cysts are frequently discovered by U.S. plant quarantine inspectors at ports of entry in small quantities of soil attached to plant parts, bulbs, packing materials, crates, and bags originating from Europe as well as from soil on travelers shoes and tires of imported used tractors. Golden nematode has been intercepted at the Mexican–U.S. border in freight cars originating in potato-growing areas of Mexico.

Despite quarantines, inspections, and other precautions, pathogens still slip through. For example, in 1983 citrus canker bacteria were intercepted more than 1000 times by APHIS.

The first hint of golden nematode in the United States came as complaints of poor plant growth from potato farmers on Long Island, New York in 1934. In 1941 an entire field was found infested, and by 1958 golden nematode was found in about 14,000 acres on Long Island. Losses often were dramatic. A lightly infested field might yield 263 bushels of potatoes; a heavily infested one yields 64 bushels.

The New York Department of Agriculture and Markets took steps to slow the spread of the nematode and promulgated specific quarantines in 1941. These quarantines prohibited sale or disposal of seed potatoes from the entire growing area of Long Island and discouraged farmers from growing other susceptible crops such as eggplant and tomato. A 1954 quarantine prohibited returning soil removed from potatoes during grading ("grader dirt") at common stations to farms because of danger of infesting clean ground. In 1955 all potatoes grown within the quarantined area had to be packaged in approved paper bags or other approved containers for movement within the United States and Canada.

The New York quarantine and Federal Quarantine 85 must have been effective because thus far the nematode has not become established elsewhere in the United States. However, there was a mild panic among potato growers in the Pacific Northwest in the 1960s when golden nematode was discovered on several farms on Vancouver Island, British Columbia. Seed potatoes from British Columbia were planted in the Pacific Northwest and there was concern that the nematode may have been introduced. The Bureau of Entomology and Plant Quarantine immediately surveyed the area of concern by sampling for the nematode in tare dirt piles at potato processing plants. None was found so either seed tubers from infested fields of Canada were not used in this region or the nematode was unable to become established.

Tracing the history of the Vancouver Island fields infested by the golden nematode indicated that the nematode had been present since the 1920s, perhaps introduced on ornamental bulbs from Europe. These fields had grown bulb crops and potatoes through the ensuing years but there was no evidence that the nematode had spread beyond the originally infested fields. For unexplained reasons the nematode was unable to survive if it was introduced.

States are subject to federal quarantines but may also impose state quarantines for protection of specific commodities and interests. Washington state has quarantines for several important agricultural crops. Some of these are summarized below:

1. Azalea flower spot quarantine, directed against *Ovulinia azalea*, prohibits importation of azalea, rhododendron, and kalmia from several southeastern states and California (except Humbolt County). Shipments must be accompanied by official certificates stating where the plants were obtained, that they were grown in a disease-free nursery, and that tops were cut back to within an inch of the soil immediately prior to shipment.
2. The onion white rot quarantine isolates three onion-growing counties in Washington and prohibits or regulates movement of *Allium* bulbs, sets, transplants, soil, tools and equipment, and livestock pastured on infested ground into the protected area.
3. The grape virus quarantine protects the expanding wine grape industry of the state from fanleaf, leaf roll, and other viral diseases by restricting importation of grape plants (but not fruit) from outside the state. Grape plants may be imported if inspected and certified to be free of fanleaf and leaf roll by a state agency using accepted indicator plants for indexing.

As might be expected, states with major valuable crop industries also have restrictive quarantines to protect such industries. California and Florida are major citrus growing states and have quarantines protecting against introduction of citrus pests and diseases such as citrus canker. Florida prohibits importation of citrus plants from other state and foreign sources except under special permit, but does not regulate importation of citrus fruit since these commodities are regulated under federal domestic quarantines. Ornamental trees such as flowering dogwood are important to Florida's large nursery industry and hence are protected by specific quarantines. Importation of dogwood trees or any plant or plant parts capable of harboring the dogwood anthracnose fungus is prohibited, except by permit issued by the Director of the Florida Division of Plant Industry. At least 16 states and any other state, territory, or country where dogwood anthracnose has been identified are quarantined. Sugarcane is another major crop in Florida and any article, including soil, capable of carrying one or more of 10 listed sugarcane pathogens is prohibited. However, sugarcane for research purposes may be admitted under special permit.

New York, a major grape-growing state, has detailed regulations concerning importation of grape stocks. Provisions are often made for introduction of materials that are not accompanied by a phytosanitary certificate or other document certifying apparent freedom from disease. Grape rootstocks and propagative material may be admitted into New York state under the following conditions:

1. *Vitis vinifera* (European varieties) nonrooted propagative material is accompanied by a statement that the mother vines were visually inspected twice during the growing season (early June and late August) and are apparently free of the fanleaf group, leaf roll, and corky bark viruses, and *Flavescence doree*, a mycoplasma.
2. Nonrooted and selfrooted vines of American varieties are accompanied by a statement that they were derived from selfrooted mother vines free from cluster abnormalities such as shot berry and cluster abortion.
3. Nonrooted and rooted propagative material of resistant rootstocks and French hybrids (fruiting varieties hybridized in Europe) are admitted under a 1-year postentry quarantine. During the postentry quarantine period the material must be indexed on Mission and St. George grape varieties and *Chenopodium quinoa*. Virus infected material is destroyed.

States forming a common agricultural area often have similar quarantines. The expansion of the wine grape industry in the Pacific Northwest resulted in grape virus quarantines in Washington (1970), Oregon (1971), and Idaho (1972). These common quarantines reduce the opportunity for infected plants to move across state lines.

State quarantines follow two distinct patterns. The simplest and most direct is where state authority prohibits or regulates movement of pest-carrying material into the state, that is, the "destination." The other is where a state recognizes the need to prevent dissemination of pests to other states, coincidentally protecting their own agricultural interests, and sets up safeguards at the "source." While a "source" quarantine is biologically sound, as demonstrated by nursery stock inspections, seed potato certification, and similar programs, it requires better organization and efficiency than "destination" quarantines. An effective "source" quarantine eliminates the need for multiple "destination" quarantines and is less of a hindrance to normal flow of commercial traffic.

States can impose urgent emergency quarantines on short notice in response to apparently threatening diseases. Such a situation led to the Washington State Department of Agriculture Emergency Order quarantine for PVY-N virus on potatoes.

Potato virus Y (PVY) is a latent virus producing no symptoms and no apparent damage to potatoes. It had not previously been of concern to potato producers and no quarantine or regulatory action had been taken against it. However, in 1990 a strain of PVY appeared in one potato seed-producing province of Canada. It was extremely virulent on tobacco, a close relative of potatoes. This strain causes severe necrosis on tobacco and could seriously harm the tobacco industry. As a result of infected seed potatoes being imported into Idaho the Washington potato industry petitioned the Washington State Department of Agriculture to impose an immediate quarantine against introduction of seed potatoes from the infested area. There is no immediate threat to commercial production but potatoes from Washington state are exported to southeastern states where tobacco is a major crop, and is threatened by PVY-N. Tobacco-producing states are concerned that some market potatoes might inadvertently be planted and constitute a danger to tobacco. Also the virus may move from potatoes to nurseries that ship tomato, pepper and eggplant transplants to southern states. The PVY-N quarantine is intended to protect the Washington potato industry from losing a significant market.

Developed agricultural countries such as Australia, Great Britain, France, and Japan have stringent quarantines. Japan with a highly regulated agricultural system has many quarantines and regulations. One prohibits importing rice because of related insect pests and diseases. Canada prohibits *Prunus* and other tree fruits from areas without virus indexing programs. Other less developed countries are not as restrictive but usually have some quarantines to protect important crops.

Quarantine requirements and restrictions of countries other than the United States are summarized for USDA inspectors in an Export Certification Manual. This provides instructions to the Plant Protection and Quarantine (PPQ) branch of APHIS for inspecting and certifying plants and plant products offered for export and issuance of Federal Phytosanitary Certificates (FPCs). These activities are authorized by the Organic Act of 1944, which was passed to enable PPQ to help exporters meet plant quarantine import requirements of foreigh countries (see CFR, title 7, part 353).

Because of a commonality of crops and quarantine interests in various regions of the world, cooperation among countries in these regions has been encouraged. The International Plant Protection Convention (IPPC) was sponsored by the Food and Agriculture Organization (FAO) of the United Nations in 1951 to provide for international cooperation in controlling plant pests and pathogens and to prevent international spread. The IPPC became effective in 1952, provides for phytosanitary export certification between member countries, and prescribes a format for the certificate. By 1980 IPPC had 79 member countries.

In accordance with IPPC requirements, eight regional plant protection organizations have been formed (Figure 6.1). The oldest of these is the European and Mediterranean Plant Protection Organization (EPPO), which predates the IPPC treaty by a year. It has about 35 member countries in Europe, North Africa, and the Near East and is probably the most organized and active of the regional groups.

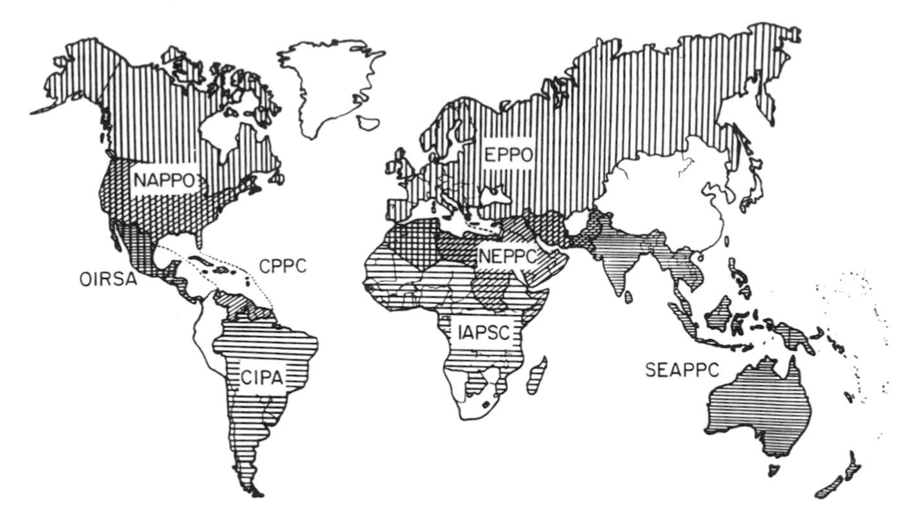

Figure 6.1. Regional plant protection organizations of the world. NAPPO, North American Plant Protection Organization; OIRSA, Organismo Internacional Regional de Sanidad Agropecuaria; CPPC, Caribbean Plant Protection Commision; CIPA, Comité Interamericano de Protección Agrícola; EPPO, European and Mediterranean Plant Protection Organization; NEPPC, Near East Plant Protection Commission; IAPSC, Interafrican Phytosanitary Council; SEAPPC, Plant Protection Committee for the South East Asia and Pacific Region. (Reprinted with permission from I. M. Smith, in Ebbels, D. L., and J. E. King (eds.), *Plant Health: The Scientific Basis for Adminstrative Control of Plant Diseases and Pests.* Blackwell, Oxford, 1979.)

EPPO has prepared two lists of pathogens that meet certain tolerance and importation requirements. These lists are revised as situations change.

List A-1 includes pests not yet introduced into the EPPO region. A zero tolerance is required from all countries for these quarantined organisms. The list includes the following selected pathogens that occur in the United States: *Xanthomonas citri* and *X. oryzae*, *Dibotryon morbosum, Endocronartium harknessii, Phymatotrichum omnivorum*, *Arceuthobium* spp., cherry rasp leaf virus, raspberry leaf curl virus, and X-disease mycoplasma.

The A-2 list contains pests present in some but not all EPPO countries. Normally a zero tolerance would be required, but member countries can select their own phytosanitary regulations according to their ecological conditions and distribution of pest. In some circumstances a stated tolerance can be specified. Some examples from the United States are *Corynebacterium insidiosum*, *Erwinia amylovora*, *Xanthomonas campestris* pv. *phaseoli, Ceratocystis ulmi*, *Glomerella gossypii*, *Phytophthora fragariae*, *Tilletia controversa,* beet curly top virus, prunus necrotic ringspot virus, and pear decline mycoplasma.

Quarantine restrictions may be one of four categories depending on degree of risk posed by the pathogen.

1. *Unrestricted.* No risk exists. The pathogen may already be established and the hazard of more virulent strains being introduced is nil. Or the host is not of sufficient economic or ecological value to justify a quarantine. The item being imported is an unlikely carrier of the pest or pathogen of concern.
2. *Restricted.* Plants in this category are low risk items and can be introduced with few restrictions. An inspection and chemical treatment may be required before phytosanitary certificates are issued. For example, Argentina requires that all alfalfa seed must be certified resistant to bacterial wilt and be fumigated with methyl bromide before it is accepted.
3. *Postentry.* These are quarantines aimed at higher risk items requiring that the plant be held or grown in an isolated location for a prescribed period. Peru requires that sugarcane "seed" be grown in postentry quarantine at designated locations for 2 years to determine if it is free of pests and disease. India requires that crops such as peanuts (*Arachis* spp.) be grown under postentry quarantine in an intermediate country *where peanuts are not grown.*
4. *Prohibited.* This is the highest risk group and includes specimens likely to harbor latent viruses, new races of bacteria or fungi, or to originate from areas with widespread diseases or pests not known to occur in the importing country. This category usually is admitted only under special conditions requiring an import permit (IP) as well as a phytosanitary certificate (PC).

An IP is a special authorization granted by the plant protection service of a country to allow entry of prohibited plants or plant products. It specifies the exact entry requirements for restricted items and is often required for importation of plant material for scientific purposes.

PCs are intended to expedite entry of plants or plant products into a foreign country. This is accomplished by certifying to the foreign plant protection agency that the shipment has been inspected and found to conform to phytosanitary import requirements of the importing country, as specified in the Export Summary for that country.

The general principles of quarantine systems from many countries have several common features that require that they

1. Specify restrictions.
2. Grant exceptions for scientific purposes.
3. Require import permits.
4. Require phytosanitary certificates and/or certificates of origin.
5. Stipulate inspection on arrival.
6. Prescribe treatment on arrival to eliminate risk.
7. Prescribe quarantine, postentry quarantine, isolation, or other safeguards.

Quarantine measures should be justified or based on certain fundamental information. For example:

1. Measures must be based on sound biological principles. A pest must pose a real threat to crops of significant importance in the country or region. Frequently, and unfortunately, quarantine action is attempted before adequate knowledge is available on which sound procedures can be based.
2. Quarantines should not be used for furtherance or hindrance of trade. Economic embargoes should not be disguised as quarantines. Some workers believe that quarantines should be filters, not barriers. Formerly, quarantine stations grew imported material and, if suspect, destroyed it. Such a station is a barrier. A filter quarantine lets the host through and holds back the pathogen, although such a process is expensive in time and money.
3. Quarantines must derive from adequate law and authority. In the United States there was no legal means of restricting movement of plant material into or within the country until the Plant Quarantine Act of 1912.
4. Quarantines should be modified as conditions change, or further information warrants, such as has been done with the white pine blister rust and citrus canker quarantines.
5. Preventing introduction and spread of pests must be considered feasible and reasonable. Quarantines may have value even if they fail in the long run. At the least they may only be holding actions that delay establishment of pests perhaps until more durable control measures are developed.
6. Effective quarantines require cooperation, both from the international community and the general public. These groups must be informed as to the purpose and importance of the quarantine restrictions.
7. Quarantine measures can be effective only if those responsible for them are well informed. Searching for a pest not known to occur in certain areas requires extensive training of inspectors, even sending them to countries where the pest is present so that they can become familiar with its appearance and variations. Administrative and enforcement agencies must be able to mollify the natural resistance of the populace to restrictions of any kind, especially those they do not understand.
8. Quarantine measures are only one facet of domestic pest management programs. Careful integration of measures is needed to achieve maximum effect. Use of seed treatments or other protective measures can further reduce establishment of potentially dangerous pathogens. It is also advantageous to incorporate resistance to a wide array of pathogens even if certain ones are not yet established in a region. Genetic resistance is an additional shield in the event that pathogens breach quarantine walls.

REFERENCES

*Anon., *EPPO Recommendations on New Quarantine Measures*. EPPO Bulletin, Special Issue, 1982.

*Ebbels, D. L., and J. E. King (eds.), *Plant Health: The Scientific Basis for Administrative Control of Plant Diseases and Pests*. Blackwell, London, 1979.

*Gussow, H. T., Plant quarantine legislation—A review and a reform. *Phytopathology* 26:465–482 (1936).

*Hewitt, W. B., and L. Chiarappa (eds.), *Plant Health and Quarantine in International Transfer of Genetic Resources*. CRC Press, Cleveland, OH, 1977.

*Kahn, R. P., Exclusion as a Plant Disease Control Strategy. *Annu. Rev. Phytopathol.* 29:219–246 (1991).

*Kahn, R. P. (ed.), *Plant Protection and Quarantine*. 3 Vols. CRC Press, Boca Raton, FL, 1989.

*Mathys, G., and E. A. Baker, An appraisal of the effectiveness of quarantines. *Annu. Rev. Phytopathol.* 18:85–101 (1980).

*McCubbin, W. A., State quarantines on interstate movement. *Plant Disease Reporter Suppl.* 195:483–500 (1950).

*McCubbin, W. A., *The Plant Quarantine Problem*. Chronica Botanica, Waltham, MA, 1954.

*Plucknett, D. L., and N. J. H. Smith, Quarantine and the exchange of crop genetic resources. *Bioscience* 39(1):16–23 (1989).

*Waterworth, H. E., and G. A. White, Plant introduction and quarantine: The need for both. *Plant Dis.* 66:87–90 (1982).

*Weber, G. A., *The Plant Quarantine and Control Administration*. The Brookings Institute, Washington, D.C., 1930.

* Publications marked with an asterisk (*) are general references not cited in the text.

7 Methods of Accomplishing Exclusion

Quarantines provide the legal framework to exclude pathogens, but just as laws alone cannot prevent illegal activies, quarantines are not completely effective by proclamation. Instead they require strategies and techniques to make them work. Quarantines were decribed in Chapter 6 and include the following options.

1. Embargoes that completely prohibit importation of specific plants or products.
2. Inspection at ports of entry.
3. Inspection at points of origin.
4. Field inspections during growing seasons (i.e., preclearance inspections).
5. Controlled entry with expressed conditions that may require some type of disinfection or disinfestation treatment.
6. Postentry quarantines which specify holding, and possibly growing, imported plants at designated and controlled locations.

The methods by which quarantines are accomplished can be simplified into three actions.

Prohibition takes the form of embargoes characterized as quarantines. Embargoes are absolute prohibitions directed at pathogens of such potentially high risk that plants or other items capable of harboring those pathogens are prohibited from entry without exception. Such absolute quarantines, or embargoes, are rare but some occur. China, for example, prohibits introduction of all corn products from the United States because of Stewart's wilt. India prohibits cacao and related plants from Africa, Sri Lanka, and the West Indies because of swollen shoot virus and other pathogens; coffee beans from Africa, South America, and Sri Lanka for various diseases (but not coffee rust), rubber from South America and the West Indies because of leaf blight, sugarcane from the South Pacific region because of several diseases, and sunflower from Argentina and Peru because of downy mildew.

Interception stops pathogens at destination points and is accomplished by inspection of subject items. If inspection detects the pathogen the item is not admitted. All countries enforcing quarantines have some manner of an inspection program. A survey of IPPC members revealed only part-time inspectors in some countries to more than 1500 in the United States. Inspectors are employed by the agricultural authority responsible for crop protection, federal, state, province, county, district, or zone. They can be full-time, academically educated professionals or nonprofessionals hired on a seasonal basis. Inspectors are trained to acquaint them with specific pests. Dealing with plant pathogens, unlike insects pests and weed seeds, poses certain inherent difficulties. For example, most plant pathogens are extremely small, even submicroscopic, as in the case of plant viruses, making visual inspection difficult or impossible in most cases. It is also difficult to distinguish between pathogenic and nonpathogenic forms of an organism. For example, macroconidia of a pathogenic *forma speciales* of *Fusarium oxysporum* is indistiguishable from macroconidia of *Fusarium oxysporum*, a ubiquitous saprophyte. Similarly, teliospores of the dwarf smut fungus of wheat, the object of a restrictive quarantine by the Peoples Republic of China, are difficult to separate from those of the more prevalent common smut fungus. The basic differences in spore morphology, a wider gelatinous sheath on the exospore with fewer and deeper reticulations on teliospores of the dwarf smut fungus, may be apparent to an expert on the smut fungi but are subtle enough to be missed by the average plant quarantine inspector, or even many plant pathologists.

Interception of plant material that may carry plant pathogens and other pests is the function of the Plant Protection and Quarantine (PPQ) branch of APHIS. PPQ has trained inspectors stationed at ports of entry around the country. These inspectors work in close cooperation with the U.S. Customs Service. Figure 7.l shows the location of ports of entry for the United States. These ports are located around the periphery of the country at major shipping centers and large cities along international boundaries.

States may also have plant inspection facilities at their borders or other appropriate locations. At least two of the states, California and Florida, have or have had drive-through inspection posts at their borders or along major highways where vehicles, sometimes including private automobiles and commercial buses, are stopped and the occupants asked to declare any plant materials in their possession. These precautions are aimed at protecting the citrus and deciduous fruit industries of those states.

Inspections can be general or specific. During a *general* inspection any plant material, soil, containers, or similar items are examined for any pest or pathogen. The inspections are intended to intercept plant pests of unknown or questionable economic importance that are not present or widely distributed within the country. *Specific* inspections consist of searches for economically important plant pests that are not present or widely distributed. Inspection of Unshu oranges admitted under provisions of Quarantine 28 and examinations for cysts of the golden nematode constitute specific inspections. Examination of cargos of wheat by Peoples Republic of China pathologists for *Tilletia controversa* teliospores also constitutes a specific inspection.

Inspections may also be *deferred* by holding material for a time and reexamining to better detect pathogens in latent or incubating conditions at the time of initial inspection. Imported rosebushes can be grown on the importer's premises but not propagated until they have been inspected for diseases such as crown gall. Fruit trees are held by the importer until inspectors visually inspect them and/or they are indexed or tested serologically.

Circumstantial inspections are based on the origin of plant material, or some other basis, such as season of the year. Every lot of a particular plant or plant part is not examined, only those coming from countries, regions, or areas where a particular pest is known.

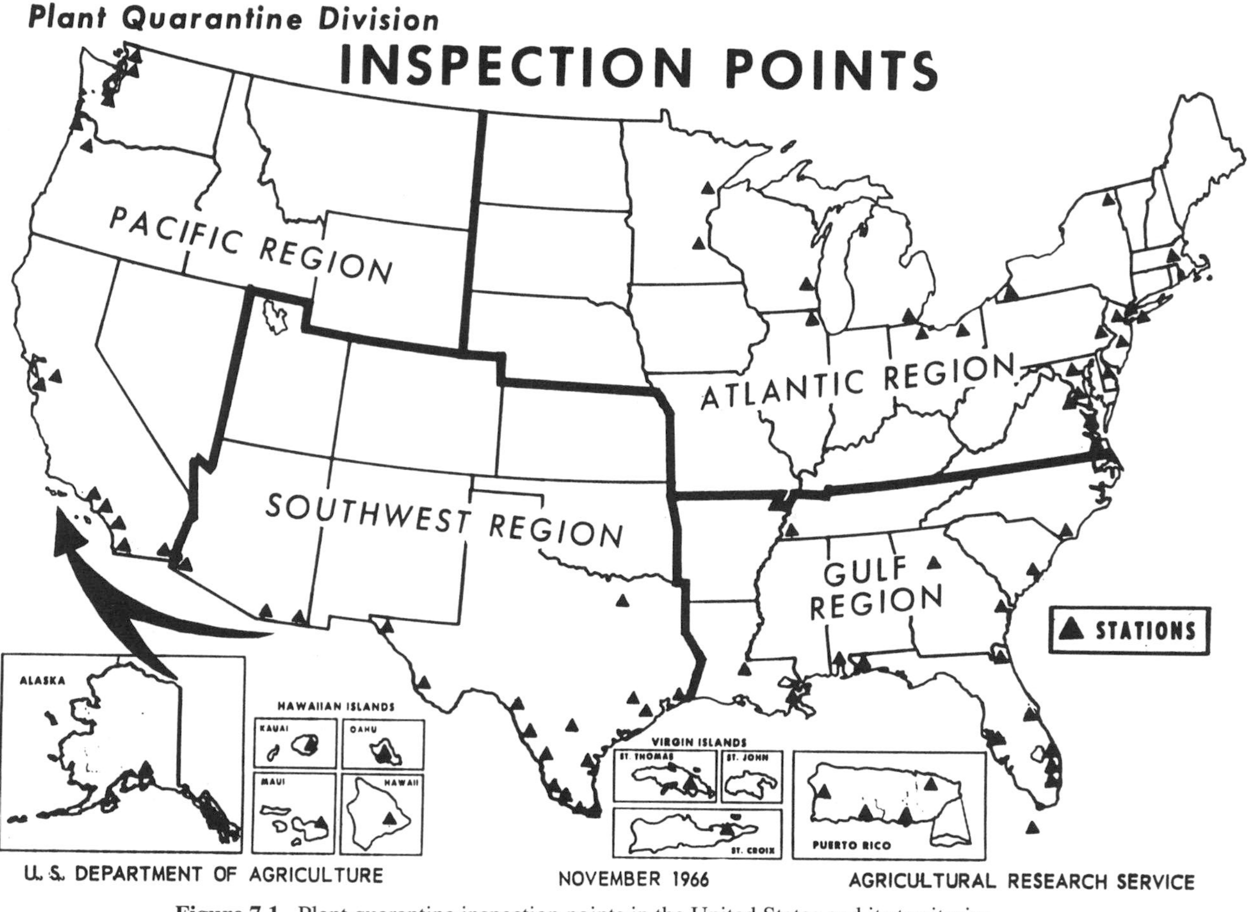

Figure 7.1. Plant quarantine inspection points in the United States and its territories.

Inspections and clearance of cargo in transit take place at all ports of entry and are mostly of a general nature. Commercial lots of plant material are inspected at one of 14 plant inspection stations with specialized facilities located at principal ports of entry.

The technology involved in quarantine inspections can be as simple as a handlens or as intricate as a scanning electron microscope. Visual examination is the primary way quarantine inspections are conducted, but culturing microorganisms on general or selective media, soil sieving, staining plant tissues, electron microscopy, serological tests of various kinds such as ELISA or agar diffusion, virus indexing, and bacteriophage assays may be used in searching for specifically designated pathogens. The techniques employed are many, often developed to detect a specific pathogen.

The techniques used to detect pathogens require time, expertise, and money. Therefore, most general inspections are based on symptoms even though symptoms alone are often unreliable for identifying specific pathogens.

Countries differ in the way low-risk plant material is inspected before being imported. For example, the United States prohibits importation of propagative stock (budwood, cuttings, transplants, seedlings, etc.) from countries known to have certain pathogens. Imports from countries where certain pathogens are not known to exist may be permitted on condition that plants be grown on the importer's premises for a prescribed period subject to official inspection. All plants must be accounted for, even those that die for unknown reasons, before the propagative material can be increased for sale. If the target pathogen, or other pest or disease, is detected the material can be destroyed under the supervision of the delegated authority, usually the federal or state plant inspection service.

The approach is different in Canada. There, representative samples are drawn from all lots of imported material and examined by indexing or by other methods at locations owned by the regulatory agency. If samples are free of the target pathogens it is assumed that the entire lot is free of infection. Meanwhile the importer is permitted to increase the material.

Postentry quarantines admit plant materials but require that they be held and grown for a period at a designated location controlled by the regulating agency to ensure freedom from pathogens and pests. These postentry quarantine stations are located in the country where the plant material ultimately will be grown. They are usually isolated from commercial agriculture areas and have closed-circuit plant growth rooms, propagating houses, glasshouses, screenhouses, and other facilities where material can be grown with minimum risk.

A more cautious postentry quarantine is the *intermediate* or *third-country* quarantine, which requires that plant material be grown in countries where the quarantined plant does not occur. An example in the previous chapter was India's requirements for the importation of peanuts for seed. The Inter-African Phytosanitary Council recommends a third-country quarantine for cacao because of witches broom fungus and swollen shoot virus, and rubber because of South American leaf blight. Export of cacao from Africa requires thorough inspection in the country of origin, followed by isolation at an intermediate quarantine station such as Mayaguez, Puerto Rico or Salvador (Bahia), Brazil, both of which are tropical locations removed from cacao-growing regions. An example of international cooperation in the conduct of intermediate quarantines is the passage of *Coffea* propagative material from Africa and Asia through USDA quarantine facilities at Glenn Dale, Maryland before it is admitted to South America for breeding purposes. Safety precautions for many tropical crops usually require intermediate quarantine in a temperate country at stations such as those at Kew, England, Wageningen, The Netherlands, and Glenn Dale, Maryland or Miami, Florida in the United States.

Elimination involves various techniques to produce or ensure pathogen-free plant material at its source, in contrast to cleaning it up at its destination. Elimination cannot be separated from prohibition and interception since the latter strategies often require elimination by preentry or postentry treatment using chemicals or heat to eliminate pathogens. Elimination is also broader in scope than prohibition and interception. The latter two strategies are almost exclusively limited to regulation of pathogens by means of quarantines, whereas elimination can exclude pathogens from individual fields (e.g., by seed treatment) as well as from countries.

Orchids are an important industry in Florida and Puerto Rico supplies many of the plants for this industry. Cultivated orchids grown in Puerto Rico for import to Florida must be grown in greenhouses, free of soil, and above ground in nurseries registered with the Puerto Rico Department of Agriculture. The plants must be inspected at least every 2 months and sprayed or dipped with a malathion solution. Wild orchids collected for export must be grown in a greenhouse for at least 6 months and inspected and treated the same as cultivated ones. Orchid plants being shipped must be certified to have been produced under the provisions described above *or* fumigated with methyl bromide at a rate of 2 pounds per 10^3 ft^3.

As a rule seed poses less of a hazard than plants, cuttings, tubers, bulbs, rhizomes, etc. Had Henry Wickham spirited *Hevea* seedlings out of Brazil in 1876 instead of seed it is likely that today leaf blight would be a serious disease in the rubber plantations of Malaysia instead of a potential hazard or threat to be feared. Nevertheless, seed can carry pathogens both externally and internally, in addition to those on fragments of crop debris mixed with seed. Production of pathogen-free seed is discussed in Chapter 17.

Elimination mostly amounts to physical removal, disinfection, or disinfestation. *Removal* separates diseased from healthy material by hand sorting, roguing diseased plants, pruning infected branches, flower clusters, or other parts, and mechanical sorting by flotation, screening, or blowing. *Disinfection* destroys or inactivates pathogens that are intimately associated with the host, specifically pathogens in seeds or plants where they are protected from the external environment. *Disinfestation* destroys or inactivates pathogens that are in nonvital associations with the host. This includes pathogens on the seed surface, in plant debris mixed with the seed, on senescent plant material, or on nonhost surfaces such as soil, machinery, implements, containers, and buildings.

Sorting and grading are important in processing bulbs of ornamental plants such as tulip, daffodil, German iris, Croft lily, gladiolus corms, rhizomes of bearded iris, orchid pseudobulbs, rooted cuttings or grafts of roses, rhododendron, and many other ornamental shrubs, grafted and budded fruit trees, banana corms, mint and asparagus roots, hop crowns, and other large fleshy vegetative organs used for propagation. Visual examinations often reveal symptoms or signs of bacterial, fungal, and nematode diseases such as Fusarium corm rot of gladiolus, Fusarium basal rot of daffodil, downy mildew in hop crowns, fungal and nematode infections in banana corms, blue mold rot of asparagus roots, and crown gall of roses and other woody plants. Relatively large and easily seen indicators of disease are effectively detected and the plants confidently rejected. Elimination by visual sorting will continue to be restricted to propagative materials of high value. It is too time consuming and costly to be used for large volume propagative items such as potato seed tubers.

Ergot sclerotia are removed from cereal grains by blowing, screening, or flotation. Smut balls are removed from grain by screening and blowing, but, unfortunately, many smut balls are broken and contaminate grain with teliospores before they can be removed. Nematode cysts can be removed from soil and plant debris by flotation and screening but inevitably some eggs can be left behind. Seeds of some parasitic plants such as dodder (*Cuscuta*) and witchweed (*Striga*) can be separated from larger seeds by screening, but often the carrier host has seeds of about the same size, for example, dodder and alfalfa, making separation impractical.

Roguing removes and destroys infected plants and is a common requirement for many preentry quarantines and certification programs. Certification of seeds and planting stock ensures that plant material is free (or relatively so) of pathogens. These programs are administered by governmental regulatory agencies that set standards and issue the necessary certificates to ensure importers that propagative material satisfies quarantine requirements. Certification programs usually involve three stages:

1. *Establishing tolerances*. A tolerance is the amount of disease allowed in particular grades or classes of material, such as breeders seed, foundation seed, registered seed, and certified seed. Tolerances vary with disease and inspection. Table 7.1 gives the tolerances allowed for three certification programs in Washington state. These rules are provided in the Washington Administrative Code and are the formal laws governing production of certified materials. There is a zero tolerance for approximately half of these diseases in all classes of stock.

2. *Field inspections or other tests*. Field inspections are conducted by personnel of the Washington State Department of Agriculture (WSDA) and financed at least in part by fees charged to owners who request certification. For example, a grower seeking to have a bean seed field certified must apply by July 1 and pays a fee of $4.00 per acre (or fraction) for each inspection. Beans require two inspections, one during the growing season and another after the plants have been windrowed but prior to threshing. Sampling for additional tests such as serological tests for viruses or bacteria costs five cents per hundred pounds. Fees for the tests are additional and depend on the type of test. Various tests are used for different pathogens on assorted crops. Plants might be grown from a seed sample (grow-out test) and examined for visual symptoms, cultured on selective media, or tested serologically.

Certification of some crops allow growers to rogue diseased plants following the first inspection and thereby satisfy the necessary tolerance for subsequent certification.

In addition to field inspections and tests other stringent requirements must be met. For example, sprinkler-irrigated bean fields are not eligible for certification against

Table 7.1. Disease Tolerances for Three Crops in Washington State

Seed Potatoes: Tuber inspection

Disease	Tolerances (%) Foundation	Certified
Bacterial ring rot	0	0
Powdery scab	0	0
Wart	0	0
Nematodes	0	0
Net necrosis associated with leaf roll virus	0.25	1.00
Deep pitted scab	1.00	3.00

Seed Potatoes: Field inspections

	Foundation		Certified	
Disease	1st	2nd/3rd	1st	2nd/3rd
Leaf roll	0.2	0.1	0.4	0.2
Mosaic and other virus diseases	1.0	0.5	2.0	1.0
Black leg and wilts	2.0	1.0	4.0	2.0
Bacterial ring rot	0.0	0.0	0.0	0.0

Strawberry Plants

Disease	Foundation	Registered	Certified 1st	Certified 2nd/3rd
Virus diseases	0	0	1.0	0.5
Red stele	0	0	0	0
Nematode	0	0	0	0
Lethal decline	0	0.5	2.0	1.0

Bean Seed

Disease	Foundation	Registered	Certified
Total seedborne diseases	0	0.5	1.5
Bacterial blights and wilts	0	0	0
Anthracnose	0	0	0
Seedborne mosaic	0	0.5	0.5

bacterial diseases. Idaho, a major bean seed producer, has had catastrophic experiences with halo blight in sprinkler-irrigated fields. As a result, bean fields were burned and could not be sown to beans for two years. Now, bean seed can not be produced in fields under sprinkler irrigation.

In Washington, fields planted for certification must have been free of halo blight the previous 2 years. Fields must also be one-quarter mile (1320 feet) from any field where any target diseases occur. Similar buffer requirements are common to many certification programs.

Certification rules also require disposal of crops that fail certification standards, especially of edible portions (seeds, tubers, bulbs, etc.). Idaho does not allow use of halo blight infected beans because of the high risk of contaminating other bean growing areas. They are destroyed by burning them. Bacterial ring rot is a serious potato disease and in Washington infected seed potatoes can be disposed of several ways. They can be processed or sold for fresh consumption, diverted outside the state at the growers expense and discretion if they were not grown in Washington, or destroyed by burning them.

Specific methods for producing pathogen-free seed and vegetative propagative material are discussed in Chapters 17 and 18. Additional examples of eliminating infected material are given in the chapters on eradication and cultural control.

3. *Issuance of phytosanitary certificates.* A phytosanitary certificate issued by the appropriate authority allows, and in most cases expedites, movement of propagative material through interstate and international trade channels. Disinfection involves pathogens that enter seeds or plants and are protected by the plant. Since disinfection is the treatment of a diseased plant it is sometimes considered therapy. Disinfestation and disinfection are similar in that both involve destruction or inactivation of pathogens by chemical or physical methods. Disinfestation involves pathogens that are merely transients on the plant or other surface.

Disinfectants generally are milder than disinfestants because they are in contact with living plant tissue and must be lethal to the pathogen but noninjurious to the host. Disinfection requires penetration into plant tissues and while heat has been the main method there now are some effective systemic chemicals such as carboxin for seedborne smut fungi. Sometimes a combination of hot water and a fungicide or nematicide is used. Some specific treatments are discussed in Chapter 8 on eradication. Both disinfectants and disinfestants must be economical and reasonably safe to the applicator.

Disinfestation utilizes heat and chemicals, or both. Many forms of heat have been tried including flaming, hot water, steam, sunlight, electric resistance, ultraviolet, ionizing radiations, ultrasound, and microwaves. Chemical disinfestation has included brining, fumigating, dusting (mainly to seed), drenches, and so forth. It has used a variety of chemicals such as formaldehyde, inorganic mercuries (mercuric and mercurous chlorides), organic mercuries, coppers, carbamates, thiram, captan, carbon disulfide, sodium and calcium hypochlorite, and acid treatment of seeds.

Barriers are the primary mechanism that makes exclusion effective. Pathogens are unable to penetrate these barriers without man's help or other disseminating agents. Major barriers are oceans, mountain ranges, jungles, and deserts where adverse conditions or absence of suitable host plants do not favor survival and establishment of pathogens. It is generally agreed that the larger the land mass covered by uniform quarantine regulations, the greater the protection of the area therein. Eight biogeographical regions have been proposed for quarantine purposes (Figure 7.2). These regions generally coincide with the eight regional plant protection organizations of the world (see Figure 6.1).

Also contributing to effectiveness of barriers is the marked reduction in concentration or effectiveness of inoculum as distance from its source increases. Early aerobiology studies showed that rust urediospore concentrations drop sharply with increases in altitude (Table 7.2). This reduces the chances of spores being caught up by and carried for long distances in high altitude wind currents. Fungal propagules and subsequent infection levels decline sharply at ground level as the distance increases from the source of inoculum. Several examples illustrated in Figure 7.3 indicate that infection or inoculum of three of the four pathogens listed drops to a minimum level less than 100 feet from the source. However, the large numbers of spores or other propagules produced by many pathogens mean that the small fraction that travels for some distance can be a significant number. Broad expanses of oceans do not always prevent the spread of pathogens by wind. Coffee rust was widely distributed through Africa and Southeast Asia in the late 1800s and early 1900s but absent from South and Central America. The sudden appearance of coffee rust in Brazil in 1970 is believed to have resulted from airborne urediospores being blown across the Atlantic Ocean (Figure 7.4). Similarly, Sigatoka disease of banana spread from Southeast Asia and the Pacific Region across the Indian Ocean to Africa and across the Atlantic Ocean to Central America by airborne spores (Figure 7.5). Sugarcane rust and peanut web blotch are believed to have reached the United States by spores of their respective pathogens being airborne across the Atlantic Ocean from Africa. Such long distance spread is probably not the norm and large natural barriers remain the major accessory to quarantines.

Other barriers also prevent or impede spread of pathogens over smaller areas. Ecological barriers prevent pathogens from becoming established. Cotton anthracnose is widely distributed through the cotton growing region of southeastern United States but absent from west of a line through central Oklahoma and eastern Texas. This line

TABLE 7.2. Number of Rust Urediospores Captured at Different Altitudes

Average Altitude in Feet	Number of Rust Urediospores
1000	355
1500	166
3000	136
12000	9
16000	2

Source: From Stakman, E. C., et al., *J. Agr. Res.* 24, No. 7, (1923).

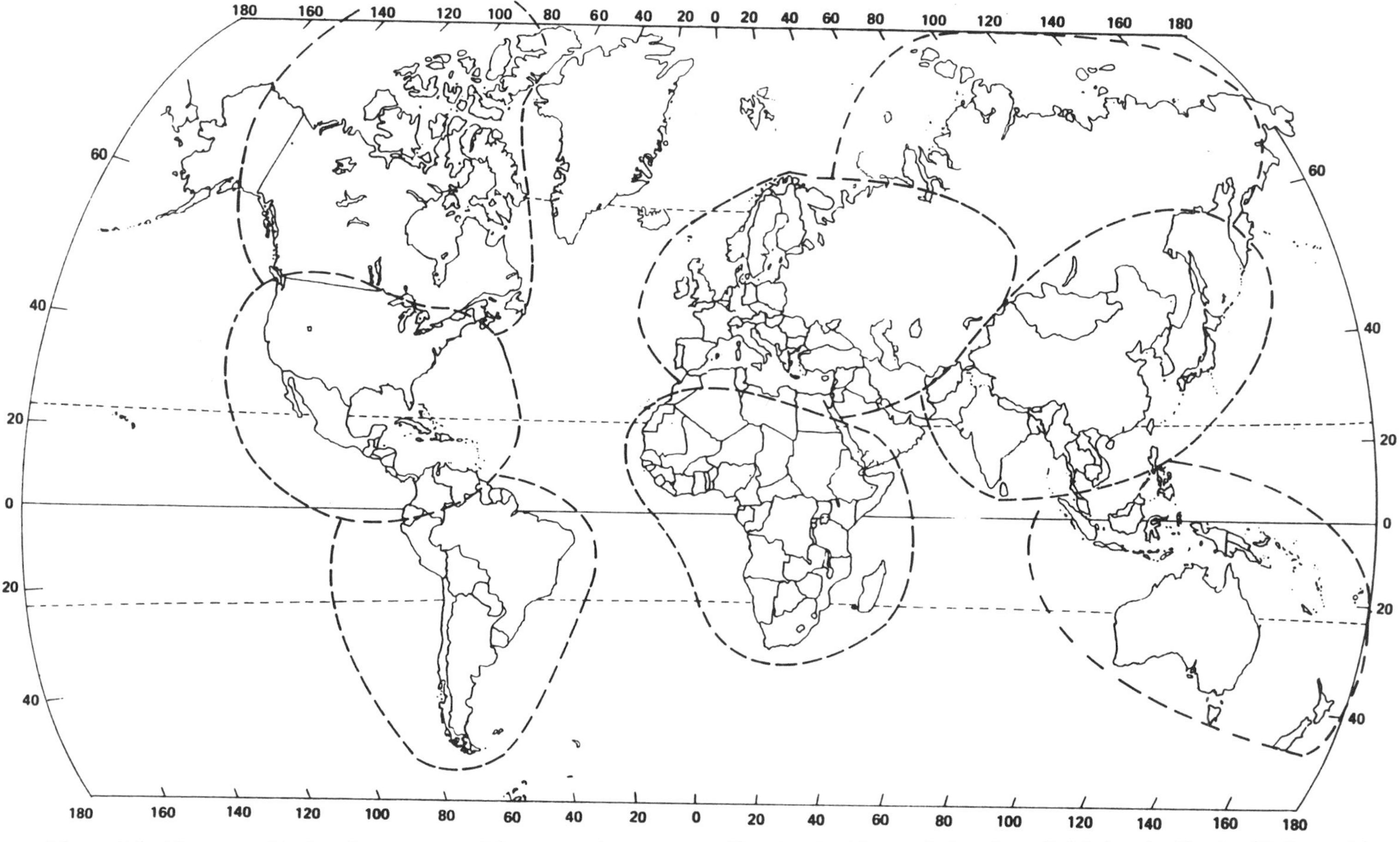

Figure 7.2. Biogeographical regions suggested for quarantine purposes. [Reprinted with permission from G. Mathys in, Hewitt, W. B., and L. Chiarrapa, (eds.), *Plant Health and Quarantine in International Transfer of Genetic Resources*, CRC Press, Cleveland, OH, 1977. Copyright CRC Press Inc., Boca Raton, FL.]

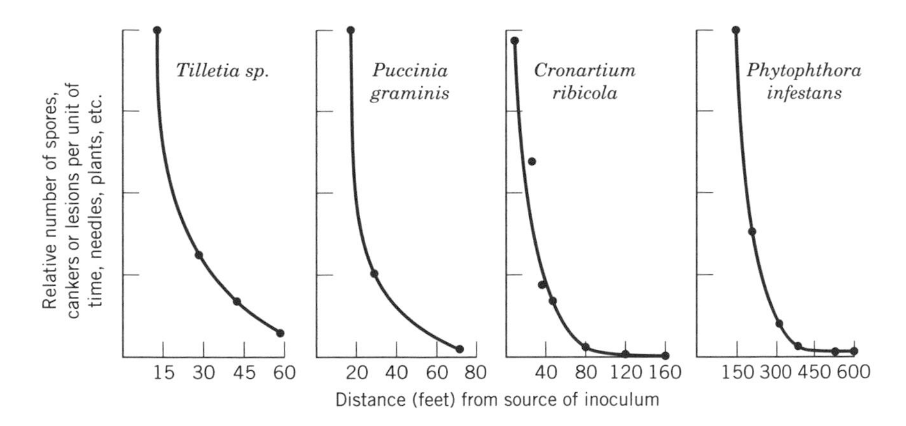

Figure 7.3. Distance of spread of several plant pathogens. (Adapted from Ingold, C. T., *Dispersal in Fungi*. Oxford University Press, 1953.)

coincides with summer rainfall. The average summer rainfall in western Texas, New Mexico, Arizona, and California is less than 6 inches while to the east summer rainfall ranges from 6 to more than 14 inches. This environmental condition determines whether or not anthracnose develops.

Soil temperature is also an important ecological barrier. Verticillium wilt is restricted largely to the northern half of the United States even though many of the plants subject to this disease are widely grown. For example, Verticillium wilts on raspberry or strawberry do not occur or are extremely rare south of 35 degrees latitude except on the Pacific Coast where a milder climate prevails. At the other extreme Texas root rot, which has a host range of more than a thousand plant species, is restricted to the southern United States and Mexico.

Broad expanses of jungles and forests form ecological barriers because absence of host plants prevent pathogens from vaulting these barriers. As openings are made in jungles and forests by settlers who then grow a variety of crops, pathogens have the opportunity to occupy these niches and progress across diminished barriers.

Artificial barriers such as trenching, screenhouses, plastic covers, glasshouses, and the like can also exclude pathogens but the usual intent is to protect a valuable plant or planting from disease rather than exclude a pathogen from a region. These barriers are more logically considered under the principle of protection and are discussed in Chapter 15.

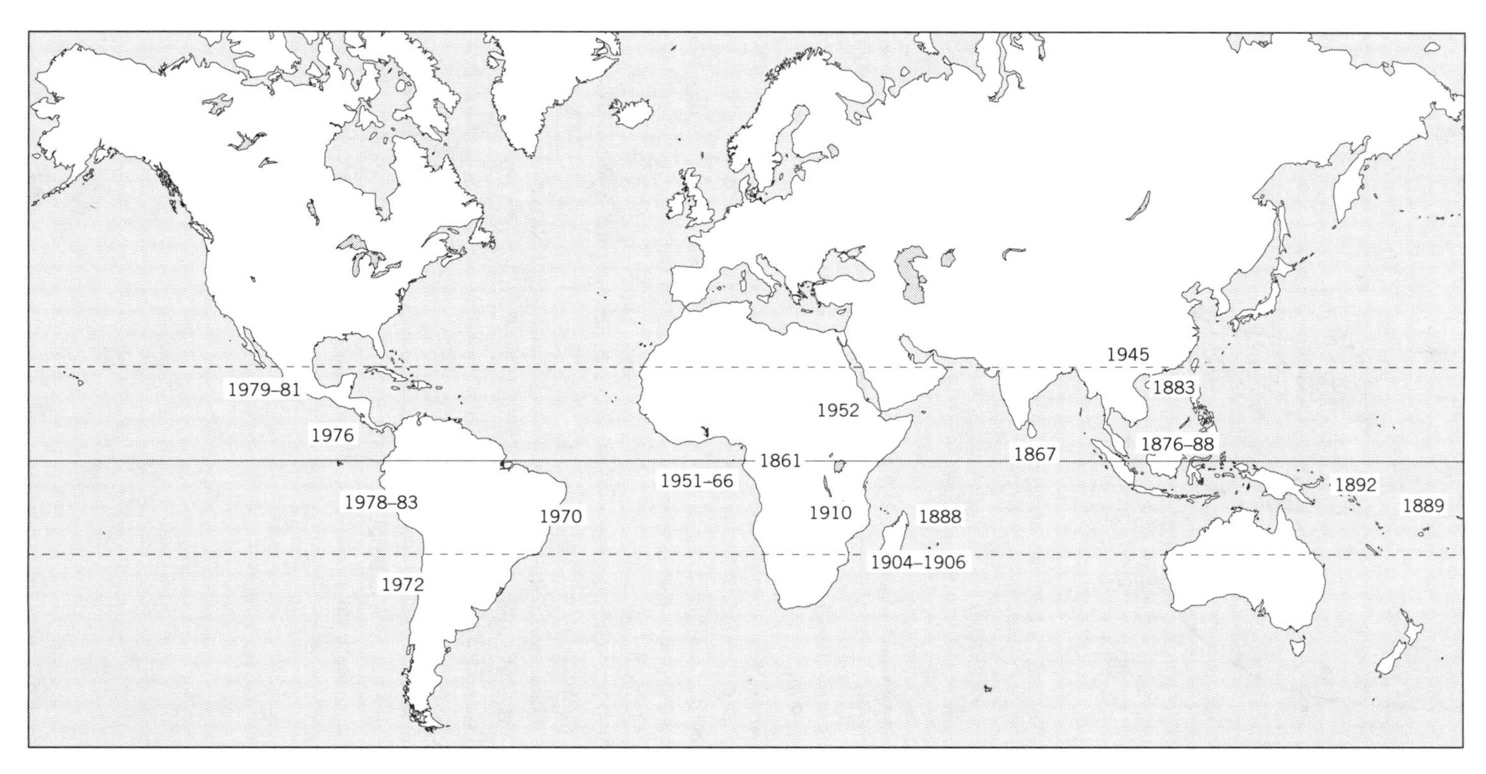

Figure 7.4. Worldwide spread of coffee rust. (Adapted from Schieber, E., and G. A. Zentmyer, *Plant Dis.* 68:89-93, 1984.)

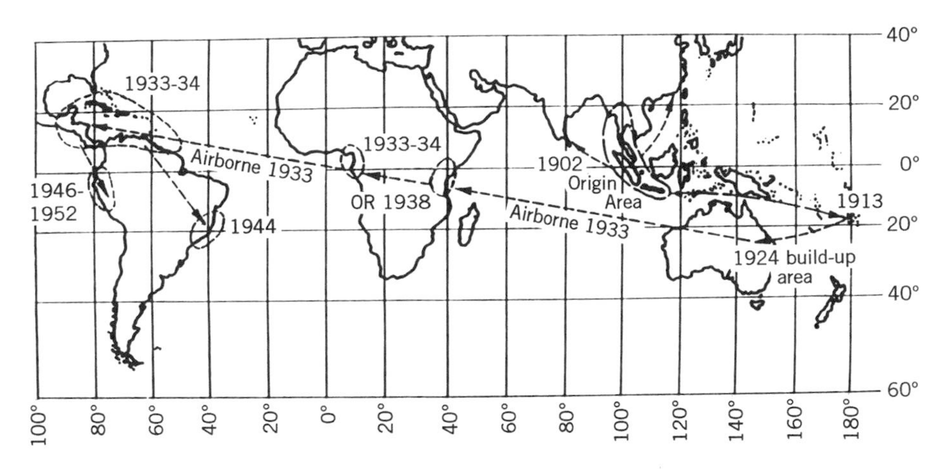

Figure 7.5. Sequential world spread of Sigatoka disease of banana. (From Stover, R. H., *Banana, Plantain and Abaca Diseases*. CAB International, 1972, with permission.)

REFERENCES

*Berg, G. H., Post entry and intermediate quaratine stations. In Hewitt, W. B., and L. Chiarappa (eds.), *Plant Health and Quarantine In International Transfer of Genetic Resources*. CRC Press, Cleveland, OH, 1977. Pp 315–326.

*Joshii, N. C., Plant Quarantine in India. *Rev. Trop. Plant Pathol.* 6:181–200 (1989). Published by Scholarly Publications, Houston, TX.

*Kahn, R. P., Plant quarantine: Principles, methodology, and suggested approaches. In Hewitt, W. B., and L. Chiarappa (eds.), *Plant Health and Quarantine in International Transfer of Genetic Resources*. CRC Press, Clevland, OH, 1977. Pp. 327–331.

*Karpati, J. F., C. Y. Schotman, and K. A. Zammarano (eds.), *International Plant Quarantine Treatment Manual*. FAO Plant Protection Paper 50, 1984.

*Limber, D. P., and P. R. Frink, The inspection of imported plants. In *1953 USDA Yearbook of Agriculture: Plant Diseases*. Pp 159–161.

*Rohwer, G. G., Plant quarantine philosophy of the United States. In Ebbels, D. L., and J. E. King (eds.), *Plant Health*. Blackwell, London, 1979. Pp 23–34.

*Shutova, N. N. (ed.), *A Handbook of Pests, Diseases, and Weeds of Quarantine Significance*. Kolos Publishers, Moscow, USSR, 1970. (Translated from Russian and Published for the USDA Agricultural Research Service by Amerind Publ. Co. Pvt. Ltd., 66 Janpath, New Delhi, 1978.)

*Weltzien, H. C., Geophytopathology. *Annu. Rev. Phytopathol.* 10:277–298 (1972).

*Publications marked with an asterisk (*) are general references not cited in the text.

8 Eradication

Eradication is the appropriate disease control action when a pathogen has breached the exclusion barrier but is not yet widely distributed or well established. Eradication aims at eliminating or reducing primary inoculum. Although it may ultimately fail, like exclusion, it can be a holding action until disease resistance or some other control is developed. Some so-called eradicative measures are really protective in their action. For example, removal of *Ribes* species, the alternate hosts of the white pine blister rust fungus, is generally included as an eradication measure. However, the pathogen is perennial in white pine, the economic host. Elimination of *Ribes* merely protects uninfected pines within a certain area but does not eliminate the pathogen already established in pines. Destroying infected pines to eradicate the pathogen was attempted in early stages of the control program. However, wide distribution of native pines and the long latent period, 3 or more years, between infection and symptom development mitigated against eradication of the pathogen.

Eradication is accomplished by one of three general means using different technologies applied in various ways.

1. *Removal* is direct destruction of a pathogen together with all or part of the host or substrate. The action and results are immediate.
2. *Elimination* is indirect destruction of a pathogen. The pathogen is not immediately removed or destroyed and host tissue may or may not be involved.
3. *Destruction* is direct destruction of a pathogen but not host or substrate. This is accomplished by chemical or physical methods of disinfection or disinfestation. These measures (e.g., seed treatment, and soil fumigation) are discussed in later chapters.

Following are examples of techniques employing removal and elimination. These methods are covered in more detail in other chapters, especially those dealing with cultural and biological methods of disease control.

REMOVAL

Of Plants or Groups of Plants

Roguing removes individual defective plants and plants harboring pathogens from an otherwise healthy population. This extreme measure sacrifices the host to kill the pathogen. Roguing is a common procedure in certification programs for propagative material. Examples include removal of leaf roll and other virus- infected potato plants from seed potato plantings, fruit trees infected with Kootenay little cherry and other viruses, various viral diseases in small fruits, citrus canker infected trees from nurseries and groves, and excising swollen shoot virus infected cacao trees. These are a few of the diseases controlled by roguing. Cutting down trees infected with dwarf mistletoe in commercial forests is a means of removing sources of pathogen seed that infects understory trees. Similarly, cutting out Dutch elm diseased trees removes them as a source of both the fungus and the disseminating bark beetle.

Of Parts or Organs

Pruning is an effective way to remove fire blight-infected branches from apple and pear trees, and canker diseases of roses, caneberries, and many trees and shrubs. Pruning removes branches diseased by white pine blister rust, pine gall rusts, dwarf mistletoes, sycamore blight, and a large number of diseases of perennial plants. Severe therapeutic pruning is used to eliminate or retard some vascular pathogens such as *Verticillium* and *Ceratocystis* in economically valuable trees.

Surgery involves removal of diseased or decayed tissue from tree trunks. It is standard in the maintenance of high value ornamental trees. Affected tissue is removed and cavities filled with concrete to provide strength to support the trees.

Scraping and scarification are methods of mechanically excising diseased tissue from tree branches and trunks. It has been used in the treatment of fire blight of pome fruits, bacterial canker of stone fruits, white pine blister rust, and other trunk cankers of economic and ornamental trees.

Hand picking leaf curl-infected peach leaves, azalea leaf galls, and galls of common smut of corn can be done on a limited scale by gardeners to eliminate inoculum of these pathogens. Harvesting asparagus removes sources of aeciospores that perpetuate the disease early in the growing season and hence reduce rust in production fields.

Contact herbicides such as dinitro compounds, or *flaming,* have been used to destroy basal spikes of hops to eliminate the initial flush of inoculum that starts downy mildew infection cycles in early spring. The same technique is used to destroy the first rust-infected shoots of spearmint to prevent initial production and dispersal of aeciospores.

Defoliating trees in rubber plantations with herbicides is the contingency plan for Malaysia in the event that South American leaf blight is detected. This extreme measure is described later in this chapter.

Of Infested Debris

Sanitation involves physical and chemical measures to remove pathogen-containing plant residues and exudates from equipment or other sources. These measures include cleaning grafting sheds to remove the rose black mold fungus, washing grain to remove smut teliospores, washing and disinfesting potato storage equipment and seed piece cutting knives to remove potato ring rot bacteria, cleaning potato storage bins to remove tuber rot pathogens, and cleaning and disinfesting contaminated poles and stakes to remove bean rust spores.

Raking and burning infected leaves of trees and shrubs with diseases such as hawthorn leaf spot, aspen leaf blight, cherry leaf spot, and a number of other fungal diseases are recommended practices for individual trees but probably are not feasible for commercial operations.

Vine pulling is a relatively new technique for removing potato stems to control Rhizoctonia infections on tubers.

Field burning is an old technique for eliminating inoculum but recent concerns about air quality, especially in populated areas, make this practice unpopular. Burning is used to reduce inoculum of brown spot needle fungus of longleaf pine, Cephalosporium stripe of wheat, Verticillium wilt of potato, and blind seed, ergot, and other diseases of grasses, to name a few.

Replacement of infested soil has been used to remove inoculum of persistent and difficult pathogens such as golden nematode, root pathogens of fruit and ornamental trees, sclerotial pathogens in flower beds, and various pathogens in nurseries and greenhouses. It is expensive and is not justified for low value commodities.

ELIMINATION

Alternate hosts are unrelated plants, such as wheat and common barberry for stem rust, required for completion of life cycles of heteroecious rust fungi. Other combinations are five-needle pines and *Ribes* species for white pine blister rust, eastern red cedar and apple for cedar apple rust, junipers and pear for trellis rust, oak trees and southern pines for fusiform rust, and many more. One host is the aecial host and supports the pycnial (if produced) and aecial stages. Aeciospores infect the telial host, which produces the uredial (repeating) stage if one occurs, and telial stage, which produces basidiospores, which, in turn, infect the aecial host and complete the life cycle. Since both hosts are required to complete the cycle, elimination of either will break the cycle and prevent infection of the other host. In practice the primary host is usually important economically and the other is called the alternate or secondary host. The relative value of the two hosts can vary. For example, to the forester white pine is the economic host and currants and gooseberries the alternate hosts. But to the jelly manufacturer currants and gooseberries are the plants of value and five-needle pines are the alternate hosts. Similarly, where trellis rust occurs, pear growers consider junipers alternate hosts while landscapers views pear trees as alternate hosts.

The term alternate host is often wrongly applied to other hosts of a pathogen, but these are not essential for completion of the life cycle and are more properly called additional or alternative hosts.

Elimination of alternative hosts is important in the control of many diseases. Some pathogens have a wide range of hosts while others have a few, common plants that serve as reservoirs of the pathogen. Many plant viruses have weed hosts. For example, some pea viruses survive in wild legumes such as vetch, and curly top and sugar beet yellows viruses can infect a large number of weed plants. Root-knot, root-lesion,and other nematodes also infect many plants including both crops and weeds. The X-disease mycoplasma of sweet cherries and peaches survives in wild chokecherry. The ergot fungus can infect many wild grasses and several rust fungi occur on cultivated and wild *Rubus* species.

Some crops or weeds serve as bridge crops for pathogens that cannot survive without a living host. The wheat streak mosaic virus is vectored by the wheat leaf curl mite, which cannot survive for long without a green host. In areas where both spring and fall-seeded wheat crops or late maturing winter wheat varieties are grown, planting and emergence of fall wheat before the other wheats mature provides a "green bridge" for viruliferous mites to move from the older wheat crop to the new crop where it can cause serious losses.

Crop rotation with nonhosts eliminates the food source and starves a pathogen. To be effective crop rotation must be clean. For example, all grass weeds for take-all, volunteer potatoes for late blight, or overwintering volunteer barley for Cephalosporium stripe of wheat must be destroyed.

Trap and decoy crops are sometimes considered to be the same, but in practice trap crops allow a pathogen to invade the plant where it cannot complete its development. Marigolds (*Tagetes* species) take the root-knot nematode but no eggs are produced. The potato powdery scab fungus can infect roots of some *Datura* species but does not produce galls or resting spores. The cabbage club root fungus also infects some nonhost plants but does not produce resting spores.

Decoy crops stimulate pathogens to germinate but resist invasion. Encysted eggs of golden nematode can remain inactive for many years. The host, however, excretes hatching factors that stimulate emergence of larvae from cysts. Nonhost hatching factors also stimulate emergence, but the plants are not suitable hosts and the larvae starve. Witchweed (*Striga*) is a phanerogamic parasite of corn and its seeds usually germinate only in the presence of a host plant. However, some nonhost plants such as soybean and cotton also stimulate germination of seeds but the seedlings are unable to produce haustoria and infect the decoy plants.

Sclerotia of the onion white rot fungus can survive without germinating in soil for as long as 18 years in the absence of *Allium* hosts, but a number of mercaptan (sulfhydryl-containing) compounds stimulate the sclerotia to germinate. Theoretically, germination in the absence of a host will result in depletion of food reserves in the sclerotia and hasten demise of the fungus. This technique is being tested in some garlic and onion-growing areas to control white rot.

Cultivation is one of several routine farming practices that can eliminate inoculum. Plowing buries inoculum such as ergot sclerotia and rust teliospores. Crop residues such as corn stover infested with the stalk rot fungus are incorporated into the soil where they decompose.

Biological control has generated great interest in the past decade because it relies on natural systems to eliminate pathogens by antagonism or other action. This topic is covered in Chapter 16.

Stump removal removes the food base required by some pathogens for infection of adjacent plants. It is used in forests and orchards to control Armillaria root rot of conifers and fruit trees and in landscaping to eliminate inoculum of the Dutch elm disease fungus.

Mechanical or chemical girdling of trees prior to felling depletes stored food from roots and eliminates inoculum of various root rot fungi.

EXAMPLES OF ERADICATION PROGRAMS

Eradication programs have been used throughout the world for a variety of diseases. Some programs were conducted long before microorganisms were known to be involved in the disease process. For example, it is reported that as early as 1660 farmers in France urged passage of laws requiring destruction of barberry bushes to protect wheat and other cereals from stem rust. If true, this was undoubtedly based on observations that rust on wheat was more severe in the vicinity of barberry bushes. Even after the causal nature of fungi, bacteria, and nematodes was known, field observations were frequently the basis for developing eradication strategies. When eastern white pine was planted extensively in Europe because of its superiority to the native pines severity of white pine blister rust was soon apparent. Foresters and other technicians noticed that the disease was absent or much less frequent on pines more than 1000 feet from currants and gooseberries. This formed the basis for *Ribes* eradication in the United States, although technically this is a protective action rather than eradicative.

The following are examples of past or present eradication programs in the United States, Canada, Central America, Africa, and Malaysia. These are examples that have had some degree of success, but there have been many more that were unsuccessful. These programs involve crops that constitute major economic assets of countries, regions, or districts and the pathogen being eradicated poses a serious risk to that economy. Eradication programs involving governmental agencies cannot be justified for mild diseases on minor crops. Nor would they receive the public support essential for their success. Most programs include intensive public relations and educational activities to gain public and legislative support.

Stem Rust

This is probably the first eradication program conducted in the United States, possibly in the world. In 1725 a law was passed in Connecticut authorizing barberry eradication measures; similar laws were passed in Maine in 1755 and Rhode Island in 1766. Wheat is a major crop grown over a wide range of environments from temperate to tropical and from arid to moist. Barberry, the aecial host, is essential for overwintering in colder regions because the teliospores from wheat produce basidiospores that infect only barberry, not wheat. In warmer areas such as the southern part of the United States the fungus survives in the uredial (repeating) stage. Although pathogenic races occur in its absence, the barberry functions in the production of new races since the pycnial stage preceding aecial development allows for genetic recombination. Two native barberry species occur in localized areas of Colorado and Virginia/West Virginia but the common barberry imported from Europe is the most important alternate host. This plant was brought to the United States by early settlers for its edible fruit and ornamental value. In hedgerows it provides shelter and food for wildlife. This simplifies the eradication process since the bushes are more likely to be in an exposed and visible area near some type of habitation.

The Plant Quarantine Act of 1912 set the stage for a countrywide barberry eradication program. Domestic quarantine 38 provided for eradication of common and other barberries and designated eradication districts, usually counties within cooperating states. The federal barberry eradication program began in 1918 and a total of 19 cooperating states eventually joined. Washington joined the program in 1923, although Idaho and Oregon were never in the program. The number of personnel was never large, consisting of an administrative corps in the USDA with a state director and a small cadre of permanent employees. In comparison, the white pine blister rust *Ribes* eradication program described later in this chapter had over 2500 employees in the Northwest alone in 1946 while the barberry eradication program had only 175, for the entire country. Temporary seasonal employees searched and destroyed barberry bushes wherever they were found. Because barberry bushes (and *Ribes* in the case of white pine blister rust) resprout if any crown tissue remains in the soil, large bushes were cut off at ground level and the cut crown sprayed or dusted with ammonium sulfamate (Ammate™). This material was as effective as hormone-type herbicides and more practical for use on individual bushes.

The discovery strategy often relied on reports of rust on wheat, whereupon an organized search was made in the area. Maps were maintained showing the location of removed barberry bushes and were used in reworking the areas. One major difficulty was the dispersal of barberry seeds by birds eating the berries and passing the seeds at distant locations from the source.

By 1955 the program had destroyed almost half a billion barberry bushes and covered over a million square miles in the 19 cooperating states. An additional function of the program was the collection of rust for race determination at the Cooperative Rust Laboratory affiliated with the University of Minnesota at St Paul. In 1955, 2000 collections were processed and 25 races identified.

The federal eradication program was discontinued in 1977, but some states continued on their own initiative. Fortunately, resistant varieties and the general reduction of barberry plants made the program unnecessary.

Apple Rusts

Apples were brought to North America by the first settlers and constituted a major food source, eventually becoming a crop of great economic value. Several rust diseases, the most important being cedar-apple rust, occurred in the United States on native crabapples and the alternate hosts, *Juniperus* species. In the 1800s apple rust became a severe problem on cultivated apples in the eastern United States. Observations and experiments in orchards of Virginia and elsewhere demonstrated that removal of junipers, primarily eastern red cedar, from the vicinity of orchards dramatically reduced incidence of rust.

West Virginia state law empowered the Crop Pest Commission to enter private property and destroy red cedars wherever they posed a risk to apple orchards. Similar compulsory state action was taken in the Hudson Valley of New York to remove junipers within 0.5 mile of apple trees. Several other states had similar restrictions on growing red cedar in the vicinity of apples. Needless-to-say, such heavy-handed governmental action was resented by many landowners whose livelihood did not depend on apples and much litigation followed. Usually the states prevailed, but often they were forced to compensate the owners for the loss of their prized junipers.

Several eastern states required that junipers (the telial host) be removed from the vicinity, usually 1 or 2 miles of apple orchards, despite the fact that basidiospores, although short-lived, could move 7 or 8 miles. In addition it was unlawful to plant junipers as ornamentals or for other purposes such as windbreaks. By 1940, nine states, including New Mexico, had laws requiring removal and restricting growing of junipers. Eradication of junipers was eventually discontinued when resistant apple varieties and effective fungicides became available.

Three different, but similar, rusts affect apples. In addition to cedar-apple rust, which infects both leaves and fruit, quince rust and hawthorn rust attack leaves of apples as well as other pomaceous hosts. Telial hosts consist of several juniper species. Their life cycles are similar with the exception that the telial galls of cedar-apple rust produce basidiospores for only 1 or occasionally 2 years, while the telial galls of quince and hawthorn rusts are perennial on juniper and may produce spores for as long as 20 years. This perennation of the telial stage further justifies eradication of junipers. Cedar removal had to be thorough and if cut out in winter or spring trees had to be burned immediately. Otherwise, galls could produce telial horns and sporulate before the galls dried out and died, thereby infecting nearby apple trees.

Pruning galls before they can swell and discharge spores offers some control for individual trees of ornamental value. However, this is usually not practical because some galls are invariably missed and the value of red cedars is so low that complete removal of the alternate host is the only realistic action.

An eradication program for control of pear trellis rust exists today around orchards in southwestern British Columbia. When rust is found on pears the Ministry of Agriculture and Fisheries surveys the surrounding area to locate the junipers. There are no susceptible native junipers in the area and the alternate hosts are either cultivated or abandoned ornamental junipers. These are removed by the government and replaced with ornamental shrubs other than junipers. There appears to be some question of the value of this program other than to placate pear growers since the rust apparently causes little damage to either pears or junipers.

White Pine Blister Rust

When this disease was discovered on currants in a planting at Geneva, New York in 1906, the affected plants were destroyed. In 1909 the first infections on pine also were found in nurseries in New York state and were destroyed. Even after the *Ribes* eradication program, which actually is a program of host (pine) protection rather than pathogen eradication, was started in 1915, survey crews trying to determine the extent of the disease eradicated diseased pines as they found them. This attempt to eliminate the pathogen was doomed from the start for several reasons. The pathogen was already established in the major pine regions of the country, the long latent period of infection makes early detection in pine difficult, and the long distance that aeciospores can travel (up to 200 miles) and infect *Ribes* means that pockets of infection can spread far and fast.

The eradication program unleashed a formidable army in the pine forests of the northern United States and California. The number of men grubbing out *Ribes* reached a peak during the depression years when almost 20,000 workers combed the woods for currant and gooseberry bushes in some seasons. This number declined markedly with the beginning of World War II and never again approached the peak numbers of the mid-1930s when 8600 persons were employed in the east and 10,000 in the west.

The blister rust control program, apart from the elimination of diseased pines, had two phases. One was the eradication of cultivated *Ribes*, primarily European black currants. This part of the program was relatively easy because cultivated currants were located in home gardens where they were readily found and identified and there were few escapes to the wild, such as occurred with common barberry described earlier. Eradication of cultivated currants was administered by Plant Protection and Quarantine personnel (Figure 8.1).

Elimination of wild *Ribes* was another matter. As one early forest pathologist quipped "that's a mighty big garden to weed out there." Early operations consisted primarily of hand pulling *Ribes* by crews working strips marked off with grocers string

Figure 8.1. Removal of European black currant bushes from a home garden by personnel of the USDA Bureau of Entomology and Plant Quarantine. (U.S. Forest Service photo.)

Figure 8.2. Blister rust control *Ribes*-eradication crew working a strip marked off with grocers string. (U.S. Forest Service photo).

(Figure 8.2). *Ribes* along streams were killed by spraying them with sodium chlorate or removed mechanically by bulldozers and replaced with grass meadows. New techniques involved gasoline-powered blowers (Figure 8.3) or helicopters spraying large areas of upland *Ribes* with hormone-type herbicides such as 2,4-D or 2,4,5-T.

Hundreds of thousands of acres were worked and reworked and millions of *Ribes* were destroyed, but in the end the program failed because it was based on erroneous premises. One was that as much as 25 feet of live *Ribes* stems was tolerable in worked areas, but it turned out that this amount produced much more inoculum than expected. The other premise was the distance that basidiospores could travel and still be viable and infective. Originally it was believed that spores were infective for no more that a thousand feet to one-quarter mile from their source. More recent research demonstrated that under some conditions of climate and topography the spores could travel a mile or more and infect pines.

Figure 8.3. Spraying 2,4,5-T on a brush field with a Buffalo turbine blower to eliminate large areas of *Ribes* at lowest cost. (U.S. Forest Service photo.)

The final death blow to the *Ribes* eradication program was the failure of cycloheximide (Actidione™) sprayed on pines to eliminate cankers. Early observations suggested that the antibiotic in a solution of diesel oil sprayed on the lower 5 feet of trunks of young pine trees translocated upward and killed cankers on the trunk, thereby saving the tree. However, more carefully conducted evaluations showed that many of the cankers were either not killed or coincidentally parasitized by *Tuberculina maxima*, a hyperparasite of the blister rust fungus.

Ribes eradication and antibiotic spray programs were abruptly cancelled in the western United States in 1977, but *Ribes* eradication continues on a limited basis in the east. The eastern program did not have to contend with *Ribes petiolare* and *Ribes bractyosum*, two stream-type *Ribes* that are as susceptible to blister rust as are European black currants. Moreover, climatic and topographic conditions are much more favorable for control in the eastern United States. In the east pine stands are smaller, more often in plantations than wild stands, and there are few deep valleys as in the west into which cool damp air can fall carrying viable basidiospores to pines.

Citrus Canker

About 1910 citrus canker, a bacterial disease, was introduced into the Gulf area of the United States on nursery stock from Japan, but was initially misidentified as scab, a fungal disease. When the true and serious nature of the disease was discovered an eradication program was started in Florida in 1915 and eventually extended to other citrus states in southeastern United States. Losses from the disease in Florida were estimated at 6.5 million dollars between 1914 and 1931. Eradication consisted of burning affected trees on the site. Properties were quarantined where the disease was found and stringent sanitation measures were taken to prevent further spread of the pathogen. By 1917, canker disease was mostly eradicated from Florida and none was found after 1926. In the process 257,745 mature trees and over 3 million in nurseries were destroyed on 414 properties in Florida. By 1934, 20 million trees had been destroyed in all states affected. Periodic surveys were conducted and occasionally infected trees were found in several states. The last canker found in the United States from that outbreak was in Texas in 1943. No additional disease was found by 1952 and in 1964 it was thought that citrus canker had been eradicated from the United States. But not for long.

Citrus canker was again found in Florida in September 1984. But this time the disease was caused by a different strain of the pathogen, one that affected nursery trees but had little effect on mature trees in groves. A Citrus Canker Disease Action Plan was developed by USDA APHIS specifying the procedures for survey, eradication, and regulatory action.

The survey was monumental in scope because of 650,000 acres of citrus, 1000 citrus nurseries, and 1300 retail outlets. The number of inspectors assigned to the survey reached 600 at times and more during seasons when conditions favored the disease. The survey also posed a risk because of the movement of contaminated vehicles and inspectors to other locations. This risk required disinfestation of vehicles, tools, hands, shoes, and clothing with chemicals such as 70% isopropyl or ethyl alcohol or 300 ppm of quaternary ammonium compounds. The surveyors were trained to recognize symptoms of the disease and field diagnoses were supplemented by laboratory methods that included immunofluorescent analysis, ELISA, or DNA probes.

By November 1986 almost 20 million nursery trees and 5000 grove trees had been destroyed, mostly by burning at approved locations. In some areas commercial citrus groves adjacent to heavily infested nurseries were sprayed with sublethal concentrations of a contact herbicide (300 ppm of diquot). This caused leaves to die within a few days and twigs to die back in a week or two but the trees resprouted in about 30 days, thereby allowing for bacteria on leaves and wood to die before new leaves developed.

The cost of the program from September 1984 to July 1986 was 25 million dollars, of which 12.3 million compensated owners for destroyed trees. These payments gained the cooperation of the citrus industry and the program was viewed with favor, at least initially. However, resistance to the program increased with time and eventually resulted in 18 lawsuits involving destruction of exposed citrus trees and/or compensation. Some of the trouble centered around debates, namely, whether the nursery form of the pathogen could damage mature citrus trees.

Initially all nurseries and groves in Florida were quarantined, but a short time later only nurseries where stock may have been exposed to the bacterium were quarantined while other intrastate movement was permitted. The following year a quarantine was placed on nursery stock movements to retail outlets and homeowners, but all quarantines were lifted as the eradication program progressed. Time will tell if citrus canker has been eradicated from Florida.

Citrus is grown in most of the tropical and subtropical parts of the world and citrus canker causes problems there as well as in the United States. Similar eradication programs have been conducted in South Africa, Australia, and New Zealand.

Kootenay Little Cherry

This viral disease was first observed in 1933 in one sweet cherry orchard in eastern British Columbia. Fifteen years later nearly every cherry tree in the Kootenay region was infected. Surveys revealed that the virus was also common in Kwanzan, Shirofugen, and other flowering cherries, some of which are used as rootstocks for sweet cherries. Major control efforts resulted from the British Columbia Plant Protection Act of 1954, which provided for a little cherry control area embracing the principal disease-free cherry growing districts of the province. Preventive measures prohibited all movement of fruit trees and fresh fruit from the Kootenay region to the control area and required removal of infected cherry trees found in surveys of the control area. These measures were complemented by compulsory eradication of all ornamental flowering cherry trees within the control area and prohibition of further plantings. As of January 1976 the disease had not become established outside of the Kootenay region.

However, in 1977 infected trees were found in the Okanogan and Kootenay valleys and a new round of eradication was undertaken. Because of the sizable sweet cherry industry in Washington state immediately south of the British Columbia areas affected, survey activities were started there. Since the symptoms on sweet cherry are apparent only at fruiting, which even then may be difficult to discern, some method of detection was needed. In the United States the APHIS surveys utilized a fluorescent staining technique on sections of leaf petioles. Canada relied on budding suspect trees onto indicator varieties such as Sam and Star, which produce a bright reddening of foliage in fall. More recently Canindex was selected as a superior indicator variety for little cherry virus.

A few infected trees were found in Washington, usually on flowering cherry roots or in proximity to flowering cherries. But thus far no evidence of widespread infection such as has occurred in the Kootenay area has been shown.

Golden Nematode

In 1946 the Plant Pest Control Division (now APHIS) of the USDA conducted surveys in potato-growing areas other than Long Island, New York, but no other infestations of the golden nematode were found. This made eradication feasible and 1500 acres of infested land was fumigated with 50% dichloropropene using 450 pounds per acre. After 1950 all newly infested fields not adjacent to previously infested areas were also fumigated. All land of each grower, not just infested fields, was considered for treating. Farmers were advised to fumigate bags and machinery with methyl bromide at 23 pounds per 10^3 ft^3 of space for 16 hours. Details of soil fumigation techniques are discussed in Chapter 10.

Quarantine regulations in Holland prohibit growing potatoes more than 1 in 3 years on clean land. Potatoes or tomatoes are banned indefinitely from fields infested with the nematode.

Use of hatching factors, i.e., compounds that stimulate emergence of larvae of some cyst nematodes such as the golden nematode, has been suggested, but this approach has never shown much success. Nevertheless it is an interesting, if unusual, form of disease control.

Swollen Shoot of Cacao

Cacao is a major crop in west central Africa and is grown to some extent in other tropical regions. Swollen shoot, a viral disease transmitted by mealybugs, is the most commercially important disease in Ghana, the Ivory Coast, Nigeria, and Sierra Leone. Millions of trees have been killed by the virus or destroyed as part of a cutting-out (grubbing) program.

Swollen shoot was discovered in Ghana in 1936, but little was done about it until 1947 when enactment of the Plant Pest and Disease Ordinance allowed the Division of Agriculture to undertake systematic control work. If the disease was deemed too advanced for effective control the area was abandoned, but cutting-out was done around the edges of such areas to contain the virus. The mealybug vector is apterous, consequently insects spread the virus as they crawl from infected trees to healthy ones.

By the end of 1957 nearly 70 million infected trees had been removed in Ghana. Inspection of grubbed areas revealed that visibly infected trees had been removed but not contact (adjacent) trees. Arrangements were made to pay farmers to remove apparently healthy contact trees. By 1963 over 113.5 million trees had been removed by the government or by farmers after marking by survey teams. Compensation to farmers was more than 11 million pounds.

The virus was discovered in Nigeria in 1944 and cutting-out became more severe than in Ghana. All cacao trees up to 30 yards from margins of infected areas were removed. Trees were removed to a distance of 5 m around disease centers with fewer than 6 obviously infected trees, 10 m from groups of 50 to 60 infected trees, and 15 m from larger centers. Missed and latently infected trees amounted to less than 5%. These were then treated as separate outbreaks in subsequent inspections.

To prevent regeneration of cut-out trees and mealybug crawlers from moving to nearby healthy trees, a 1-m section was removed from surface roots and piled several paces (10–15 feet) away from surrounding vegetation.

One of the lessons learned from this program and applicable to other eradication attempts is the futility of trying to cut out infection centers with more than 50% infection. Once the canopy has closed in there is no barrier to mealybug movement.

Banana Diseases

Several viral and soilborne diseases of banana are eradicated by destroying diseased plantings even though well-established ones are hard to destroy. Bulldozing is the cheapest way to destroy large areas of bananas. Successive crops of suckers are removed with brush cutters. Small areas can be treated with 2,4-D but this method, while effective, is slow and repeated applications are usually necessary.

Eradication of bacterial wilt (Moko disease) requires early detection based on inspections. Infected and adjacent healthy plants are killed with herbicides. Buffer zones vary from 15 to 30 feet depending on the bacterial strain. Some strains are mechanically transmitted by insects and require wider buffer zones than others. Weeds in buffer zones also are killed with herbicides because they can serve as alternative hosts. After bananas are killed areas are fallowed for 6 to 12 months to allow bacteria in soil and plant debris to die.

Bunchtop virus control also requires early detection and prompt destruction of infected plants. Australia requires destruction of diseased plants both in production areas and abandoned plantations.

South American Leaf Blight of Rubber

The adage that introduced Chapter 2, "desperate diseases require desperate remedies", is appropriate to the contingency eradication plan for South American leaf blight in Malaysian rubber plantations.

Until 1900 practically all of the world's rubber came from Brazil and Peru, but today only 1% comes from South America, 92% from Asia, and 7% from Africa. This is due to the devastation wrought by the leaf blight fungus and the basis for stringent quarantine regulations on importation of plant material from South America. The fungus was described in 1904 but the disease was not serious on wild trees in native forests. However, when trees were concentrated in plantations the fungus spread from wild *Hevea* species into plantations and severe losses occurred. One planting of 2000 acres was destroyed by the disease in 1933; another of 16,000 acres was partly saved by crown budding with partially resistant material. Costa Rica reported in 1941 that all but 10% of 36,000 trees were lost to leaf blight. The Ford Motor Company developed a 7900 acre rubber plantation in Brazil, but in 1933 leaf blight destroyed one-fourth of the plantation.

Rubber plantations in Malaysia originated from seeds Henry Wickham carried from Brazil in 1875. Because the pathogen is not seed transmitted and the Wickham seeds were germinated at Kew, England, the *Hevea brasiliensis* clones that established Asian rubber plantations were free of leaf blight. However, this narrow genetic base is highly susceptible. In the absence of resistance a contingency plan was devised in the event that leaf blight appears in Malaysian rubber plantations.

If the fungus is found, the affected area and a one quarter mile surrounding zone will be sprayed by air with a 5% solution of the herbicide 2,4,5-T in 3 gallons of diesel oil per acre to defoliate the rubber trees, the same material that was sprayed on the jungles of Vietnam in the 1970s war. A day later 1% pentachlorophenol in 3 gallons of diesel oil per acre is applied by air as an eradicant fungicide.

REFERENCES

*Anon., *Control of Plant-Parasitic Nematodes*. National Academy of Sciences, Washington, D.C., 1968.

*Benedict, W. V., History of white pine blister rust control—a personal account. USDA FS-355, 1981.

*Duffus, J. E., Role of weeds in the incidence of virus diseases. *Annu. Rev. Phytopathol.* 9:319–340 (1971).

*Entwistle, P. F., *Pests of Cocoa*. Longman, London, 1972.

*Fuller, E. H., Plant life and the law of man. IV. Barberry, currant and gooseberry, and cedar control. *Bot. Rev.* IX (8):483–592 (1943).

*Holliday, P., *Fungus Diseases of Tropical Crops*. Cambridge University Press, Cambridge, 1980.

*Hutchinson, F. W., Defoliation of *Hevea brasiliensis* by aerial spraying. *J. Rubber Res. Inst. Malaya* 15:241–274 (1958).

*Mai, W. F., and B. Lear, The Golden Nematode. *New York (Cornell) Agr. Exp. Sta. Ext. Bull.* 87:1–32 (1959).

*Martin, J. P., E. V. Abbott, and C. G. Hughes (eds.), *Sugar-Cane Diseases of the World*, Vol. I. Elsevier, New York, 1961.

*Roelfs, A. P., Effects of Barberry Eradication on Stem Rust in the United States. *Plant Dis.* 66:177–181 (1982).

*Schoulties, C. L., et al., Citrus canker in Florida. *Plant Dis.* 71:388–395 (1987).

*Stover, R. H., *Banana, Plantain and Abaca Diseases*. Commonwealth Mycological Institute, Kew, Surrey, England, 1972.

* Publications marked with as asterisk (*) are general references not cited in the text.

9 Seed Treatment and Eradicant Chemicals

Eradication of pathogens with chemicals is accomplished by treating true seed, disinfesting contaminated nonhost surfaces, treating vegetative propagative material, soil fumigation, application of eradicant fungicides to growing plants, and chemotherapy. Limited mention is made of seed treatment, chemotherapy, and eradicant fungicides in other chapters and soil fumigation is a specialized application and is treated in Chapter 10.

Eradication with chemicals eliminates pathogens by disinfection or disinfestation. Disinfection reverses or neutralizes established infections. Elimination of loose smut from wheat or barley seed by the systemic fungicide carboxin and fungicidal control of powdery mildews are examples of disinfection.

Disinfestation eliminates pathogens from plant surfaces before infection. Many seed treatments are eradicants, but they can also serve as protectants since they prevent infection by pathogens in surrounding soil. Most foliar sprays are protective and applied prior to arrival of inoculum but spraying dormant peach trees with lime sulfur or other eradicant fungicide kills fungal spores on the buds before they can infect newly exposed leaves and is an example of disinfestation.

Eradicating pathogens from propagative material, either true seeds or vegetative material such as sugarcane setts, potato tubers, or sweetpotato roots, aids exclusion, but the basic principle is to eliminate pathogens from plant organs. Exclusion is a secondary benefit.

The three major food crops in the world, wheat, rice, and corn (maize), are all propagated by seed. Potato, sweetpotato, and cassava are the next most important, but they are propagated vegetatively. Although lower in total production than the above mentioned crops, peas, beans, lentils, chickpeas, faba bean, soybean, and other large seeded legumes provide a major source of protein in many regions of the world. All of these crops are subject to seedborne pathogens and measures to eradicate them from seeds, tubers, roots, and cuttings are important to agriculture throughout the world.

SEED TREATMENT

Seed treatment is probably the simplest and least costly in time, material, and money of any chemical or cultural control measure. Seeding is an essential farming practice and using treated seed costs little.

Seeds were soaked in various materials in ancient times as palliatives against various pests. Two of the commonly used materials, wine and urine, happen to contain excellent disinfectants, namely, ethanol and ammonia.

The accidental discovery that salt (sea) water could control common smut spores on wheat seed led to routine soaking of wheat seed in brine solutions. There followed a procession of other compounds used for seed treatment, including mercuric chloride, arsenic, lye, and copper sulfate. Most applications of these materials were probably happenstance based either on empirical observations or speculation. The first scientific basis for chemical seed treatment was the work of Benedict Prévost, a French scientist who, in 1807, observed that traces of copper sulfate inhibited germination of wheat smut spores. However, it was not until 1854 that use of copper sulfate seed treatment for cereal grains became popular.

An undesirable effect of copper sulfate, mercuric chloride, and sodium chloride is phytotoxicity because of their high water solubility. This trait is apparent in Bordeaux Mixture discussed in Chapter 12. Copper sulfate reacts with lime to form a relatively insoluble compound that is much less phytotoxic than soluble copper sulfate. These insoluble coppers are sometimes called fixed or neutral coppers. In 1873 copper sulfate-treated seed was immersed in milk of lime, presumably to hasten drying of the seed. This process made the treatment safer and continues today in the form of copper carbonates.

In the search for less injurious seed treatments salicylic acid was found to be very effective but too expensive. Liver of sulfur (potassium polysulfide) was satisfactory and became Ceres™ powder, one of the first proprietary materials for treating seed.

Rice is a major food crop in tropical and subtropical regions and is grown in many countries, but the leading producers are Asia, Egypt, Italy, and the United States. Rice is subject to many serious diseases, some of which are seedborne, and seed treatment is the main method used to reduce losses from seedborne diseases. Disinfestation of seed is the main control measure for blast, Helminthosporium leaf spot, Bakanae disease, bacterial leaf blight, bacterial grain rot, and white tip nematode. Mercury fungicides are no longer used in Japan and have been replaced by benzimidazole fungicides, hypochlorite, antibiotics, carbamates, captan, and copper fungicides.

Advantages of Seed Treatment

Seed treatment for disease control has several advantages over other chemical and many nonchemical control measures.

1. Seed treatment is *easily applied* but must be done properly. The seed must be completely and uniformly coated with the chemical. Stickers enhance adhesion of fungicides to seed and include natural materials such as milk casein, wheat flour, blood albumin, gelatin, oils, gums, resins, and fine clays, or synthetic stickers such as methyl cellulose and latex.
2. Seed treatments are *inexpensive* compared to foliar sprays or soil drenches. For example, fungicidal seed treatment of wheat seed costs about $1.00 a bushel and a bushel of wheat plants about an acre. If a disease is not checked by a seed treatment it might have to be controlled later at a cost of 15 to 20 dollars an acre.

3. Seed treatments are very *effective* since every plant (seed) is protected. This stops development so that further cycles do not occur.
4. Seed treatments can *supplement* other control measures and are frequently required in some quarantines. Alfalfa seed exported to Canada from the United States must be treated with the fungicide thiram. This fungicide eliminates the Verticillium wilt fungus that may be carried in plant debris along with seed. Similarly, canola (edible oil rape) is treated with benomyl, iprodione, or carbathiin (the Canadian common name for carboxin) to prevent movement of virulent forms of blackleg to areas where they do not occur.

As a supplement to quarantines the USSR State Service for Plant Quarantine required disinfection of wheat seed to destroy wheat gall nematodes by treating seed with a mixture of formalin and ethylmercury chloride. Control of wheat gall nematode helps control yellow slime and twist diseases of wheat since the nematode may be a prime disseminating agent of the pathogens. Corn seed is treated with mercuric chloride or thiram to destroy *Diplodia zeae* inoculum.

Seed treatment is often encouraged even for crops where good disease resistance is available to prevent or delay development of new pathogenic races. For example, in the Pacific Northwest common smut of wheat is controlled largely by resistant varieties. However, a new variety has only 3 to 8 years before new pathogenic races of smut appear to attack that variety. Seed treatment extends the life of resistant varieties by slowing emergence of new pathogenic races.

Seed treatment can also eliminate minor pathogens from seed that could otherwise weaken developing plants and make them more vulnerable to other pathogens or pests. Seed held in storage for several years can produce weak seedlings that are vulnerable to attack by opportunistic pathogens such as *Pythium*. Treating seed with fungicides enables these weaker seedlings to survive and develop into vigorous plants.

Most seed-treatment fungicides kill or inhibit pathogens on seed surfaces but few kill pathogens embedded in the seed coat. Fewer fungicides eradicate pathogens beneath the seed coat. Many seed-treatment fungicides, especially those prior to the 1960s, were broad spectrum fungicides. The more recently developed materials are lower in phytotoxicity, narrower in range of activity, and more specific in their action.

The first organic mercury compound, chlorophenyl mercury, containing 18.8% metallic mercury, was developed in 1915 by Farbenfabriken Bayer, and named Uspulum™. The use of mercury fungicides increased rapidly after World War I because of a shortage of copper sulfate. After World War II mercuries came into common use because of their wide range of activity and relatively low cost. Many liquid formulations were available and were much easier to apply to seed than wettable powders. The following includes many of the mercury fungicides used from 1915 to 1966. Toward the end of this period the use of organic mercury fungicides began to decline because of high mammalian toxicity.

Germisan™, cresylmercuric cyanide, with 16.1% mercury, was marketed in 1920, and Semesan™, hydroxymercury chlorophenol, with 19% mercury, in 1924. Following these compounds mercury content was lowered considerably to reduce seed injury. Ceresan™ (European formulation), phenylmercuric acetate, had 1.5% mercury, as did Agrosan G™, tolylmercuric acetate, and Ceresan™ (American formulation), ethylmercuric chloride. A new European formulation of Ceresan™, methoxymethylmercuric silicate, had 2.5% mercury. In 1933 New Improved Ceresan™, ethylmercuric phosphate, with 3.2% mercury replaced earlier formulations until 1945 when Ceresan M™, *N*-(ethylmercury)-*p*-toluenesulfonanilide, also with 3.2% mercury, replaced them for several reasons including the advantage of being applied as a slurry.

Other organic mercuries included phenylmercury ammonium acetate, phenylmercury urea, methylmercury 8-hydroxyquinolinate, and phenylmercury formamide. Panogen™, methylmercuric dicyanodiamide, with metallic mercury contents of 1.5 to 4.2% depending on formulation, was the last of the mercury formulations widely used to treat seed. Its decline in 1969 came as a result of an unfortunate poisoning.

In the mid-1960s residues of mercury were found in pheasants in several areas as a result of the birds eating treated seed spilled alongside roadways or in fields. Hunting seasons were curtailed in Montana and Alberta, Canada. Hunters in some other states were warned about the possibility of mercury residues in pheasants and cautioned against eating too much of the wild meat.

About this time mercury residues were detected in beef carcasses. Evidently some ranchers mixed surplus treated seed grain with cattle feed. Then in the summer of 1969 a farmer in Alamagordo, New Mexico, obtained some waste grain contaminated with a mercury fungicide or fungicides. The grain was fed, along with garbage, to a number of hogs. After a month, one of the hogs became ill and was butchered. The meat was eaten by seven of nine family members from September through December. In December 1969, three of the family who had eaten the tainted pork became permanently injured from mercury poisoning.

This tragic incident was reported on nationwide television in February 1970, and on March 9, 1970, the United States Department of Agriculture suspended the registration of all fungicides containing alkylmercuries. This event not only resulted in the discontinuance of most mercury fungicides in the United States but, moreover, it caused plant disease chemicals to fall under the same cloud of suspicion and scrutiny suffered by insecticides and herbicides.

Mercury fungicides can be placed in one of three chemical groups:

1. Alkylmercuries—mercury compounds combined with methyl, ethyl, or other hydrocarbon chains.
2. Phenylmercuries—mercury combined with a phenol group.
3. Miscellaneous mercuries—mercury in ring compounds other than phenol.

Of the three groups the methylated mercuries tend to be more toxic than other forms. About the time of the Panogen incident, concern surfaced about mercury content in various fresh water and marine fishes in several parts of the world. The mercury levels in fish were associated with methylmercury compounds derived from industrial uses of mercury such as for slime control in pulp and paper mills, where mercury discharged into lakes, rivers, and estuaries became methylated under anaerobic conditions. Even before the decline of mercury seed treatments a few nonmercury eradicant seed treatment chemicals were available, but none had either the range of activity or the low cost of the mercuries and hence could not compete with them for cereal seed treatment.

The organomercury fungicides still play a major role in seed treatment in many parts of the world. As recent as 1983 mercury fungicides comprised up to 98% of the seed treatments in some countries of Northern Europe, but not at all in North America, Australia, Africa, South America, and southern Europe.

Although the benefits of seed disinfection are well known there are several possible disadvantages and risks. The most frequent is toxicity to seeds or developing plants. Germination or sprouting may be delayed or otherwise impaired and thin stands or weak plants can result.

Disinfection can remove beneficial organisms from seeds and allow facultative pathogens to develop without opposition. In one study in Washington wheat seed disinfested with ethanol and hypochlorite had unusually high infection by *Fusarium graminearum*. Similar observations have been made with other pathogens including *Helminthosporium sativum*. Presumably the disinfestants removed antagonistic bacteria from seed surfaces and allowed the pathogens to develop unchecked.

Sometimes one pathogen is controlled but another unexpectedly comes into prominence because it is unaffected by the chemical treatment. In Japan a disease complex caused by pathogenic fungi on rice was correlated with the ban on mercury fungicides. Newer fungicides, especially antibiotics, are ineffective against a broad range of rice pathogens. Antibiotic treatment of potato seed tubers to eliminate bacterial decay often results in greater incidence of Fusarium seed piece decay.

In Asia, the United States, and other parts of the world, paddy rice farming often involves production of fish in the flooded fields. Use of chemicals in rice paddies is a concern because rice and fish are rotated or coproduced. Benzimidazoles, thiram, and the mercuries are toxic to fish or leave toxic residues. Hypochlorite seed treatments have the distinct advantage of being decomposed by sunlight and further inactivated by dilution in water and do not pose a residue problem.

Thiram, tetramethyl thiuram disulfide, a product of the rubber industry, was developed as a fungicide in 1931. It is still an effective eradicant and protectant but has no significant effect on smut fungi, a primary reason for seed treatment of cereals. Chloranil, tetrachloro-*p*-benzoquinone, appeared in 1940 for seed treatment of vegetables and other crops. Captan, *N*-(trichloromethylthio)-4-cyclohexene-1,2-dicarboximide, appeared in 1952 and serves the same purpose as thiram and chloranil.

In 1945, hexachlorobenzene (HCB) came into use as a seed treatment to control common smuts of wheat. However, it has no effect on other seedborne fungi. Its main advantage was that it controlled smut spores on seed and those in soil, which are a major source of seedling infection in wheat in the Pacific Northwest. HCB has a narrow range compared to the similar pentachloronitrobenzene (PCNB), also used as a seed eradicant as well as a soil fungicide. HCB was discontinued in the 1970s and largely replaced by carboxin, 5,6-dihydro-2-methyl-*N*-phenyl-1,4-oxathiin-3-carboxamide, for smut control on cereals. Carboxin, like HCB, has a narrow spectrum largely restricted to a few smut fungi, but has the advantage of being systemic. It is often mixed with thiram, captan, or maneb to increase its activity.

Ethirimol, 5-butyl-2-ethylamino-4-hydroxy-6-methoxyl-pyrimidine, is an effective seed treatment for powdery mildew of cereals. Fenfuram, 2-methyl-furan-3-carboxanilide, similar to carboxin, was a seed treatment for internally borne loose smuts of cereals. Furmecyclox, (*N*-cyclohexyl-*N*-methoxy)-2,5-dimethyl-3-furancarboxamide, was another carboxin-like fungicide. Both fenfuram and furmecyclox have been discontinued.

The nonmercury fungicides currently used or used in the past for seed treatment eradicate fungi from seed surfaces but function mostly as protectants. The greatest use is on wheat, rice, corn, and potatoes. These fungicides and their active ingredients are listed in Table 9.1. They are discussed further in the chapters on fungicides.

Fungicide applications to seed can be either wet or dry depending on formulations available. Soaking seed in an aqueous solution requires draining and drying before the seed is packaged and stored. Wet seed can heat up, freeze, or sprout. Wet application now mean a slurry or "quick-wet" method. Dry applications are easier to apply and cause less damage to seed.

Dry applications entail mixing seed and fungicide in a rotating drum or inclined auger cylinder. Usually a sticker such as methyl cellulose (Methocel™) is applied to the seed before the chemical is added to assure adhesion of the fungicide.

The slurry method applies the fungicide in a thick water suspension mixed with the seed in a special treater. This often consists of a spinning disk on which the slurry is fed and then discharged in a fine mist onto the seeds falling in a curtain around the disk. Liquid formulations are also applied in this manner in the "quick-wet" method.

In addition to uniform coverage and good adhesion there are several requirements for fungicidal seed treatments. These include low hazard to the operator, no damage

TABLE 9.l. Nonmercury Fungicides Used Worldwide for Seed Treatment of Rice, Wheat, and Potatoes

Common Name	Chemical Name
Amobam	Diammonium ethylene bisdithiocarbamate
Benodin	2,5-Dimethyl furan-3-carbonic acid anilide
Benomyl	Methyl 1-(butylcarbamoyl)-2-benzimidazolecarbamate
Biteranol	β-([1,1'-Biphenyl]-4-yloxy)-α-(1,1-dimethylethyl)-1*H*-1,2,4-triazole-1-ethanol
Boric acid	H_3BO_3
Calcium hypochlorite	$Ca(OCl)_2$
Captafol	*cis*-*N*-[(1,1,2,2-Tetrachloroethyl)thio]-4-cyclohexene-1,2-dicarboximide
Captan	*cis*-*N*-Trichloromethylthio-4-cyclohexene-1,2-dicarboximide
Carbendazim	Methylbenzimidazol-2-yl carbamate

TABLE 9.l. Continued.

Common Name	Chemical Name
Carboxin	5,6-Dihydro-2-methyl-*N*-phenyl-1,4-oxathiin-3-carboxamide
Chloroneb	1,4-Dichloro-2,5-dimethoxybenzene
Copper hydroxide	$Cu(OH)_2$
Copper oxychloride	$3Cu(OH)_2.CuCl_2$
Copper 8-quinolinolate	$Cu(OC_9H_6N)_2$
Etherimol	5-Butyl-2-ethylamino-4-hydroxy-6-methyl pyrimidine
Fenaminosulf	Sodium [4-(dimethylamino)phenyl] diazene sulfonate
Fenarimol	3-(2-Chorophenyl)-3-(4-chlorophenyl-5-pyrinidine methanol
Fenfuram	2-Methyl-furan-3-carboxanilide
Formaldehyde	HCHO
Fuberidazole	2-(2'-Furyl)-1*H*-benzimidazole
Furmecyclox	(*N*-Cyclohexyl-*N*-methoxy)-2,5-dimethyl-3-furan carboxamide
Guazatine	Mixture of polyamine reaction products
Hexaclorobenzene	C_6Cl_6
Imazalil	1-[2-(2,4-Dichlorophenyl)-2-(2-propenyloxy)ethyl]-*H*-imidazole
Iprodione	3-(3,5-Dichlorophenyl)-*N*-(1-methylethyl)-2,4-dioxo-1-imidazolidinecarboxamide
Mancozeb	Coordination product of zinc ion and maneb
Maneb	Manganese ethylene bisdithiocarbamate
Metalaxyl	*N*-(2,6-Dimethylphenyl)-*N*-(methoxyacetyl)-alanine methyl ester
Methfuroxam	2,4,5-Trimethyl-3-furanilide
Methylarsonate	$Fe_2(CH_3AsO_3)_3$
Metiram	Tris{ammine-[ethylene bis(dithiocarbamate)] zinc(II)}[tetrahydro-1,2,4,7-dithiadiazocine-3,8-dithione] polymer
Nabam	Disodium ethylene 1,2-bisdithiocarbamate
Nuarimol	α-2-Chlorophenyl)-α-(4-fluorophenyl)-5-pyrimidine methanol

TABLE 9.l. Continued.

Common Name	Chemical Name
Organo-tin	Triphenyl tin-acetate, -chloride, -hydroxide
Propineb	Zinc ethylene bisdithiocarbamate polymer
Prochloraz	1-*N*-Propyl-*N*-[2(2-(2,4,6-(trichlorophenoxy)ethyl] carbamoylimidazole
Pyrocarbolid	5,6-Dihydro-2-methyl-*N*-phenyl-4*H*-pipran-3-carboxamide
8-Quinolinol	$(C_9H_7NO)_2$
Quintozene	Pentachloronitrobenzene
Sisthane	α-*N*-Butyl-α-phenyl-1*H*-imidazole-1-propane-nitrile
Sodium hypochlorite	NaOCl
TCMTB	2-(Thiocyanomethylthio)benzothiazole
Tecnazene	Tetrachloronitrobenzene
Thiabendazole	2-(4'-Thiazolyl)-benzimidazole
Thiophanate methyl	Dimethyl-[(1,2-phenylene)bis(imino-carbonothioyl)] biscarbamate
Thiram	Tetramethylthiuram disulfide
Tolchlofos methyl	*O*-2,6-Dichloro-4-methylphenyl-*O*,*O*-dimethyl phosphorothioate
TPN	Tetrachloroisophthalonitrile
Triadimenol	β-(4-Chlorophenoxy)-α-(1,1-dimethylethyl)-1*H*-1,2-triazole-1-ethanol
Tridemorph	Mixture of C_{11}-C_{14}-4-alkyl-2,6-dimethylmorpholine homologues
Triforine	*N*,*N*'-1,4-Piperazinediylbis(2,2,2-trichloro-ethylidene)-bis[formamide]
Zineb	Zinc ethylene bisdithiocarbamate

to the environment, speedy application, convenience, low cost, versatility to adapt to different seed species and sizes, formulations, and so on. Several types of application machines have been described and these are of two basic types. One involves mechanical mixing of the seed with the chemical by rotation, stirring, augering, or some other means. Another meters the chemical onto spinning disks, cones, or brushes from which it is dispersed in a fine mist to a passing curtain or band of seeds. Figure 9.1 shows a diagram of a revolving drum type applicator and Figure 9.2 is a diagram of a spinning disk applicator. Some of these machines are batch mixers in which weighed amounts of seed are coated with measured amounts of chemical. Others are

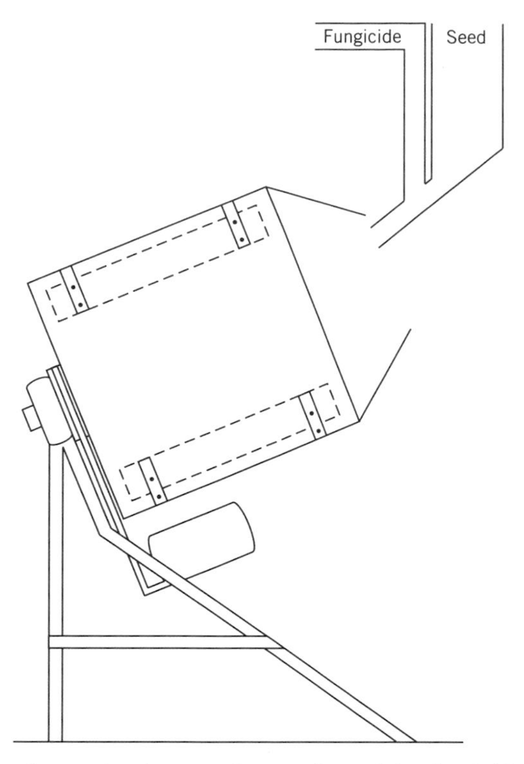

Figure 9.1. Diagram of a rotating drum seed treater for applying fungicidal powders to seeds.

continuous flow mixers in which seed moves continuously through a mixing chamber where chemical is metered based on speed of the operation. A wide variety of machines are in use throughout the world and names vary, even for the same machine. Chemicals can also be applied to seed by other means. Pelleting involves building up layers of adhesives and chemicals on the seed in successive applications. Coatings or dips involve dipping seeds or vegetative parts in suspensions containing chemicals. Repeated passes through different solutions can apply more that one pesticide to the same seed. Steeping is used primarily for treatment of rice and involves soaking seed for several hours in solutions of pesticide.

Mercury fungicides are somewhat volatile. Their vapor action facilitates uniform coverage of the seed since the fungicide continues to move in the vapor phase. The volatility also means that some fungicide is lost from the seed when seed is stored in open containers. This is not a problem with seedborne pathogens since fungicides have time to eradicate the surface organisms. However, there may be loss of toxicant due to vaporization from treated seed when applied for protection against soilborne pathogens.

Pelleting is a technique by which layers of clay and fiber containing fungicides (or other chemicals) are built up on seeds. The pelleting process produces seeds of uniform size and shape to facilitate machine planting and allows loading-up (high concentrations) of fungicides. Pelleting has little value for eradicant fungicides but allows for

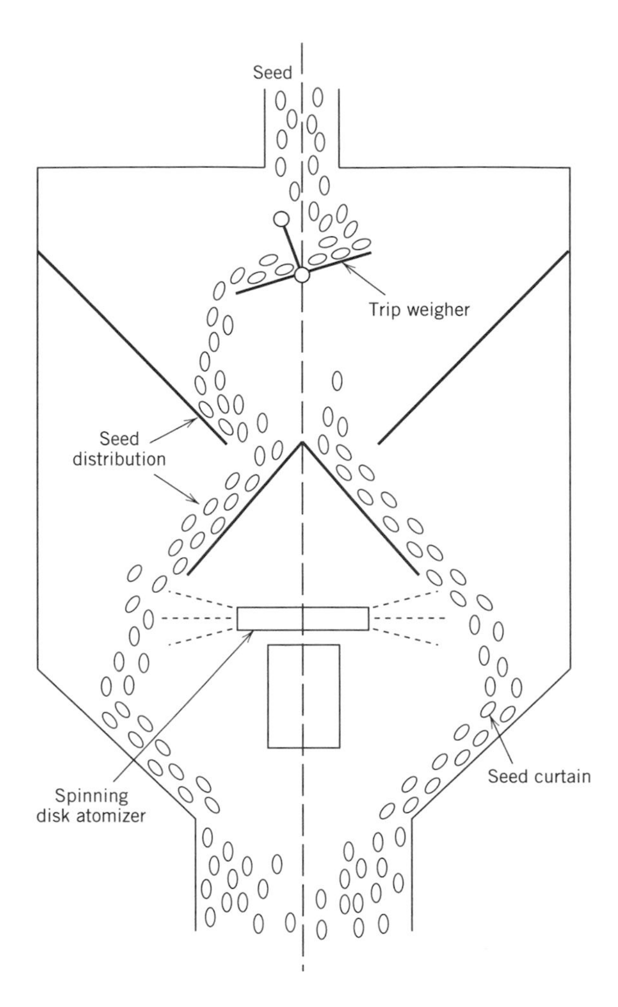

Figure 9.2. Diagram of a spinning disk seed treater for applying liquids and slurries to seeds.

maximum amounts of protectants to adhere to seeds. Pelleted onion seeds containing thiram at a rate of 1 pound of formulated fungicide per pound of seed reduced smut in onion seedlings to half or less that of untreated seed but there was no difference between two different pelleting methods. Pelleting can interfere with normal germination and emergence of seedlings if the pelleting material inhibits water uptake during the germination process. It is also too costly for widespread use. A thiram soak is recommended in several countries for elimination of the black leg fungus from crucifer seeds. The same soak is recommended for Septoria leaf spot of celery and Alternaria leaf spot of cabbage.

Ascochyta infections of several large seeded legumes (peas, faba bean, and lentils) are not readily controlled by surface applications of fungicides, but soaking seed in a benomyl/thiram mixture for 8 to 12 hours can markedly reduce infection.

ERADICANTS OTHER THAN SEED TREATMENTS

Many organic mercury compounds were used to treat vegetative propagative materials such as tubers, rhizomes, bulbs, and corms. These compounds included Mersolite™, phenylmercury acetate; Merthiolate™, sodium ethylmercury thiosalicylate; Sanoseed™, ethyl mercury chloride; Corona P.D.™, mercury bromine phenol; Semesan™, hydroxymercurichlorophenol; Semesan Bel™, a mixture of Semesan™ plus hydroxymercurinitrophenol; and Puratized N-5-E™, phenylmercury triethanol ammonium lactate. Some of these fungicides were short lived, being discontinued after a few years on the market. Many brand names of fungicides appear in the literature that are, or were, used only in Europe or countries other than the United States.

Disinfectant dips are used for certain vegetatively propagated materials. Sugarcane setts, for example, are dipped in solutions of organomercury fungicides containing 0.015% mercury for control of pineapple disease. This process was developed in South Africa and is now extensively used in many sugarcane growing regions. Sugarcane smut now is controlled by soaking setts in 0.1% mercuric chloride or 1% formalin for 5 minutes. Apple budwood dipped in streptomycin sulfate solution (2 g/liter) for 20 minutes or mercuric chloride (1 g/liter) for 2 hours eliminates fire blight bacteria.

Dipping transplants or seedlings in systemic nematicides eradicates nematodes from plants used for propagation. Root-knot nematodes have been eliminated from rose and strawberry plants by dipping them in fenamiphos or ethoprophos for 30 minutes.

Potato seed tubers are disinfected to control several latent fungi like *Oospora pustulans* (the cause of skin spot), *Helminthosporium solani* (silver scurf), and *Rhizoctonia solani*, (black scurf). These pathogens were usually eliminated by dips in organomercury fungicides when these materials were still available. Dips in thiabendazole (TBZ) effectively reduces the first two fungi but has no effect on *Rhizoctonia*. Sweetpotato roots are treated to eradicate the black rot fungus by dipping in TBZ or ferbam. In Russia, potato seed tubers are immersed for 1 hour in a 2% boric acid solution to eliminate powdery scab and common scab and in a 1% dip for *Rhizoctonia* control. This treatment is also reported to control potato blackleg.

Prior to 1961 rice seed was soaked (steeped) in a 25 ppm solution of organomercuries for 6–12 hours to control blast. Mercury fungicides were banned in Japan in 1961 and Blasticidin S has since been used for this purpose. Some writers claim that this is the first antibiotic to be used commercially for control of a plant disease, but both streptomycin and cycloheximide were used for this purpose well before 1961.

Banana fruit rots can be a serious problem because of the long time in transit; shipments taking more than 10 days are especially vulnerable to fruit rots. The problem was much more severe when bananas were shipped in bunches instead of boxes as they are now. Benzimidazole fungicides have essentially eliminated fruit rots and most, if not all, bananas are now treated by dipping or spraying them to control crown, stem, and finger rots.

A few antibiotics are used to control plant diseases, all of which are eradicative in their action. All antibiotics are complex chemical molecules. Streptomycin was the first antibiotic used in plant pathology. Although it can eradicate bacteria from plant tissue it is used primarily as a protectant to control diseases such as fire blight of pear.

Oxytetracycline (Terramycin™) is used as a chemotherapeutant for several mycoplasma diseases of tree fruits. Most antibiotics are antibacterial, but several used in plant disease control are antifungal. Kasugamycin is an antifungal antibiotic used in Japan to control rice blast and several vegetable diseases. Cycloheximide (Actidione™) is a byproduct of the streptomycin process and was first used to control cherry leaf spot. Candicidin is an antifungal antibiotic used as a postharvest dip to destroy spores of brown rot fungi on peaches before fruit is stored.

A number of foliar fungicides have eradicative action and destroy established infections but are generally considered protectant fungicides. This is particularly true of powdery mildew fungicides that are usually applied after the disease has developed. Powdery mildews are a special case, however, since the mass of the fungus is on the outside of the plant where it is exposed and vulnerable to fungicidal action. One undesirable feature of many eradicative fungicides is the plant injury they can cause. It is probably a phytotoxic effect that provides the eradicative action. Many of the foliar applied materials kill a band of healthy tissue around sites of infection and thereby perform a type of chemical excision of lesions.

One of the oldest fungicides, lime-sulfur, is an eradicant when applied as a dormant spray, but it can cause considerable plant injury if applied to green tissue at the same concentration. Dichlone, 2,3-dichloro-1,4-naphthoquinone, was used to control many foliar diseases of fruits and vegetables but crops showed varying degrees of sensitivity. For example, in controlling Coryneum blight of stone fruits the same rate of dichlone after blossoming that caused little or no damage to cherries and peaches caused severe fruit drop in apricots. Similarly, apricots are called sulfur-shy plants since they can be easily damaged by sulfur fungicides whereas other tree fruits are unaffected.

Dodine, *N*-dodecylguanidine acetate, has long been a standard orchard fungicide for control of apple scab and peach leaf curl, and diseases of ornamentals such as sycamore blight. It acts by desiccating scab lesions or destroying fungal spores on plant surfaces. Other eradicative fungicides include procymidone, *N*-(3',5'-dichlorophenyl)-1,2-dimethylcyclopropane-1,2-dicarboximide; piperalin, 3-(2-methylpiperidino)propyl-3,4-dichlorobenzoate, an eradicant fungicide effective primarily against powdery mildews; edifenphos, *O*-ethyl-*S,S*-diphenyldithiophosphate, an excellent eradicative fungicide for rice blast; tridemorph, *N*-tridecyl-2,6-dimethylmorpholine, and aldimorph, 4-*N*-dodecyl-2,6-dimethylmorpholine, both listed as eradicative or curative foliar fungicides; and several triazole compounds, propiconazole, penconazole, fenbuconazole, tetraconazole, terbuconazole, and hexaconazole, all listed as curative fungicides.

The phytotoxic nature of some chemicals is used to eradicate plant pathogens by either direct destruction or chemical excision of diseased tissue. Calcium cyanamide or other high nitrogen compounds have been used to destroy mummified fruit (sclerotia) on the ground in orchards and blueberry plantings, thereby eliminating overwintering sources of primary inoculum of brown rot and mummyberry fungi. Used otherwise these chemicals would cause severe plant burning if applied on trees or bushes. Sprinkler rot of pears, peaches, and other tree fruits results when irrigation water contaminated with sporangia and zoospores of the pathogen is sprinkled onto developing fruit. Early orchardists eradicated this inoculum in irrigation water by placing burlap bags containing large crystals of bluestone, copper sulfate, in the distribution boxes where lateral irrigation lines emerge. Enough copper dissolved into the water to kill the fungus spores.

Elgetol™, sodium dinitro-*o*-cresylate, was used to cure crown gall on fruit trees. The bark of affected tissues was scraped away and the area painted with the Elgetol solution. Within a few days the crown gall tissue is killed and healthy callus tissue forms to produce healthy wood. If the gall is extensive and extends more than half the circumference of the stem the technique is not practical because the plant will be killed by girdling. A similar technique has recently been used with Bacticin™ and Gallex™, mixtures of 2,4-xylenol and *m*-cresol.

Bacterial canker on cherries and other stone fruits has been eradicated by scraping the bark around the edge of cankers and applying copper sulfate paint, which kills bacteria in affected tissues and some surrounding healthy bark.

As part of the white pine blister rust control program the antifungal antibiotic cycloheximide in an organic solvent was sprayed onto scarified cankers (Figure 9.3). This method worked if the canker was not too far advanced; otherwise the tree trunk would be girdled by the chemically excised tissue.

Figure 9.3. Spraying scarified trunk of a white pine sapling with cycloheximide/diesel solution. (U.S. Forest Service photo.)

Fumigation is usually considered to be too severe to use on living plant parts but the FAO Plant Quarantine Treatment Manual cited in Chapter 7 estimates that 95% of all plants moving in international commerce can tolerate methyl bromide at dosages required to kill insects and certain pathogens. Fungicidal dips of benzimidazoles, captafol, formaldehyde, trichlorophenol, or carbamates are used for the few genera in which all species are intolerant of methyl bromide.

Methyl bromide fumigation is a based on concentration of the gas (*C*) in grams per cubic meter over time (*T*). It is usually carried out at normal atmospheric pressure under tarpaulins or in enclosed chambers. For bulb nematode in onions $C \times T$ is 800 to 1000 so that approximately 40 g of methyl bromide per cubic meter of volume (2.5 lb/1000 ft^3) is used at 10–20°C for 24 hours.

Elimination of coffee rust spores from green coffee beans requires 48 g of methyl bromide for 6 hours at 25–35^{3}C. The dose of fumigant to eradicate sugar beet cyst nematodes depends on temperature. At 10–15°C 256 g for 8 hours is required, at 16–32°C 128 g for 4 hours, and above 32°C only 64 g for 2 hours is needed.

A number of inorganic and organic compounds have been used to eliminate or reduce postharvest decay in fruits and vegetables. Sulfur dioxide is used to eliminate wild yeasts from grapes prior to crushing to enhance normal fermenation. In stored grapes, however, retreatment is needed because some spores survive initial fumigation. Treatments may include applications of sodium bisulfite, $HNaO_3S$, which breaks down in storage to release SO_2. Sulfur dust applied to peaches before storage releases hydrogen sulfide and prevents brown rot.

Nitrogen trichloride is used as a fumigant to control blue and green molds on stored citrus fruits. It also reduces decay in tomatoes and peppers but causes scarring and bleaching in vegetables. Ethylene and propylene oxides reduce mold on unsulfured prunes, figs, dates, and raisins.

Formaldehyde gas has been used to fumigate empty sweetpotato houses. Mixing 3 pints of formalin and 23 ounces of potassium permanganate in water for each 1000 ft^3 of space causes rapid release of formalehyde gas, which is highly effective in eliminating spores of *Rhizopus nigricans* and *Ceratocystis fimbriata*, the causes of soft rot and black rot of sweetpotato roots. A similar formaldehyde fumigation was used to control Botrytis rot of grapes.

Several phenolic compounds are used as postharvest disinfectants for citrus fruits and vegetables. SOPP (sodium-o-phenyl phenate) and OPP (o-phenyl phenol) disinfect produce and equipment. Diphenyl-impregnated wraps prevent citrus blue and green molds and other fungi. Diphenyl vapors have also been applied to other fruits and vegetables but, unlike citrus, some fruits and vegetables acquire off-flavors or other undesirable effects such as browning. Iodine used in wraps or liners for wrapping or packing citrus and grapes has proven less effective than diphenyl for reducing decay.

Carbon dioxide is routinely used in controlling atmospheres in which apples and other fruits are stored. It apparently does not act as a fumigant; instead CO_2 inhibits development of decay organisms by greatly reducing their respiration rate. Increasing CO_2 concentration to 5–7% reduces various banana fruit rots in storage. However, some soft fruits such as peaches, apricots, and strawberries are injured by carbon dioxide. Ozone tried in a similar way killed exposed spores but had little effect on decay of apples.

DISINFESTATION OF NONHOST SURFACES

A number of chemicals are used to disinfest equipment and other nonhost surfaces. Technically these chemicals are bona fide *disinfestants*, not *disinfectants*; nevertheless the terms are used interchangeably. Corrosive sublimate (mercuric chloride) is very effective but is toxic to humans and corrosive to metals. Standard household bleaches, calcium and sodium hypochlorite, are also effective disinfestants. Formalin solution was a common disinfestant for nonhost surfaces, seeds, and other plant materials but its potential as a carcinogen has curtailed its use.

Mercuric chloride was long used to disinfest potato seed cutting knives and pruning tools used to excise branches infected with fire blight but high mammalian toxicity resulted in its disuse. Formalin has been used to disinfest tool surfaces such as machetes and shovels to prevent dissemination of bacterial wilt of banana. The usual concentration was 1 pint of formalin in 3 pints of water.

Quaternary ammonium compounds are surface active agents that function as bactericides, fungicides, and general disinfestants. They are used in food industries such as milk processing to sanitize equipment and in seed potato handling operations to eradicate potato ring rot bacteria from machinery. Alkyl dimethyl ammonium chloride is sold under the trade name Roccal™. Other compounds are alkyl dimethyl benzylammonium chloride, octyl docyl dimethylammonium chloride, and dimethyl benzylammonium chloride. Copper-8-quinolinolate and 8-quinolinol are also disinfestants used for cleaning of potato handling equipment.

CHEMOTHERAPY AS AN ERADICATION TECHNIQUE

Chemotherapy treats the host plant rather than the pathogen directly. Some chemicals such as fosetyl-Al induce resistance in plants and therefore implement a strategy under the principle of resistance as defined by Whetzel. Chemotherapy received a great deal of attention in the 1950s and 1960s but became successful only after systemic fungicides and bactericides were developed.

Pear decline and X-disease of peaches are diseases caused by mycoplasma-like organisms (MLOs). Formerly these diseases were thought to be caused by viruses. However, they responded to tetracycline antibiotics and are thus thought to be more like bacteria. Solutions of oxytetracycline injected at multiple points around the trunks of affected trees appear to cure them but treatment results only in remission of symptoms so that treatment must be repeated every few years.

Dutch elm disease is treated in a similar way by injecting one of the benzimidazole fungicides, usually TBZ (thiabendazole), into the trunk or flare roots of affected elm trees. As with the MLO diseases no cure is effected but only a remission of symptoms so that treatment must be repeated on a 3-year cycle. These examples do not qualify as eradication in the strict sense but are mentioned here because they are often cited as eradicative measures.

REFERENCES

*Anon., *Farm Chemicals Handbook*. Meister, Willoughby, OH, 1991.

*de Ong, E. R., *Chemistry and Uses of Pesticides*. Reinhold, New York, 1956.

*Hansen, E. W., E. D. Hansing, and W. T. Schroeder, Seed treatment for control of diseases. In *USDA Yearbook of Agriculture: Seeds*. U.S. Government Printing Office, Washington, D.C., 1961. Pp. 272-280.

*Jeffs, K. A. (ed.), *Seed Treatment*. The Lavenham Press Ltd., Lavenham, Suffolk, England, 1986.

*Leukel, R. W., Treating seeds to prevent diseases. In *USDA Yearbook of Agriculture: Plant Diseases*. U.S. Government Printing Office, Washington, D.C., 1953. Pp. 135-145.

*Maloy, O. C., *Pesticides for Plant Disease Control*. Washington State University Coop. Ext. Ser. Multilith 3644, 1972.

*Martin, H., *The Scientific Principles of Crop Protection*. Edward Arnold, London, 1973.

*Martin, T., *Application to Seeds and Soil*. British Crop Protection Council Publication, Monograph 39, 1988.

*Purdy, L. H., J. E. Harmond, and G. B. Welch, Special processing and treatment of seeds. In *USDA Yearbook of Agriculture: Seeds*. U.S. Government Printing Office, Washington, D.C., 1961. Pp. 322–329.

*Sharvelle, E. G., *The Nature and Uses of Modern Fungicides*. Burgess, Minneapolis, 1960.

*Sharvelle, E. G., *Plant Disease Control*. AVI, Westport, CT, 1979.

*Publications marked with an asterisk (*) are general references not cited in the text.

10 Soil Fumigation

Soil fumigation is possibly the most effective and demanding of the various chemical methods of plant disease control. It can be done quickly, is thorough, and lasting, if done properly. The process demands attention to detail because its success is largely influenced by chemical, physical, and biological conditions. Soil must be prepared properly, special knowledge and equipment are needed, and careful postfumigation operations are required. Soil fumigation is also one of the most costly disease control methods. As early as 1936, fumigation with carbon disulfide cost $700 per acre and with chloropicrin about $500. Some fumigation techniques presently can cost more than $1000 an acre. Obviously, only high value crops justify such high costs.

Soil fumigation is inherently feasible, but not exclusive, for control of soilborne pathogens, especially root rot and vascular wilt fungi and nematodes. However, it is not the tactic of choice for controlling every soilborne pathogen. Disease resistance is simpler and cheaper. Crop rotation and other cultural practices are preferred over fumigation for the same reasons. Few foliar diseases are amenable to control by soil fumigation since plant residues in which foliar pathogens are embedded depress effective soil fumigation. However, it is possible for fungi that pass dormancy unprotected in or on soil. Sclerotia of *Claviceps*, *Sclerotinia*, *Typhula*, and similar fungi survive in soil and later produce spores that infect aerial portions of plants. Soil fumigation is a key strategy in controlling several soilborne viral diseases by eliminating nematode vectors, usually *Xiphenema* species. Fumigation also controls fungi such as *Olpidium brassicae* and *Polymyxa graminis*, vectors of lettuce big vein and soilborne wheat streak mosaic viruses, respectively. Because of cost of treatment, however, fumigation for these vectors is probably limited to nurseries and transplant beds.

Fumigation is the application of a chemical, usually as a liquid or solid, to a matrix or space where it volatilizes to the gaseous form. Soil matrices consist of soil particles forming channels or pores, water around soil particles and in the channels, organic matter, plant residues, and microorganisms. Fumigants move through the soil pores as vapor and dissolve in the water film surrounding soil particles. Several physical characteristics of the soil greatly influence movement and retention of fumigants. A proper balance of air space or porosity, moisture content, soil particle size, and organic matter is important for effective fumigation.

DEVELOPMENT OF SOIL FUMIGANTS

Carbon disulfide (CS_2), an inorganic compound, was used as early as 1896 to eradicate grape phylloxera. Most early use of soil fumigants was to control soil insects such as phylloxera. Most soil fumigants applied for disease control are directed at nematodes, which are more like insects than other pathogens. Julius Kühn (considered the father of modern plant pathology) applied carbon disulfide to control sugar beet cyst nematodes in the 1870s. This was the first use of a soil fumigant for disease control. Carbon disulfide has since been used extensively for soil fumigation and is still used. Its chief disadvantage is its low flash point (-30°C) compared with gasoline, which is slightly more flammable with a flash point of -45°C. Flash point is the temperature at which vapors ignite. Another inorganic fumigant once advocated for control of nematodes and soilborne fungi is hydrocyanic acid (HCN), but its high mammalian toxicity discouraged its use.

Formaldehyde (HCHO) is the simplest organic compound to be used in soil fumigation. Since formaldehyde is a gas it usually is applied as formalin, an aqueous solution of about 40% formaldehyde. It was first used for soil fumigation about 1906 as a 0.5 to 1% drench at a rate of 1.5 gallons of solution per square foot to control damping-off in conifer nurseries. The large volume required for field application limited it to use as a row treatment diluted 1 part in 128 parts water (about 0.3% formaldehyde) to control onion smut. Because of its inherently noxious fumes when applied as a liquid it was sometimes adsorbed onto an inert carrier such as charcoal, ground oat hulls, infusorial earth, or sawdust for use as a dust. Aside from an unpleasant odor, formaldehyde has been implicated as a carcinogen and eliminated for soil fumigation or other agricultural uses.

Chloropicrin, trichloronitromethane, or tear gas, came into general use as a fumigant after World War I when large quantities manufactured as tear gas were declared surplus. Three million pounds a month were being manufactured in the United States by 1919. Its use as a soil fumigant was delayed because of lack of a method for confining the gas. Eventually glue-coated paper was used as a soil cover. Chloropicrin was first used to control root-knot nematode and damping-off of vegetable seedlings in the mid-1930s. From that time to the present improved application techniques have made chloropicrin one of the most effective soil fumigants. One report (Schmitt, 1949) rated chloropicrin as the best fungicide and herbicide of nine compounds considered to be most effective of 600 volatile chemicals tested. Its disadvantages are high cost and lachrymal properties that make it disagreeable to handle. However, the lachrymal feature serves as a warning agent and safety factor.

The next advance in soil fumigation was the chance discovery of D-D mixture in 1943 by an entomologist in Hawaii. D-D™, a byproduct of allyl alcohol manufacture, is a mixture of 1,3-dichloropropene and 1,2-dichloropropane. For almost 50 years it was the leading chemical for controlling plant parasitic nematodes, especially root-knot nematodes. Several subsequent soil fumigants are modifications of some component of D-D.

The most recent major achievement in soil fumigation was development in 1956 of metam (or metham) sodium, sodium methyldithiocarbamate, sold under the trade name Vapam™. Metam sodium is not a fumigant but in moist soil releases methyl isothiocyanate (MIT), which is a fumigant. MIT was marketed a few years later as a proprietary fumigant under the trade name Vorlex™. After World Wars I and II a number of soil fumigants other than chloropicrin were developed for control of plant diseases. Most are halogenated hydrocarbons with chains of 1 to 3 carbon atoms. Some of these compounds had limited use and were discontinued. Others found wide application and are still used.

Methyl bromide (bromomethane) is a colorless, odorless, highly toxic gas that has a very low boiling point (BP) and proportionately high vapor pressure (VP) (see Table 10.1). These physical properties require that methyl bromide be dispensed as a gas or dissolved in solvents such as propylene dichloride, carbon tetrachloride, or xylene, to stabilize it and allow application in liquid form. Commercial formulations usually contain 2% chloropicrin as a lachrymal warning agent. Methyl bromide was first used in 1940 to control fungal pathogens, insects, nematodes, and weed seeds in seed beds, hot beds, and potting soil.

Ethylene dibromide (EDB) came into use in 1947 and possibly earlier but was largely consumed during World War II for making tetraethyl lead antiknock additive for gasoline. EDB has a high boiling point and low vapor pressure and can be applied with little or no seal of the soil surface. Although EDB, like D-D, has some fungicidal activity, it is used primarily as an insecticide and nematicide.

CBP-55 was a mixture of chlorinated and brominated hydrocarbons containing 55% chlorobromopropene that had good activity against soilborne fungi. It received some attention in the 1950s. Although it needed no seal it was unpleasant to handle, phytotoxic, lachrymal, and expensive and hence failed to develop commercial use.

TABLE 10.1. Physical Properties of Soil Fumigants Recently or Currently in Use

Chemical	Boiling Point (°C)	Vapor Pressure at 20°C (mm Hg)	Solubility in H_2O at 20°C (g/100 ml)	Water/Air Distribution at 20°C (Henry's Law constant)
H_2O	100	17.3	—	—
Ethanol	78.5	44.5	Infinite	—
Methyl bromide	4.5	1420	1.6	4.1
Carbon disulfide	46.3	298	0.217	1.8
Chloropicrin	112.4	20	0.195	10.8
1,3-D	104	21	0.275	20.2
EDB	132	7.7	0.337	42.7
MIT	119	21	0.76	88.0
DBCP	196	0.58	0.123	163.8

Dibromochloropropane (DBCP) is effective against nematodes and as effective as D-D against golden nematode. It was sold under several trade names and was still widely used in 1980 throughout the world. It has the advantage of low phytotoxicity and can be applied around living plants without causing injury. However, it has a high boiling point, low vapor pressure, and moves largely in soil water rather than as a vapor. Its water to air distribution ratio is about 40 times that of methyl bromide. This has resulted in contamination of ground water and use discontinued.

In addition to the important soil fumigants mentioned, others have been tested and shown promise but because they were no better than existing fumigants, cost more, or created application or environmental problems they were never developed commercially. Allyl alcohol and allyl bromide have good fungicidal properties, comparing well with chloropicrin for control of *Verticillium* but do not diffuse upward as well. The tetra and pentachloroethanes and dichloroisopropylether are potent nematicides but are very phytotoxic and persist in soil. Other compounds such as 1,2-dichloropropane, 1,1-dichloro-1-nitroethane, and tetrachloroethylene are simply poor nematicides. Since fumigants are used more often for nematode control than for other pathogens, lack of nematicidal activity virtually eliminated them from further consideration.

BIOLOGICAL ACTIVITY OF SOIL FUMIGANTS

The biological spectrum of soil fumigants largely determines their eventual acceptance. In general, the phycomycetes such as *Pythium* and *Phytophthora* are more vulnerable to soil fumigation than other fungi, and vegetative mycelium is easier to kill than chlamydospores or sclerotia. Table 10.2 shows the biological activity of common fumigants used presently or in the past and their general rates of application.

Methyl bromide and chloropicrin are general biocides that kill a wide range of organisms including fungi, nematodes, insects, and weed seeds. They are often marketed as mixtures under several trade names.

Methyl bromide is the most effective of soil fumigants because of its ability to rapidly and completely penetrate soil, but an impervious seal must be placed over the soil surface. It kills even durable resting structures such as sclerotia of *Sclerotinia sclerotiorum*. It is the standard soil treatment for Verticillium wilt of strawberries in California but costs more than $1000 an acre because treated soil must be covered with a polyethylene sheet. Methyl bromide is sold under several trade names including Dowfume MC-2™, Meth-O-Gas™, and Brom-O-Sol™ and is used to fumigate nursery and greenhouse soils. The toxic action of methyl bromide to microorganisms results from methylation of sulfhydryl enzymes.

Chloropicrin is much more toxic to soil insects and nematodes than carbon disulfide. It also is highly toxic to soil fungi and weed seeds. It has the distinct advantage over methyl bromide of being less volatile and therefore easier to apply because it remains liquid at working temperatures, and is less hazardous because it is lachrymal and gives ample warning of its presence. It is marketed under several trade names including Larvacide™, Picfume™, and Chlor-O-Pic™ and is often used to compare with newer fumigants because of its broad spectrum of activity. Chloropicrin

TABLE 10.2. Biological Activity of Common Soil Fumigants

Fumigant	Chemical Structure	Activity[a]	Rate (Depends on Target Organism)
Methyl bromide	CH_3Br	F, N, I, H	180–240 lb/acre
Chloropicrin	CCl_3NO_3	F, N, I, H	200–400 lb/acre
1,3-Dichloro propene	$Cl–CH_2–CH=CHCl$	F, N, I, H	18–60 gal/acre
Ethylene dibromide	$BrCH_2–CH_2Br$	N, I	4.5–18 gal/acre
Dibromochloro propane	$BrCH_2–BrCH–CH_2Cl$	N, I	10–25 gal/acre
Methyl isothiocyanate	$CH_3N=C=S$	F, N, I, H	20–40 gal/acre
Metam sodium	CH_3NHCS_2Na	F, N, I, H	50–100 gal/acre

[a] F, fungicide; N, nematicide; I, insecticide; H, herbicide.

is frequently used in combinations with other fumigants such as methyl bromide or 1,3-dichloropropene to enhance overall activity and reduce required rate of each fumigant. These mixtures are widely used to control many soil pests including nematodes, insects, weed seeds, and in particular fungi causing root rots and vascular wilts. Some mixtures of fumigants such as methyl bromide and chloropicrin are better than when either fumigant is used alone. Methyl bromide penetrates root-knot nematode galls and root tissue more readily than chloropicrin but chloropicrin is a better fungicide than methyl bromide.

1,3-Dichloropropene, the main active component from D-D mixture, was first marketed as Telone™. Telone II™ contains 94% 1,3-D, and has replaced the discontinued D-D mixture. It is a good nematicide and insecticide but poor fungicide, and is sometimes mixed with chloropicrin or methylisothiocyanate to enhance its fungicidal action. 1,3-D, like D-D, has low volatility and the soil surface needs only moderate compaction as a seal. Although its fungicidal activity is less than chloropicrin 1,3-D controls *Pythium* and some other phycomycetes. Because of its relatively low cost, its nematicidal effectiveness, and relative ease of application, 1,3-D is perhaps the most widely used soil fumigant in recent times. Even 25 years ago (1967) an estimated 179 million acres were treated with 1,3-D in the United States alone, compared with 11 million acres treated with methyl bromide. The relative amount used may be even higher today.

Ethylene dibromide (EDB), once sold as Dowfume-45™, Dowfume-85™, and other trade names, was used extensively as a nematicide and insecticide. Although generally having little or no activity against fungi it has been used to control several important soilborne diseases including Fusarium wilt of cotton, black rot of sweetpotato, and brown rot and black shank of tobacco. EDB solidifies at 9°C and is ineffective in cool soils. Therefore, it is used almost exclusively in warm climates. EDB has a very low vapor pressure (about a third that of water) but a fairly high solubility in water and the chemical moves mainly in soil water rather than through air spaces in soil. As with DBCP this property has resulted in groundwater contamination and disuse of EDB as a soil fumigant.

Dibromochloropropane (DBCP), sold as Nemagon™ and Fumazone™ among several trade names, found wide use during its relatively short life in the United States. It was as effective as D-D against the golden nematode, had low phytotoxicity, and could be applied around living plants for nematode control. However, bromine-sensitive plants could be injured. It was first used in 1956 and by 1967 was used on about 121 million acres in the United States. DBCP properties are similar to those of EDB and like the latter was used mainly in warmer areas for control of nematodes in citrus, grapes, and other crops. Although less soluble in water than EDB it has a much higher water to air distribution ratio so that it persists in soil longer and, like EDB, moves into groundwater. Because of this and human health problems DBCP was discontinued in the United States in the late 1970s.

Methyl isothiocyanate (MIT) has a broad range of activity against nematodes, insects, fungi, and weed seeds. Its high melting point (35°C) requires that it be mixed with 1,3-D or other C_3 hydrocarbons to facilitate application. One such mixture is sold as Vorlex™. MIT is more often applied as highly water-soluble metam sodium, a liquid, or as dazomet (Mylone™ is one trade name), a solid. Metam sodium and dazomet yield MIT on reaction with water. Their high water solubility permits application as drenches or through sprinkler irrigation systems. These fumigants are especially effective against difficult to control soilborne fungi such as *Verticillium.*

PHYSICAL CHARACTERISTICS OF SOIL FUMIGANTS

Physical properties of chemicals that most influence their effectiveness as soil fumigants are volatility and solubility. *Volatility* is reflected by boiling point and vapor pressure and directly controls conversion from liquid or solid phase to vapors. Volatility thus contributes to movement of soil fumigants through air passages in soil. *Solubility* in water is measured in two ways: (1) direct solubility in weight of chemical per volume of water (e.g., grams/100 ml, which can be read directly as percent solubility) and (2) distribution of the chemical in soil water and soil air. Table 10.1 gives the physical properties of common soil fumigants and includes water and ethanol for comparison.

There is a general inverse relationship between the boiling point of a compound and its vapor pressure: the lower the boiling point the higher the vapor pressure. This regulates the rate at which a chemical leaves the liquid phase and is important in determining movement of chemicals in soil. Length of the carbon chain and degree of substitution with halogens such as bromine or chlorine directly influence boiling point. Figure 10.1 shows graphically the boiling points of most soil fumigants and their parent compounds. As length of carbon chains increase so do boiling points. This graph also demonstrates that if a chain has more than four carbons the boiling point is probably so high that vapor pressures and diffusion will be too low and too slow for effective soil fumigation. Addition of halogens to the molecule has a similar effect on boiling points. Iodine is included in Figure 10.1 to further demonstrate the halogen effect.

Solubility is important because most soil microorganisms, especially nematodes, live in water films surrounding soil particles and the toxicant must dissolve in soil water

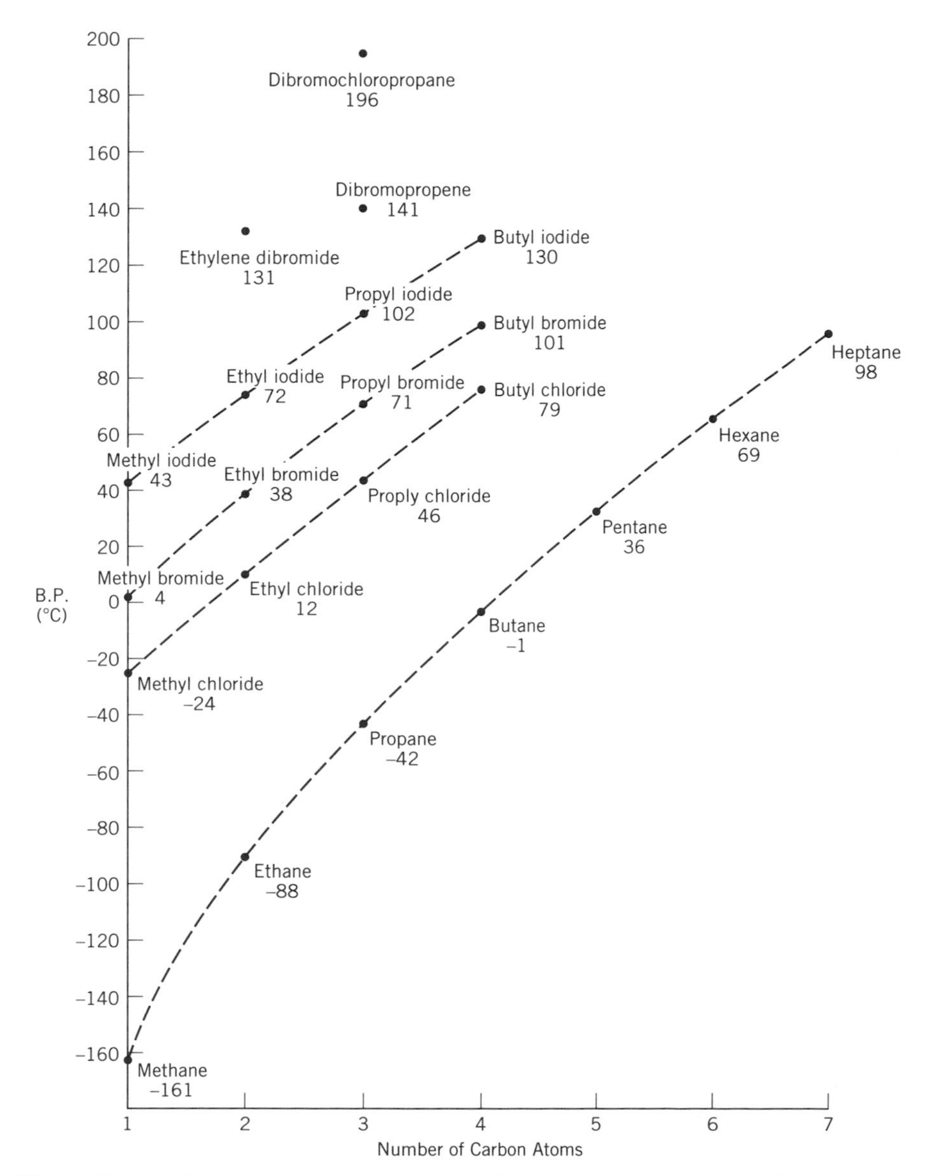

Figure 10.1. Boiling points (rounded to nearest whole numbers) of halogenated hydrocarbons and their parent compounds.

to kill targeted organisms. But solubility alone does not explain the action of soil fumigants. Since all these fumigants are relatively insoluble (less than 1%, except methyl bromide), the equilibrium between amount of chemical dissolved in soil water and in soil air is important. This ratio is expressed as Henry's Law constant. The lower the constant the greater the tendency to move to the vapor phase and, conversely, the higher the number the greater the capacity to remain dissolved in water. Fumigants with ratios lower than 10 leave the soil so rapidly that they are lost within hours or days, unless an impervious seal covers the soil. Fumigants with ratios of 10 to 100 remain in soil longer and do not require tarping but only compaction of soil surfaces. Those with ratios above 100 can remain in the soil for months. DBCP is the most persistent and methyl bromide the least persistent of the common fumigants. Table 10.1 gives Henry's Law constants and illustrates the importance of the water/air distribution to persistence. DBCP is retained in solution 40 times greater than is methyl bromide. Carbon disulfide has a lower ratio than methyl bromide, but its overall solubility in water and its vapor pressure are also lower by about one-fifth or less. These features dampen the effect of the low water/air distribution ratio.

FUMIGANT EFFICACY

A soil fumigant is usually selected for its relative toxicity to soilborne pathogens. Toxicity to targeted organisms must be weighed against hazard to applicators and ease of application. Methyl bromide is an excellent general biocide, killing a wide range of organisms. It is also hazardous because it is colorless, odorless, and has high mammalian toxicity (LD_{50} = <1 mg/kg). Methyl bromide's low boiling point and high vapor pressure dictate that it is applied as a gas and, except when the fumigant is placed 18 inches or deeper, requires that the soil surface be covered with an impervious seal to prevent rapid escape. On the other hand, chloropicrin has almost as broad a range of activity as methyl bromide and is much less hazardous because its lachrymal property gives ample warning of its presence. Its higher boiling point and lower vapor pressure make it easier to apply as a liquid and requires only compaction of the soil surface to seal it.

The amount (rate) of fumigant applied depends on whether nematodes or fungi are to be controlled, soil texture, and temperature at time of application. Except for cyst nematodes, nematode control is usually obtained with the lower recommended rates while fungal structures like sclerotia require the highest rates. Higher rates also may be needed for nematodes deeper in soil, larger populations, soils with high organic matter, or low temperatures.

Organisms must be exposed to toxicants for a minimum length of time in order to kill them. Exposure time is related to toxicant concentration, a relationship expressed in terms of fumigant concentration (*C*) and length of exposure (*T*) ($C \times T$ or *CT*). Determining *CT*s for soil fumigants is difficult because concentration decreases with increasing distance from point of application. There also are differences in response to fumigants by different organisms (e.g., fungi and nematodes) or even by different stages of the same organism (e.g., eggs, larvae, adults, and cysts of nematodes).

Among fungi in general, phycomycetes are the most susceptible to fumigants and sclerotial fungi most resistant. The *CT* interaction is not always predictable. Sclerotia of *Sclerotium rolfsii* and *Sclerotinia sclerotiorum* are more resistant to methyl bromide when exposed to low concentrations for long periods than when treated with high concentrations for short periods. Soil organisms should be exposed to the fumigant for several days to several weeks, depending on temperature, moisture, and other factors discussed previously, before disrupting surface seals to aerate the soil, in order to get maximum benefits from soil fumigation.

FACTORS AFFECTING SOIL FUMIGATION

Soil characteristics are as important as the fumigant to success of soil fumigation. Soil texture, structure, moisture, temperature, organic matter, and plant residues have a direct influence on fumigant action. Preparing soil for fumigation requires several critical steps. The soil should be ripped to a depth of 2 feet and then worked by plowing followed by disking, harrowing, or rototilling to the depth to be fumigated. Fumigants will not diffuse below the plow sole or other compacted layer. Fumigation depth depends on the targeted pest and generally is 10–12 inches with the fumigant placed at 8–10 inches. However, it can be deeper. For example, 18 inches is a common depth for fumigating with 1,3-D for root-knot nematode in central Washington and can be as deep as 36 inches for some citrus nematodes in sandy soils.

Soil Texture

Mineral soils range in texture from sands, which contain 85% or more sand, to clays, which contain more than 40% clay, with many variations between these two extremes. Silt is a soil component intermediate in size between sand and clay. Different soils consist of varying amounts of sand, silt, and clay. Sandy loams contain less than 20% clay, 30% silt and more than 52% sand. Silt loams contain more than 50% silt and 12–27% clay. Clay loams contain 27–40% clay and 20–45% sand. Mineral soils contain less than 20% organic matter. Organic soils have a high (>20%) organic matter content. Muck soils contain 20 to 50% organic matter and peat soils more than 50%.

Course textured soils generally are easy to fumigate while fine textured soils like clays and clay loams are more difficult. The clay fraction, especially colloidal clays, adsorbs large amounts of fumigant. For example, the dry sand fraction of a soil adsorbs only 0.18% of its weight of chloropicrin, the silt fraction 0.8%, the clay fraction 3.2%, but the colloidal clay fraction 10%. Organic matter adsorbs even larger amounts. High clay content also reduces size of air passages and retards diffusion of fumigant through soil. Thus heavy soils require higher rates of chemicals to accomplish adequate fumigation than do lighter soils.

Soil Structure (Porosity)

Soil consists of soil particles honeycombed with interconnecting air channels that eventually lead to the soil surface. A film of water surrounds soil particles (Figure 10.2)

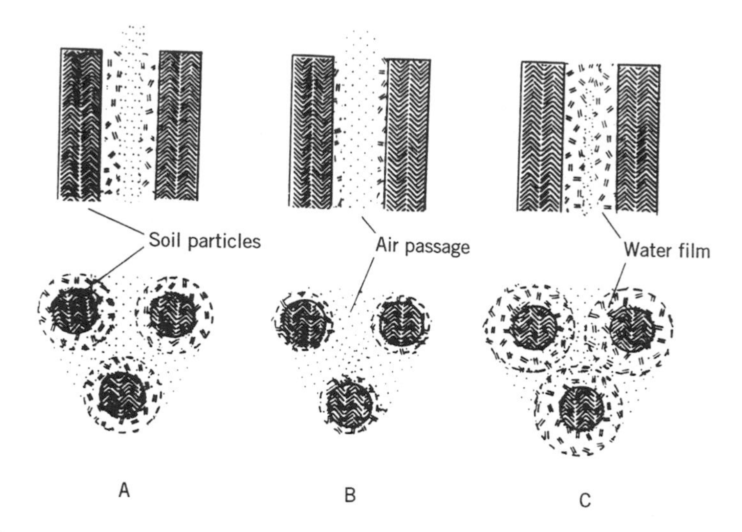

Figure 10.2. Profile (top) and plane (bottom) sections of soil tubes showing soil particles, water films, and air passages. (A) Soil moisture correct for fumigation. (B) Soil too dry. (C) Soil too wet.

producing a continuous layer of water lining these channels. Fumigants moves through the air passages and dissolve in the water film. Too many or too large air channels allow fumigants to diffuse through soil too rapidly to be effective. Moisture films should not be too thick or movement of fumigants will be impeded nor too thin so that they move too fast.

Soil Moisture

Soil moisture affects soil fumigation two ways. First it absorbs fumigant from the soil air so that the toxicant acts on soilborne organisms. Thickness of the water film is important for retaining gas in soil since the gas in soil air is in equilibrium with gas in soil water. Therefore, the greater the volume of air in proportion to the volume of water (i.e., a thin water film) the easier the gas can leave the soil and is lost. Solubility compensates to some extent for a low equilibrium ratio. For example, methyl bromide is about 8 times as soluble as chloropicrin so there is 8 times as much in the water film. However, methyl bromide has a vapor pressure 70 times that of chloropicrin, which more than negates the higher solubility.

The second function of water is the physical effect on size of air passages. An optimum amount of soil moisture provides both an adequate water film and ample air space (Figure 10.2A). Dry soils have thin water layers and large air passages (Figure 10.2B). Too wet and the air passages fill with water (Figure 10.2C). Soils drier than the wilting point (-15 bars) have little or no water around soil particles and fumigant can be adsorbed directly to particle surfaces. Fumigants move best in moist soils with a water potential between -0.6 and -15 bars (soils with a water potential of -0.1 to -0.3 bars are at field capacity, those at -15 bars are close to the permanent wilting point).

Soils wetter than -0.6 bars have blocked air passages and fumigants are impeded as well as diluted. High soil moisture in fine textured soils also reduces downward movement of fumigants. Soil moisture usually increases with soil depth making it difficult to effectively control deeply distributed pests such as those of deep rooted perennials such as fruit trees.

Soil moisture expressed as percentage is usually based on soil weight and varies with soil texture, bulk density, and organic matter. It is simpler to express soil moisture in terms of field capacity, which is the maximum amount of water a soil will hold. Optimum soil moisture content for fumigation is from 50 to 75% of field capacity. Field capacities range from about 7% available water in sandy soils to 35% in clays and 40% in peats. Soil fumigation is usually not recommended in clay and peat soils. A simple method to determine adequate soil moisture for soil fumigation is to squeeze a handful of soil into a ball. If it holds together without dripping water (too wet) or does not crumble (too dry) the soil moisture is suitable for fumigation.

Soil Temperature

Effectiveness of soil fumigation generally increases with increasing temperatures and most fumigation is done when soils are relatively warm. Temperature directly affects fumigant volatility, which determines movement through soil. Soil temperature also determines the postfumigation period before planting to ensure elimination of phytotoxic residues.

Volatile fumigants such as methyl bromide and chloropicrin are not very effective in soils below 10°C. This can be offset to some extent by increasing the dosage or length of exposure. Most effective fumigation takes place in warm soils. Sometimes temperature determines which of two effective fumigants is used. For example, greenhouse growers often use chloropicrin in summer when soil temperatures are high and use methyl bromide in winter when temperatures are low and time is limited. Fumigants such as 1,3-D and MIT give good results when applied in fall or winter at temperatures above 5°C. This may be due to seasonal differences in location of nematodes in soil or because the fumigant remains a longer time in soil. Fumigants persist longer in cool soils because of slower volatilization, hydrolysis, and decomposition.

Some soil fumigants are said to work well over a wide range of temperatures, but at low temperatures fumigation may take too long and leave phytotoxic residues. At high temperatures fumigants escape too rapidly to provide adequate exposure of targeted organsisms. Best results are usually obtained in the range of 10 to 27°C.

Organic Matter

Organic matter, sometimes called humus, is one component of mineral and organic soils and consists largely of decomposed plant and animal materials that have lost their distinct identity. In mineral soils it binds soil particles and improves soil structure. Organic matter content varies considerably and can be absent or lacking in very sandy soils or comprise more than 50% of peat soils. Soil organic matter usually ranges from 1 to 6%.

Because organic matter adsorbs large amounts of fumigant soil fumigation is less effective when organic matter is high. Dosages must by increased to obtain control comparable to that in soils with little or no organic matter. If 200 pounds of methyl bromide per acre controls nematodes in sandy loams at a given temperature, 500 pounds are required for the same degree of control in peat soils.

Plant Residues

Plant residues are portions of plants (leaves, stems, roots) remaining in soil after harvest or other farming operations. These may consist of plant roots left after pulling an orchard or vineyard, stubble after harvest of cereal grains, stover after picking corn, chaff from threshing grain, or various plants plowed into the soil. Eventually residues decompose and become humus, but as residues the material is still identifiable. Residues, like organic matter, can adsorb large quantities of fumigant. It requires up to 10 times as much EDB and 3 times as much 1,3-D to kill root-knot nematodes when plant residues are added to soil. It takes 8 times as much 1,3-D to kill root-knot nematodes inside grape and fig roots as it does in soil, presumably because the chemical is absorbed by root tissues before reaching the nematodes.

Residues such as wheat straw, corn stalks, potato haulms, and the like can also cause rapid loss of fumigants from soil by creating channels (chimneys) to the soil surface. Crop residues should be well incorporated into soil by rototilling or disking. Use of a mold-board plow may simply deposit residues in a layer at the plow sole where it is slow to decompose and can be returned to the soil surface with the next plowing. Incorporated residues should decompose for several weeks or months before fumigating.

APPLICATION OF SOIL FUMIGANTS

Methods

Soil fumigation is inherently expensive. Cost of chemical is usually the greatest expense and application costs are additional. With the exception of the plow sole method or application of MIT through irrigation systems, soil fumigation requires operations additional to those that are routine such as seedbed preparation, fertilization, and cultivation. A great deal of technological development has been necessary to ensure accurate placement of fumigants, precise dosages, and adequate sealing of soil surfaces.

Some early fumigants such as carbon disulfide and chloropicrin were injected into soil at fixed points with hand-operated injectors. These devices hold a quart to a gallon of fumigant and have a hollow probe that is pushed into the soil to prescribed depths, usually 6–8 inches, at intervals on a grid over the soil surface (Figure 10.3). The fumigant is injected by pushing down on the handle of the injector, which activates a plunger. Dosage is regulated by length of stroke of the plunger. Because this technique is labor intensive and time consuming it is used only for small plots of high value crops. The method was first used to inject carbon disulfide around roots of grapevines to control phylloxera.

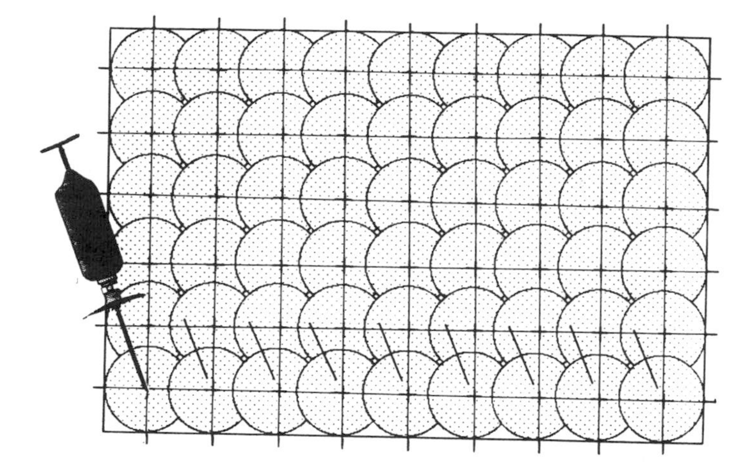

Figure 10.3. Soil surface marked with grid pattern for injection of soil fumigant with a hand applicator.

The most common method of applying soil fumigants is through chisel (tooth or knife) injectors. Chisel injectors are long steel shanks attached to the tractor toolbar. A tube is welded to the trailing edge of the shank so that chemical flowing through the tube is deposited in a stream as the chisel moves through the loosened soil. Distance between fumigant bands is regulated by spacing the chisels on the toolbar and application depth determined by the length of chisel in the soil. The application tool can be a straight or curved shank with or without a cultivator foot attached, the object being to place a uniform and steady stream of fumigant in the track of the chisel. Application can be by gravity flow or by systems equipped with pumps to maintain steady pressures of 20 to 30 pounds (Figure 10.4). A reservoir attached to a manifold feeds fumigant through lateral tubes to individual injectors. Rate of delivery is determined by size of the orifice through which the chemical flows and speed that chisels travel through the soil. Calibration is necessary to determine the volume of chemical that flows through the application tubes in a specific period of time. The amount dispensed is usually measured in milliliters per shank per minute, and the speed of travel in feet per minute. The amount of liquid deposited in 43,560 linear feet (the number of feet in an acre on a 12-inch spacing) can then be calculated or extrapolated for other chisel spacings. The amount applied per acre can be determined by converting the number of milliliters to gallons (3785 ml = 1 gallon). Conversion factors are available to calculate amounts in liters and hectares for international units of measurement.

Chisel applications place streams of fumigant in parallel lines at a given depth in soil, usually 10-12 inches, deeper in some situations, and spaced about 12 inches apart. Fumigants diffuse more upward than laterally or downward in soil and it is important that lateral diffusion is sufficient for individual application zones to overlap (Figure 10.5). This ensures that no gaps exist where pathogens escape contact with the fumigant. Subsurface blades can be used to deliver fumigant at several points across the width of the blade (Figure 10.6). The mechanics are the same as those of chisel applications.

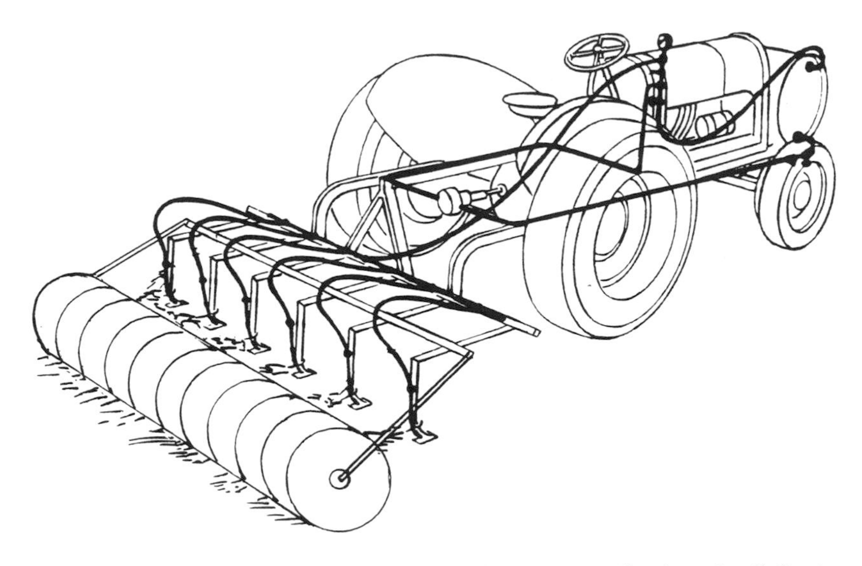

Figure 10.4. Diagram of chisel injectors set up of pressure application of soil fumigant. Compacting rollers are behind the chisels. Note that safety equipment has been omitted.

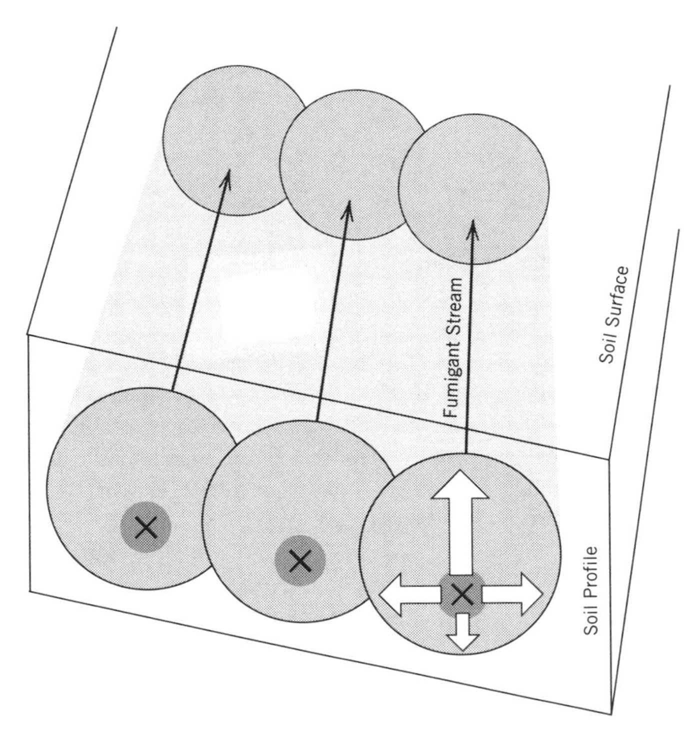

Figure 10.5. Diffusion pattern of fumigant applied in a stream by chisel injectors at 8-inch depth and 12-inch spacing.

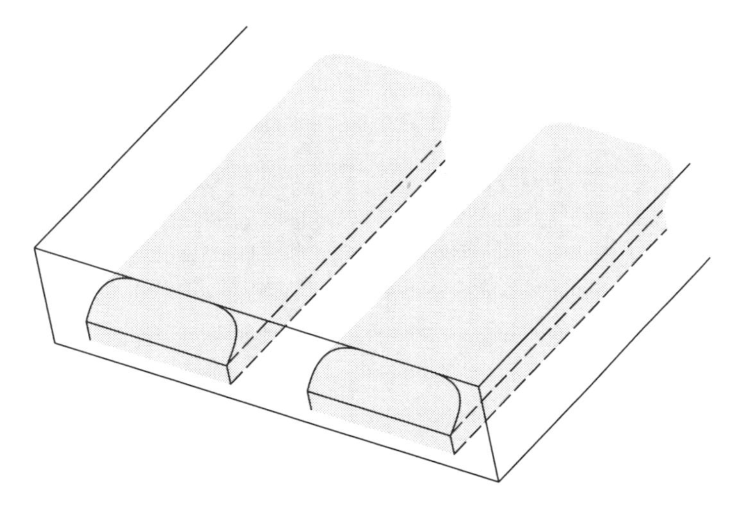

Figure 10.6. Diffusion pattern of fumigant applied in a band by blade applicators at a depth of 8 inches.

Plow sole applications have the advantage of applying fumigant during routine farming operations but have the disadvantage of turning infested soil to the surface where fumigation may be inadequate. Spray nozzles are adjusted to spray fumigant into the furrow exposed by a plow. The next plow in the gang rolls the plowed soil onto the treated soil while that furrow is sprayed and so on (Figure 10.7). A cultipacker should follow the plow to spread and compact the soil and reduce loss of fumigant. Plow sole applications may not be recommended in some regions.

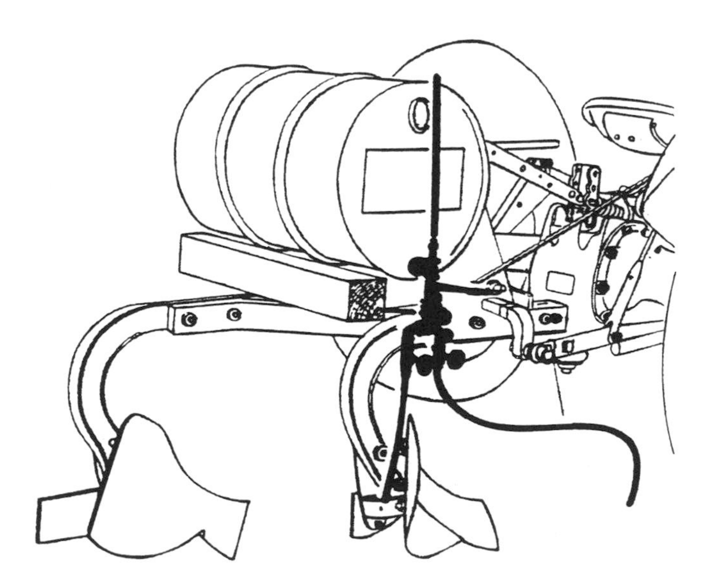

Figure 10.7. Gravity flow equipment for applying soil fumigant to the plow sole.

Surface applications drench the soil with aqueous solutions of fumigants like formaldehyde or MIT releasers. A specific amount of active fumigant is applied per volume of soil irrespective of the total volume of liquid. Methyl bromide can also be applied as a surface application, similar to space fumigation, usually to small plots such as nursery seed beds or batches of potting soil. Methyl bromide is available in tanks with pressure regulators or cannisters containing one or one and a half pounds of the gas. The soil to be treated is covered by a polyethylene tarp sealed along the edge with soil (Figure 10.8). The gas is released through tubes leading to an evaporation pan under the tarp, or cannisters are placed under the tarp and punctured.

Application Equipment

Application equipment is very diverse. It can be as simple as using a sprinkling can to pour a solution of formaldehyde or metam sodium onto the soil or as complex as a pressure-operated injection system. Hand injectors are still used for small plots in some parts of the world. Gravity flow systems for injections are probably the simplest system for volume application. Applicators for granules such as dazomet are simply adaptations of similar equipment used to apply fertilizer or insecticide granules. Water-soluble fumigants like sodium metham or MIT can be metered and applied through irrigation sprinkler systems. Whatever system is used it must be calibrated so that the proper amount of fumigant is delivered to the point where it will be most effective.

Application Procedures

Most companies that sell fumigants provide instruction on how to apply their products. These guides detail procedures the company considers necessary to obtain optimum results. There are numerous procedures, all of them essential for successful application of soil fumigants.

Soil Preparation

Soil must be worked to seed bed condition by deep plowing followed by rototilling or disking to loosen the soil at least as deep as the fumigant is intended to penetrate. Most

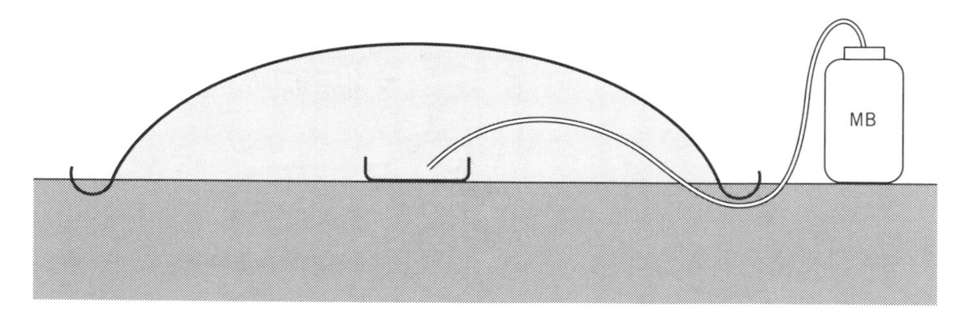

Figure 10.8. Application of methyl bromide gas from cannisters to the air space over soil covered and sealed by a polyethylene tarp.

soil organisms are in the upper foot of soil so that working soil 12 inches deep is usually sufficient. However, fumigation several feet deep may be needed to prepare for planting deep rooted perennials, such as treatments of soil for apple replant problems. Some means of breaking up the soil to these depths is required, sometimes by digging and working the soil with a backhoe or very long chisels.

Another requirement is decomposition of plant residues and elimination of as much organic material as possible. Organic matter irreversibly binds organic chemicals and renders soil fumigants ineffective. Reducing plant residues usually requires cultivation to break up the material and distribute it through the soil where microorganisms can decompose it. At least a month, perhaps a season, is needed for plant residues to decompose.

Addition of water to soil can enhance fumigation several ways. It provides the moisture needed for effective fumigation in dry soils and lowers soil temperatures by evaporative cooling. Moisture also activates dormant microorganisms and makes them more sensitive to fumigants. Excess soil moisture may have to be drained away to obtain satisfactory fumigation. In some regions sudangrass is grown in clay loam soils to remove excess water. This method can reduce the amount of soil moisture by half at a depth of 6.5 feet (13.6% in fallow soil compared with 6.6% moisture after planting sudangrass).

Dosage Rates

The rates of soil fumigants required to kill specific pathogens depend on the type and condition of that organism, relative toxicity of the fumigant, edaphic factors, temperature, and moisture. Rates are usually suggested by the manufacturer to cover a range of conditions and situations.

Depth and Width of Application

Although nematodes occur to depths of 6 feet, most soil microorganisms are in the upper 12 inches of soil. This is the region usually fumigated. Since fumigants diffuse upward more readily than downward the fumigant must be placed below the midpoint of the zone to be treated (Figure 10.5). For a 12-inch treatment the fumigant is usually applied at an 8–10 inch depth.

Spacing the applications is as important as depth. The effective lateral range of chloropicrin is about 6 inches from the point of injection so that 12 inches is the maximum spacing for chloropicrin injections to ensure that the killing zones are in contact (Figure 10.5). This way there will be an overlap of fumigant from adjacent injections so that no gaps occur.

Application Patterns

Broadcast fumigation is aimed at treating the entire planting area but is especially costly since up to half the area treated is not occupied with plants. Cost of fumigation can be reduced in row crops by treating only the band of soil in which the crop is to be seeded leaving untreated strips between the rows. Spot treatment is adequate where individual high value and productive plants such as citrus, grapevines, and figs are to be planted.

Side dressing places fumigants that are low in phytotoxicity in the vicinity of live roots. These materials are more effective against insects and nematodes that migrate in and out of roots than against fungi, which remain in the host. EDB was used at low rates (2–4 gallons per acre) as a side dressing for sugar beets, tobacco, and pineapple.

The easiest and cheapest method of applying soil fumigants is through an established irrigation system. The fumigant is metered into the irrigation water at low concentrations and applied over a prolonged period. Additional water is added to move the fumigant to the desired depth in the soil.

Sealing the Soil Surface

All fumigants benefit from a surface seal, even if it is just slight compression of the soil. Highly volatile fumigants such as methyl bromide require an impervious cover. Several covers including glue-coated paper, asphalt, latex, and waxes have been used, but polyethylene films are now commonly used. Despite their higher cost polyethylene films 25 μm thick are generally used for machine application to minimize loss of fumigant since films less than 25 μm thick are permeable to the gas. The film is rolled out immediately behind application of the fumigant (Figure 10.9). Edges of adjacent sheets are glued together on the next pass across the field to provide one continuous cover over the treated surface. Equipment is available to remove the polyethylene film but it is damaged in the process.

Thicker covers are used for bulk treated soil or spot fumigation to improve fumigation results. For example, 49 g of methyl bromide–chloropicrin mixture (67–33%) under a 100-μm tarp gave the same results as 98 g under a 25-μm tarp. However, the thicker the tarp the greater the cost.

Less volatile fumigants such as 1,3-D and EDB require only compaction of the soil surface to retard diffusion of the gas from soil. Different devices are used ranging from simply dragging a board across the soil surface to power-driven packing rollers that smear the soil surface as well as compact it. Light irrigation of the soil surface provides an adequate seal for some fumigants.

Postfumigation Treatment

Following fumigation it is important that phytotoxic levels of fumigant be eliminated from the soil. This usually requires waiting 2–3 weeks after the end of the prescribed fumigation exposure, which is generally 1–2 weeks. However, low temperatures and high soil moisture prolong the waiting period but escape of residual gas can be accelerated by working the soil with a cultivator. Since most fumigants have a distinctive odor, a common way to determine if soil is free of fumigant is to simply smell the soil. If no odor of fumigant is detected no additional aeration is necessary and planting may proceed.

Soil amendments such as fertilizers, stabilizers, or modifiers such as gypsum or lime, or organic additives (mulches, composts, manure, peat, sawdust, or other

Figure 10.9. Applying a soil fumigant and sealing the soil with a polyethylene tarp.

materials) should be added after fumigation, otherwise they interfere with the fumigation process.

PROBLEMS WITH SOIL FUMIGATION

Soil fumigation requires specialized equipment that is costly. If the procedure is unsuccessful the monetary loss can be substantial. For these reasons, most growers rely on custom applicators to apply soil fumigants.

Most soil fumigants have some phytotoxic properties and plant injury occurs if treated soil is not purged of residual fumigant. Plants not killed outright may be so weakened that they do not produce a profitable crop.

Also, toxic residues may be left when fumigants decompose. Most soil fumigants are halogenated hydrocarbons and some breakdown products, especially bromine, are injurious to sensitive plants such as carnation, onion, and sugar beet. Decomposition results mainly from microbial activity and addition of organic matter appears to accelerate decomposition of EDB, 1,3-D, metam sodium, and MIT. 1,3-D and DBCP are chemically hydrolyzed to 3-chloroallyl alcohol and then converted to 3-chloroacrylic acid. The bromine and/or chlorine is removed and intermediate products converted to CO_2 and water. Losses to decomposition range from 2 to 3% per day in sandy soils to 25% per day in clay soils. DBCP converts to n-propanol, chlorine, and bromine and EDB converts to ethylene and bromine. Some soil bacteria use small amounts of 1,3-D as an energy source but, in general, soil enrichment with fumigants does not result in increases of fumigant-decomposing microorgisms.

Availability of some nutrients can be altered by soil fumigation. One of the most pronounced is the plant growth response resulting from release of nitrogen when microorganisms are killed. The nitrogen flush is the usual explanation for improved plant growth after fumigation when no pathogen or other pest can be associated with a plant problem.

Elimination of a known pathogen usually explains the plant growth response but it may result from elimination of less than obvious pathogens. Increases of wheat yields following soil fumigation or use of the selective fungicide metalaxyl suggest that *Pythium* can weaken wheat plants enough to reduce yields without producing obvious symptoms of disease. Fumigation may also eliminate weeds and soil insects that in the aggregate reduce plant growth but go undetected.

Temporary elimination of beneficial organisms may be the greatest disadvantage of soil fumigation even though fumigation does not sterilize the soil. Eventual recolonization occurs but certain beneficial organisms are especially sensitive to soil fumigants. One such group is the nitrifying bacteria and ammonia accumulates in fumigated soil when the nitrification process is inhibited, resulting in ammonium toxicity or nitrogen starvation.

Detrimental responses to fumigation in avocado, citrus, cotton, peach, soybean, white clover, and hardwood trees have been associated with elimination of endotrophic mycorrhizae. Among fumigants that caused the reported stunting are chloropicrin, EDB, D-D, propylene oxide, MIT, methyl bromide, carbon disulfide, ethylene dichloride, and metam sodium. However, DBCP and EDB were not reported to adversely affect root infections or chlamydospore development by vesicular–arbuscular (VA) mycorrhizal fungi. Another detrimental effect of soil fumigation is reduced availability of phosphorus to plants. This apparently results from elimination of mycorrhizal fungi, which are important in phosporus transformations in soil.

Endotrophic mycorrhizal fungi are usually phycomycetes, commonly *Glomus* species, and they are especially sensitive to soil fumigants. *Glomus* species are 2–9 times more sensitive to methyl bromide fumigation than pathogenic fungi. It is unlikely that fumigant dosages can be reduced enough to permit mycorrhizal fungi to escape and still kill soilborne pathogens. Detrimental effects in plant growth resulting from elimination of mycorrhizae frequently occur in forest nurseries where soil fumigation is routinely practiced. Seedlings must be inoculated with appropriate mycorrhizal fungi or they will fail to grow when outplanted in soils that lack these beneficial organisms. It now appears that both endo- and ectotrophic mycorrhizae are much more common in plants than was once thought and they play an important role in plant health. Ectotrophic mycorrhizae enhance nutrient uptake, reduce drought stress, produce essential growth substances, and protect roots from pathogenic organisms.

REFERENCES

*Anon., *Plunder Underground.* The Dow Chemical Company, Midland,
MI, 1967.

*Anon., *Control of Plant Parasitic Nematodes.* National Academy of Sciences, Washington, D.C., 1968.

*Brown, R. H., and B. R. Kerry (eds.), *Principles and Practices of Nematode Control in Crops.* Academic Press, New York, 1987.

*Goring, C. A. I., Physical agents of soil in relation to the action of soil fungicides. *Annu. Rev. Phytopathol.* 5:285–318 (1967).

*Goring, C. A. I., and J. W. Hamaker (eds.), *Organic Chemicals in the Soil Environment*, 2 Vols. Marcel Dekker, New York, 1972.

*Hedden, O. K., J. D. Wilson, and J. P. Sleesman, *Equipment for Applying Soil Pesticides.* USDA Agr. Res. Ser. Agric. Handbook No. 297, 1967.

*Martin, J. P., and P. F. Pratt, Fumigants, fungicides and the soil. *J. Agric. Food Chem.* 6:345–348 (1958).

*Menge, J. A., Effect of soil fumigants and fungicides on vesicular-arbuscular fungi. *Phytopathology* 72:1125–1132 (1982).

*Mulder, D. (ed.), *Soil Disinfestation.* Elsevier, New York, 1979.

*Munnecke, D. E., and S. D. Van Gundy, Movement of fumigants in soil, dosages, responses, and differential effects. *Annu. Rev. Phytopathol.* 17:405–429 (1979).

*Newhall, A. G., Disinfestation of soil by heat, flooding and fumigation. *Bot. Rev.* 21:189-250 (1955).

Schmitt, C. G., Comparison of volatile soil fungicides. *Phytopathology* 39:21 (Abst.)(1949).

* Publications marked with an asterisk (*) are general references not cited in the text.

11 Eradication by Physical Means

Heat is second only to chemicals for eradicating pathogens from soil, seed, vegetative parts, or nonliving surfaces. Heat can be in the form of thermal radiation from a heating element, fire used in field burning or flaming, radiations from the sun (solarization), molecular excitation from ionizing radiations or microwaves, live or pressurized steam, or hot water. Except for field burning and flaming, the use of heat as an eradicative measure is limited to small areas such as greenhouses, nurseries, and transplant seedbeds. Soil solarization is also limited to small areas because the soil is covered with a plastic film.

Heat results from electromagnetic radiations. These may be visible and invisible wavelengths from fire or the sun, invisible short wavelengths of X-rays, or longer wavelength microwaves. In the electromagnetic spectrum gamma rays, X-rays, ultraviolet, visible, infrared, and ultrasonic (microwaves) have been used to eradicate pathogens by producing heat or similar energy. Electromagnetic energy is sometimes referred to as Hertzian waves from the system of describing them based on their frequency and wavelength. The frequency or number of vibrations per second is called a Hertz (Hz). The Hertzian value of a particular wave is inversely related to the distance between peaks of two adjoining waves. The length of a single wave may be as short as 0.01 nm (nanometer = 1 billionth of a meter) for gamma rays to as long as 10 m for ultrasonic frequencies. The product of wavelength times frequency always equals the velocity (speed) of light or 299,727,738 m/sec (rounded to 300,000,000 to simplify calculations). Thus, X-rays have a frequency of 30 billion billion (3×10^{18}) Hz. At the other end of the spectrum ultrasonic waves (microwaves) with frequencies of 1–2 billion Hz have a wavelenth of about 1/3 m.

Figure 11.1 shows the electromagnetic spectrum with the wavelengths and frequencies for different types of radiations. Note the inverse relation betwen frequency and wavelength. Gamma rays have the same characterisitics as X-rays but have more energy and are more penetrating. They differ primarily in origin. Gamma rays result from nuclear disintegrations and are emitted by certain radioisotopes. X-Rays are produced by electrons striking a metal target in a vacuum tube and therefore are extranuclear radiations.

Thermal radiation is in the range from about 0.4 μm (the long wave fringe of UV) to 1 mm (the end of infrared).

Fire was the first form of heat used to eliminate pathogens. It was recommended by the Romans as early as 70 B.C. to sterilize soil, although it is unlikely that the concept of sterilization was recognized at that time. More likely the rationale for fire was for

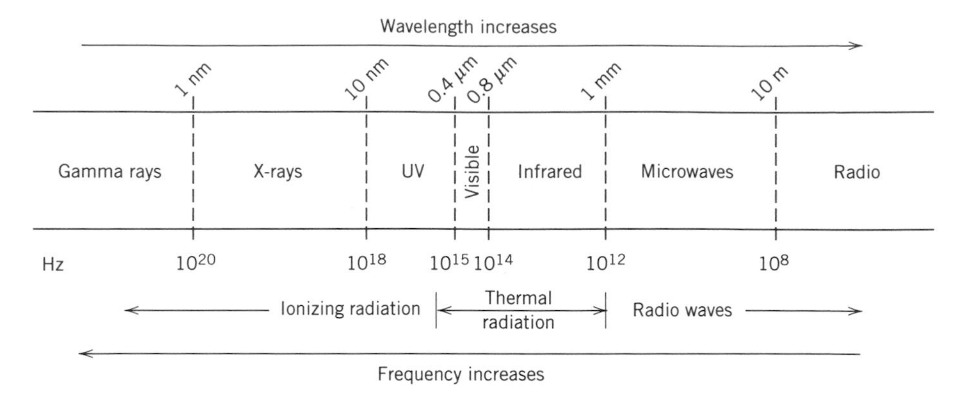

Figure 11.1. Wavelengths and frequencies of components of the electromagnetic spectrum. Note that the units are not proportional.

a ritual of purification. No other method was proposed until 1767 when hot water treatment was suggested as a treatment for millet smut, but nothing was done for another century. In 1883, J.L. Jensen, in Denmark, developed a dry heat method for eliminating the late blight fungus from potato tubers, but the method had little practical value because potato plants could still become infected via airborne spores. In 1888, Jensen described a hot water treatment for the control of loose smut of wheat and barley that had great practical value since seedborne inoculum is the only source of infection. This treatment consisted of soaking seed 15 minutes in water heated to 55°C, then spreading the seed in thin layers to cool and dry. The following year Jensen's hot water treatment was used in the United States and soon became the standard seed treatment for loose smut of cereals. Steam was first used to control soil pests in 1893 and became the most satisfactory method of controlling pests in greenhouse soil.

Many pathogenic organisms are more sensitive to heat than seeds, making heat treatments possible not only for soil and other nonliving materials but also for many seeds. Vegetative materials (bulbs, corms, roots, etc.) are less tolerant and require longer treatment at lower temperatures. Figure 11.2 illustrates the sensitiveness of various pathogens and pests to temperature. About the lowest temperature that will kill the most sensitive organisms is 45°C.

Heat treatment is fast, usually requiring only a few hours, compared with days or weeks for soil fumigation. It is relatively inexpensive, the main cost being the cost of electricity or other energy source. It leaves no chemical residues such as bromine that can cause plant injury. Heat, however, is not always effective. Sometimes pathogens recolonize sterilized soils and, having no competition from other organisms, develop unchecked, causing severe damage. Heat treatment, whether electric sterilization, steaming, or flaming, destroys organic matter in soil. Field burning pollutes air, creates environmental, social, and political controversy, and danger. Postharvest burning grass seed fields has resulted in temporary closures of interstate highways because of reduced visibility from smoke. As populations encroach on agricultural areas, environmental concerns have become more prevalent.

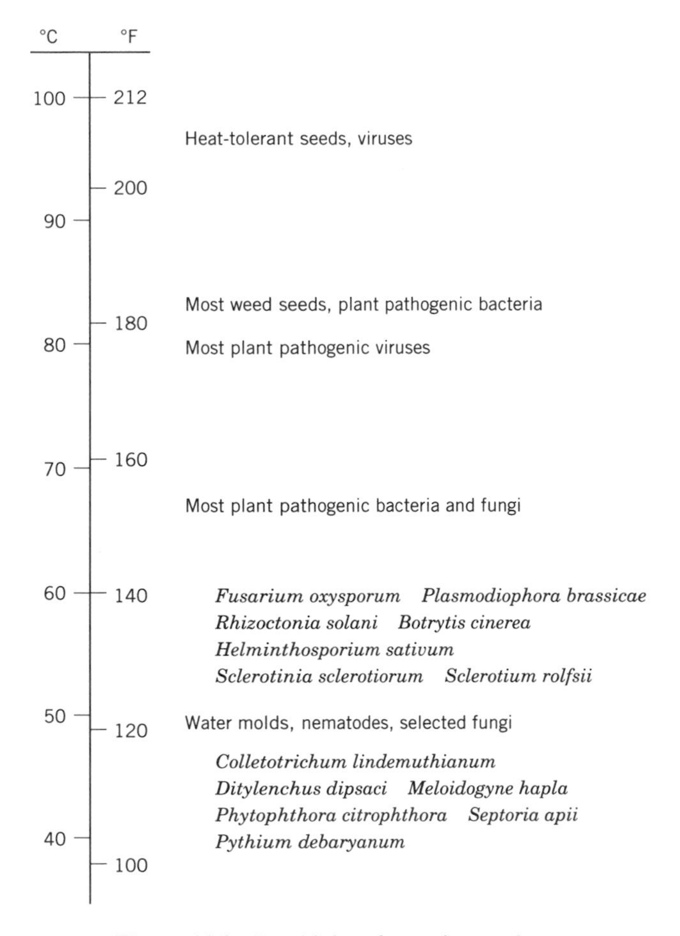

Figure 11.2. Sensitivity of organisms to heat.

Sterilization refers to complete elimination of organisms from soil and other nonliving materials. Ideally, only harmful organisms would be destroyed leaving beneficial bacteria and fungi to compete with pathogens reintroduced into the soil. Partial sterilization, called *pasteurization*, is achieved by heating substrates to 62°C. This temperature kills the more sensitive organisms and eliminates most plant pathogenic nematodes, bacteria, and fungi, especially water molds such as *Pythium* species, a major cause of damping-off in greenhouse soils. Pasteurization had its beginning with "gentle" heating of wine by Louis Pasteur (whose name is given to the process) to preserve wine and prevent it from souring.

Moist heat is more efficient in eliminating pathogens because water conducts heat better than air. Steam helps break hydrogen bonds thereby promoting rapid denaturation of proteins and affecting membrane lipids. Steam requires fewer heat units (BTUs) than dry heat to raise temperatures of soil or other substrates. Live (free flowing) steam or hot water also has the built-in safety feature of not going over 100°C.

Moist heat distributes more uniformly than dry heat because insulating air pockets are less likely to occur when moisture is present. The major disadvantage of moist heat, especially steam under pressure which can raise temperatures above 100°C, is that it can break down organic matter. This releases metallic ions, such as manganese, which can be toxic to plants. Moist heat sterilization can alter the granular structure of soil (puddling) or cause other harm from excess condensation.

PRESSURIZED STEAM

Steam sterilization kills all organisms in soil and is the usual standard with which other disinfesting treatments are compared. Steam applied under pressure to sterilize soil and soil containers (pots, flats, etc.) requires special equipment. Large mattress sterilizers or commercial canning pressure cookers are commonly used although small quantities (< 1 ft^3) can be sterilized in standard laboratory autoclaves. Steam at 15 pounds pressure per square inch reaches a temperature of 121°C, but temperature decreases when either pressure decreases (Table 11.1) or air content of steam increases (Figure 11.3).

FREE-FLOWING STEAM

Steam is usually applied to a mass of soil as live or free-flowing steam. Various techniques have been developed to apply and distribute steam through soil. Baker (1957) and others have described these methods in detail and list advantages and disadvantages of each. The earliest method used in a commercial greenhouse in 1893 consisted of three perforated steam pipes on the bottom of a wooden bin 4.5 feet deep holding 480 ft^3 of soil. This method was soon modified so that the pipes were buried at various locations in greenhouse beds and steam brought to the soil rather than soil being brought to the steam. In both methods steam under boiler pressure was applied for 1–2 hours until soil at 10–20 inches reached a temperature of 66–100°C. It should

TABLE 11.1. Relationship between Pressure and Temperature of Pure Steam

Pressure (lb/in^2.)	Temperature	
	°C	°F
5	108	226
10	116	240
15	121	250
20	127	260
25	131	267
30	134	274

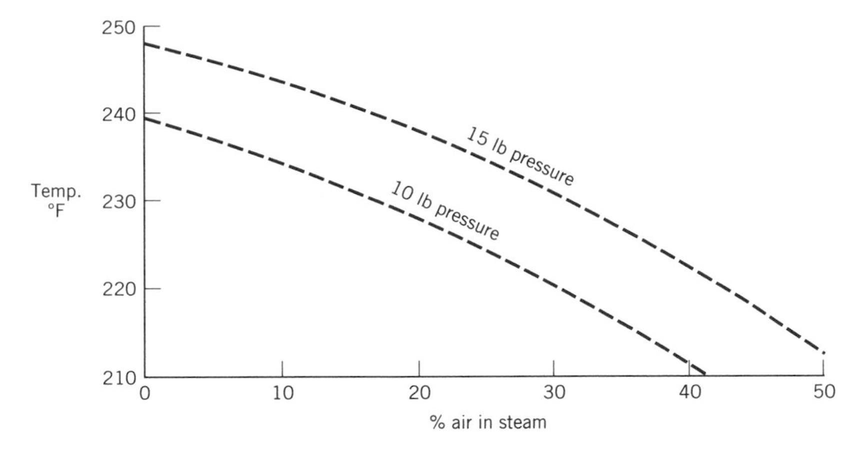

Figure 11.3. Temperature of steam mixed with various amounts of air.

be noted that the pressure referred to is the pressure of steam leaving the boiler, not pressure of the steam in the soil.

The *buried perforated pipe system* uses 1- to 2-inch-diameter metal pipes 30 to 150 feet long perforated with holes 1/8 to 3/16 inch diameter spaced 8 to 16 inches apart. The holes are staggered about 20° from the bottom of the pipe. Number and length of pipes depend on size and shape of the beds to be treated and capacity of the boiler. Depth of steaming depends on the crops to be grown, pests being controlled, and the soil. About 2.5 ft^3 of loam soil can be steamed per boiler horsepower per hour. Clay soils require more exposure time and sandy soils less. The buried pipe system effectively controls root-knot nematodes and *Fusarium* and *Verticillium*.

The *inverted pan system* developed in the 1890s is a more rapid system of applying steam to soil. In 1904 the procedure was recommended by the USDA for sterilizing tobacco seedbeds in Connecticut. A large iron pan is inverted over the soil to be treated, the edges of the pan are pushed into the soil, and steam is piped in through an opening in the side of the pan. Pans range from 30 to 100 ft^2 and require some system of tracks, hoists, or rollers to move them.

The buried pipe and inverted pan systems have short lives because of rapid corrosion of metal components and have been replaced by the *buried tile system*. Clay drain tile 3–4 inches in diameter is laid 12–16 inches deep in rows 18–20 inches apart and steam introduced under pressure through headers (Figure 11.4). This system treats about 1.5 ft^2 per boiler horsepower per hour and the area treated is limited by size of the boiler. Although the area treated is smaller than that possible with the buried pipe or inverted pan methods, the buried tile system lasts much longer and remains serviceable for 15–20 years.

Devices have been developed to force live steam into soil as the apparatus is moved along a path or track. The steam rake is a rake-like device with hollow tines 6–8 inches long that project into the soil. As the rake is pulled through soil steam is forced into it. Similarly, the steam plow injects steam into the furrow created as the plow moves through the soil. These systems are used primarily in greenhouses and nursery beds and

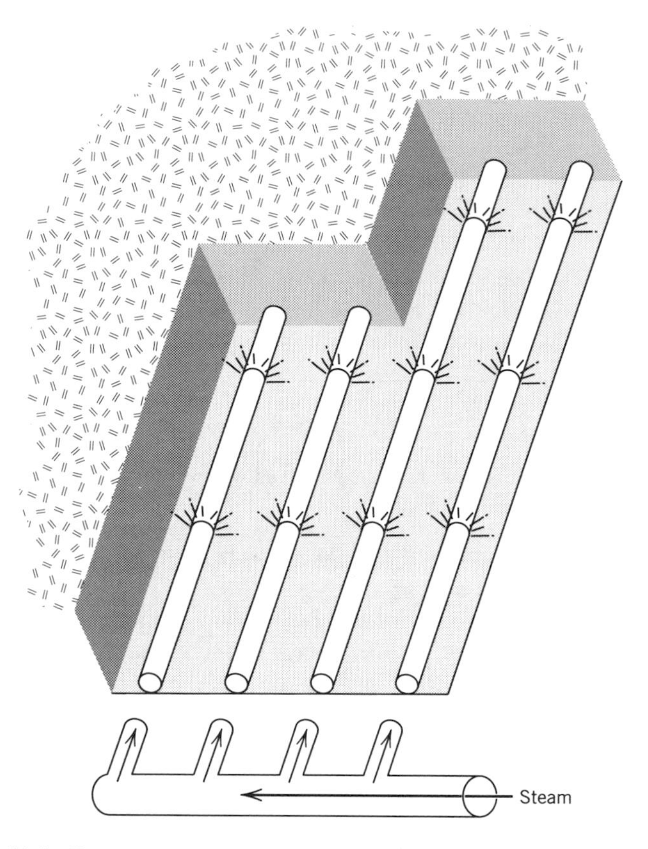

Figure 11.4 Cutaway diagram of the buried clay tile steam sterilization system.

require some mechanical device such as a winch and cable to pull the apparatus. A plastic tarp or other impervious sheet is trailed behind the steaming apparatus to hold the steam in soil for a prescribed length of time.

AERATED STEAM

The concept of mixing air with steam to conserve steam and minimize undesirable effects of steam sterilization began in England in the 1950s. The method was intensively researched by workers in California. Air mixed with free flowing steam has the same effect as air mixed with steam under pressure—it cools it, so the higher the air content of the mixture the lower the temperature. Since temperatures much lower than 100°C will kill most soilborne pathogens (Figure 11.2), especially important damping-off and root rot fungi like *Rhizoctonia solani, Pythium* species, *Fusarium roseum*, and *Colletotrichum coccoides*, air/steam mixtures are equally effective and more economical than pure steam. Lower temperatures have both physical and biological advantages over the higher temperature (100°C) of pure steam. There is less breakdown of organic matter and subsequent release of toxic elements, and soil structure apparently is not affected. Many saprophytic microorganisms survive aerated steam treatment so creation of a biological vacuum is avoided. This minimizes the potential for recontamination and colonization of treated soil by pathogens.

Several systems have been developed to mix air with steam before applying the mixture to soil, the important feature being that a sufficient volume of air be moved, pressure being relatively unimportant. A blower that forces a jet of air into steam entering a mixing chamber is the most efficient system. Generally the blower is of the paddlewheel (squirrel cage) design, rather than a propeller type fan. Air is introduced into the steam by other systems such as piston or rotary pumps, steam flowing past an intake orifice, or simply air and steam pipes emptying into a common chamber.

Amount of air in proportion to amount of steam determines the temperature of a mixture. Pure steam is 100°C, a 1:1 (air:steam) mixture by volume is 82°C, 2:1 about 71°C, and 4:1 approximately 60°C.

The thermal mass of steam (pure or aerated) moves through soil as a unit, raising the temperature of the soil along a moving front. Soil must be sufficiently porous so that the air mass pushed by the steam moves through the soil uniformly, otherwise portions of the soil will not reach desired temperatures because of air locks. Soil must be loose and friable, not compacted. As in soil fumigation, excess moisture restricts or blocks air passages and impedes movement of steam through soil. High moisture content also requires more heat to produce desired temperatures because water absorbs heat.

Aerated steaming is usually carried out in large stationary or portable bins or boxes, at least 2 feet deep, with air and steam introduced through a perforated plenum at the base. Shallower depths result in volcano-like blowouts releasing the steam into the atmosphere instead of it being retained in the soil. Final temperature is determined by length of steaming period. However, it should be emphasized that temperture and not time is the critical factor in successful sterilization. When the desired temperature is reached steam flow is adjusted to maintain the temperature for 30 minutes and then shut off. Thirty minutes cooling time is allowed before the bin is emptied.

SUPERHEATED STEAM

Steam can be heated to temperatures much higher than 100°C by passing steam through coils or pipes heated by electricity, propane, oil, or other energy source. Superheated steam is used primarily for cleaning machinery. To prevent disease it is used to clean tillage, harvesting, or postharvest machinery that carry pathogens or infested crop debris and soil on their surfaces. Other than for sanitizing purposes, superheated steam is not used to control plant diseases.

HOT WATER

Since Jensen first developed the hot water treatment to eliminate smut from cereal seeds hot water has had many additional applications. This treatment has been used to eliminate many seedborne pathogens and others in or on vegetative organs. The maximum temperature used is about 55°C, somewhat lower for sensitive material. There are some undesirable features. Considerable labor and time are required to treat and dry seed or other material. Temperatures must be high enough to kill pathogens and low enough to prevent damage to seed. Development of systemic fungicides such as carboxin allows eradication of some seedborne fungi without resorting to hot water treatment but few such chemicals are available.

Hot water treatment is the usual, often the only, control measure for seedborne diseases in a wide variety of seeds and vegetative propagative organs. It controls many fungi and bacteria on or in true seeds and controls endoparasitic nematodes in vegetative structures such as rootstocks, corms, tubers, bulbs, and rhizomes. Hot water treatment is a general practice for flowering bulbs because infected bulbs may show no evidence of nematodes or other pathogens. Treatment thereby prevents introducing pathogens to uninfested areas. The standard treatment is to soak bulbs for 4 hours at 43°C or 3 hours at 44°C. Formerly it was 3 hours at 43°C, but some races of the bulb nematode can tolerate this time-temperature combination.

Root-knot and other nematodes are eliminated from grape and citrus seedlings treated for 15 minutes at 50°C or 5 minutes at 51–52°C. Treatment at the higher temperatures requires immediate cooling to minimize injury to plantstocks. Mint roots treated at 50°C for 10 minutes gave complete control of root-knot nematodes and 100% plant survival.

Banana corms are routinely soaked in hot water at 55°C for 25 minutes to eliminate burrowing nematodes. Sugarcane setts are treated for 30 minutes at 52–53°C to eliminate several viruses and to control anthracnose, a fungal disease. Shallot cloves are treated for 2 hours at 43°C, 1.5 hour at 45°C, or 1 hour at 46°C to control damping-off fungi, especially *Pythium* and *Rhizoctonia*.

Air trapped in crevices in bulbs and corms can interfere with hot water treatment because of the insulating effect of air. These organs are presoaked in water for about 2 hours to displace air. For the same reason banana corms are peeled and insulating outer layers of tissue are removed before treating.

Cabbage and cauliflower seeds are soaked for 20 to 25 minutes at 50°C to control black rot, black leg, and bacterial leaf spot. Other seedborne diseases controlled by hot water are *Septoria* leaf spot of celery, browning of flax, and bacterial leaf spot of carrot. Tomato seed is treated for 25 minutes at 50°C to control bacterial canker and bacterial spot. Pepper is also susceptible to bacterial spot but is damaged by heat and this method is not used for pepper seed. Likewise, hot water treatment is not used to control lettuce mosaic, a seedborne viral disease, because lettuce seed is killed by heat as quickly as the virus.

The large seeds of beans, peas, corn, and cucurbits are rarely, it ever, treated with hot water. The long time required to heat the seed mass permits excessive water imbibition, which results in seed decay or impaired germination.

Figure 11.2 shows that nematodes and water molds are killed at the lower temperatures (about 45°C) used for heat treatment. Even heat tolerant fungi cannot withstand temperatures above 60°C for prolonged periods but this is also about the highest temperature that plant material can tolerate. Temperatures below 40°C for prolonged periods of time are used in thermotherapy of plant material (see Chapter 18).

Hot water treatment was used to control decay in green tomatoes as early as 1920 but used infrequently since then. However, recent trends toward less use or availability of chemicals has increased use of hot water treatments to control postharvest diseases of fruits and vegetables. Temperatures range from 45 to 63°C for 0.5 to 20 minutes. Hot air at 43–54°C for 15–30 minutes has also been used but is less effective and causes more injury than moist heat. Dipping peaches and nectarines in water at 54°C for 2.5 minutes destroys surface mycelium of *Monilinia* and *Rhizopus*. Stem rot of melons is controlled by hot water at 54°C for 0.5 minute and hot air-steam at 40°C for 1 hour controls decay in figs, raspberries, and strawberries. Hot water treatment of guava fruit at 50°C for 30 minutes reduced the amount of fungal fruit rot from 7 to 1 or 2%. Control with heat is less satisfactory than postharvest chemicals but at least gives some control. The most suitable application is for produce sold soon after treatment and not stored, since there is rapid deterioration, discoloration, spotting, and general reduced storage life.

COMBINATIONS OF HEAT AND CHEMICALS

Some combinations of fungicides or nematicides and hot water permit use of lower water temperatures and lower rates of pesticides for effective eradication. One of the earliest combinations of hot water and a fungicide was the use of a 0.5% formalin solution at 43°C for treating ornamental bulbs to eliminate *Fusarium* and other decay fungi, and root-lesion and bulb nematodes. A mixture of 1000 ppm (0.1%) benomyl in hot water at 55°C has been used to eliminate *Fusarium* species from asparagus seed.

DRY HEAT

Dry heat, in the form of hot air from an oven, was suggested for control of loose smut of cereals as early as 1832 but Jensen found that even 7 hours at 51–54°C was ineffective. Dry heat is not as effective as moist heat, requires higher temperatures, has some undesirable effects, and is generally restricted to pasteurization of soil.

High voltage electricity as a nonthermal means of treating soil is impractical, especially for field application, because of the large amount of energy required. It has been estimated that it would take more than 10,000 horsepower (about 7500 kW of power) to treat a swath 6 feet wide and 2 feet deep at 2 miles per hour. Electricity is more commonly used to create heat through resistance, the heat being transferred to soil by thermal conductance.

Baking or burning is undesirable because soil is dried and organic matter destroyed, rendering it less suitable for plant growth. Furthermore, organisms are more resistant to dry heat than to moist heat. Most so-called dry heat methods actually utilize moist heat because the heat applied is transferred to moisture in soil.

Sterilizing or pasteurizing soil by dry or moist heat entails heating soil uniformly to 70–80°C, maintaining that temperature for 0.5 hour, then cooling it as rapidly as possible. Prolonged heating adversely affects soil structure and humus content, resulting in flocculation. One of the major defects in heating soil by electricity is that temperatures are too high close to heating elements and too low away from them and the required slow heating over several hours is harmful to soil. However, dry heat provides a means of sterilizing soil when steam is unavailable.

Electric resistance using the soil as the heating element by passing a current directly through it has been attempted but is extremely dangerous and inefficient. Electric bulk soil pasteurizers are of two types. One uses electric heating elements buried in soil to generate heat from soil moisture. This system is simple, safe and inexpensive but produces uneven temperatures. The second type uses an electric induction grid in which current is passed through insulated elements inside iron pipes buried in the soil. This system is costly because of rapid deterioration of insulation and pipes.

Several continuous bulk soil pasteurizers have been developed in which a layer of soil moves over a surface heated by electricity or oil or gas flame. Most of these are rotating drums heated on the outside much in the way that asphalt–gravel mixes are prepared.

Flash-flame pasteurizers are similar to gas or oil space heaters in which a jet of flame is directed into one end of a tubular container and soil is fed into the other end. The soil moves in the direction of the flame and is heated as it passes through the flame. These devices are portable and can be used where electricity is unavailable but the volume of soil treated is small.

FLAMING FOR DISEASE CONTROL

Aside from the ancient references to the use of fire in agriculture, there is historical evidence that fire maintains forests and grasslands in a good state of health. One of the earliest and perhaps best known examples of disease control by fire is brown spot needle disease of longleaf pine. This disease tends to keep longleaf pine seedlings in the "grass" stage rather than allowing internodal elongation. Burning by early ranchers in southeastern United States renovated grazing lands by destroying scrub oak and other ground vegetation, which then promoted forage plants. Coincidentally brown spot inoculum on diseased pine needles was also destroyed and young longleaf pines developed into healthy saplings. Effectiveness for brown spot control notwithstanding, use of fire has always been controversial.

Partial or complete control of a number of plant diseases has been achieved by burning crop residues after harvest. In this regard grass seed production, especially of fescues and ryegrass, has received the most attention. Blind seed, grass seed nematode, ergot, and other diseases are adequately controlled in Oregon by field burning. However, as residential populations increase near grass seed production areas political pressure is increased to stop field burning. Attempts are being made to develop systems whereby burning can be rapid and relatively smoke-free. This usually involves some type of enclosed system where air, fuel, usually propane, and crop residues are combined to develop an intense flame and rapid incineration. Part of the argument for burning grass seed fields is stimulation of flower buds and increased seed set and growers are very reluctant to abandon this multipurpose practice that simultaneously eliminates unwanted crop residues and controls pathogens and pests.

Cephalosporium stripe of wheat occurs in several regions of the world including Japan, Great Britain, and the United States where it can cause severe losses. The fungus invades the vascular system of the plant extending to the heads so that considerable inoculum remains in straw after harvest. Burning markedly reduces the amount of infection in subsequent wheat crops (Table 11.2). However, burning wheat stubble is contrary to longstanding conservation policy and considered a destructive practice. Moreover, burning alone will not control Cephalosporium stripe. Crop rotation and tolerant varieties are preferred for management of this disease. Unfortunately, some farmers use disease control as justification for burning heavy straw residues even when the disease is insignificant.

Propane or oil fueled flamers have been used to destroy inoculum of Verticillium wilt of potatoes, mint rust, hop downy mildew, and other diseases. Flaming is generally considered economical unless there is a drastic increase in fuel prices as has happened during petroleum shortages of one kind or another. In all of these diseases the basic strategy is to destroy primary inoculum before secondary disease cycles start. Examples of burning or flaming for control of plant disease are given in Chapter 15.

A unique use of fire for disease control involves cauterization of bacterial cankers on fruit trees. A propane burner is used to cauterize cankers on apricot, peach, and cherry trees by heating the bark for 5–20 seconds until the inner bark begins to crackle and char. The method is considered better than cutting and/or chemical excising because it permits rapid wound reaction, which forms an effective barrier against reinfection.

In some cases burning can have undesirable effects and increase disease severity. Hot slash fires in conifer forests stimulate spore germination and growth of *Rhizina undulata,* which infects and kills planted or natural seedlings. Ground fires in forests also cause basal scars on less fire-tolerant trees and create entry wounds for decay fungi. Fire also can stimulate germination of seeds of undesirable species such as currants and gooseberries, which serve as alternate hosts for white pine blister rust. However, this effect of fire can be turned to good use in eradication of alternate and alternative hosts. Seeds are stimulated to germinate and the developing plants more vulnerable to chemical or physical eradication.

TABLE 11.2. Effect of Residue Management on the Incidence of Cephalosporium Stripe in Kansas

Residue	Cephalosporium Stripe Infection (%)			
Disposal Method	1980	1981	1982	3 year average
Burn and disk	18	17	3	13
Plow	39	30	4	24
Disk	42	34	13	30
Chop and disk	53	40	18	37
Direct drill	55	54	29	46

Source: Adapted from Bockus, W.W., J.P. O'Connor, and P.J. Raymond, *Plant Dis*. 67:1323-1324, 1983, with permission.

SOLARIZATION

Heat from the sun has long been used for partial sterilization of soil in India and Egypt by leaving soil fallow during the hot season of the year. In soils along the Nile in Egypt soil temperatures at a depth of 15 cm can range from 31°C in May to 36°C in August. Today solar heat is captured by covering soil with a sealed, thin, clear polyethylene sheet (plastic mulching). The heat adsorbed during the day is held under the tarp at night. The thinnest possible film gives the best results because it is cheaper and has better transmission of radiant heat. Temperatures have reached 45–55°C at 5 cm (2 inches) and 39–45°C at 20 cm (8 inches). These temperatures are well into the lethal range for many soilborne pathogens (Figure 11.2). The method has been applied in Israel, countries along the Mediterranean region, and in California. It probably will be restricted to regions where sufficient sunlight and high temperatures prevail.

Soil should be kept moist during treatment to enhance the thermal sensitivity of fungal structures and improve heat conduction. Some degree of control has been achieved for Verticillium and Fusarium wilts, southern blight, pink root, and other soilborne diseases.

Burning straw on the soil surface is a modified form of solarization. Increased heat adsorption by blackened surfaces elevated soil temperatures sufficiently without tarping to reduce *Pythium* populations by 75% in the top 2 inches of soil. This reduction increased wheat yields 15–20% in experimental plots.

PHYSICAL METHODS OTHER THAN DIRECT APPLICATION OF HEAT

In addition to traditional methods of applying heat to soil or plant parts for eradication of pathogens several somewhat unique methods have been tried. These involve nonthermal radiation of some form even though heat is generated in the soil or plant tissue. These methods include short wavelengths of the electromagnetic spectrum (gammma rays, X-rays, and ultraviolet rays) and wavelengths beyond infrared such as microwaves and ultrasonics or other high-frequency radio waves.

Development of any new technology is invariably followed by diverse attempts to use it. Hence it is not surprising that plant pathologists have tried to control plant diseases with X-rays, gamma rays, and microwaves soon after these physical forces were widely applied to other commercial uses.

X-Rays and Gamma Rays

X-Rays and gamma rays have been used to detect cavities and decay in wood utility poles, heart rot in forest and shade trees, and hollow heart in potato tubers. However, because of damaging effects such as chromosome breakage, ionizing radiations have not been very successful in controlling plant diseases. Their greatest contribution in this regard is sterilizing plant materials to be used in culture media such as carnation leaves for use in culturing and identifying teleomorphs of *Fusarium* species, or extending storage life of fruits and vegetables.

X-Rays were shown to inhibit crown gall formation on plants as early as 1917, and similar effects obtained in 1922 with buried tubes of radium. Later studies using high dosages of cobalt-60 gamma radiation on crown gall of tomato indicated that the effect of radiations was on cell division of the host rather than on the causal bacterium. Therefore the effect should be considered therapeutic rather than eradicative.

Neither high nor low-voltage X-rays or gamma rays from cobalt-60 or iodine-131 kill potato ring rot bacteria contaminating burlap bags and these radiations appear to have no practical value for this use. Gamma ray dosages required to kill bacterial and fungal pathogens in carnation cuttings kills the plants before the pathogens. Nor does prolonged and repeated exposure to X-rays have a measurable effect on teliospores of *Puccinia graminis*.

Ionizing radiations have been used more successfully in preserving fruits and vegetables. Generally the source of radiation is cobalt-60 because this isotope produces high energy (i.e., penetrating) gamma radiation and unlike X-rays does not require an external power source. However, as with most ionizing radiations, thick shields are necessary to protect nontarget organisms, including humans, from injurious effects of radiation.

Gamma rays in the range of 4–16 krads (a chest X-ray uses an exposure of 20 to 30 mrads or about two-millionths this dosage) delay spoilage and extend storage life of onions, potatoes, and tomatoes but higher dosages predispose potatoes to rot and affects firmness and weight of tomatoes. Strawberry storage losses were reduced by 200 krad treatment but 85 krad exposure of apples had no effect on either tissue breakdown or fungal decay. Higher levels, however, resulted in greater breakdown of apple tissue. Sensitivity to ionizing radiation is related to water content of tissues, probably because of production of hydroxide, peroxide, and ozone radicals, which interfere with enzyme activity in tissue.

Ultraviolet

Ultraviolet (UV) is ionizing radiation of longer wavelength than gamma or X-rays and is usually produced by mercury vapor ("black light") lamps. UV is not a penetrating radiation and is useful only for sterilizing surfaces. However, attempts to use it to disinfest burlap bags contaminated with potato ring rot bacteria were not successful.

Microwaves

Microwaves include high-frequency radiowaves just beyond the infrared region of the spectrum. In the 1920s it was observed that animal tissues exposed to short-wave radio frequencies had higher temperatures than unexposed tissues. Medical uses were developed for microwaves in the frequency range of 200 to 3000 MHz. These high-frequency vibrations interact with tissue to produce heat and rate of heating is generally proportional to water content. This has been the basis for using microwaves to control plant diseases.

Early attempts to pasteurize small quantities of soil for disease control were unsuccessful because heterogeneity of soil results in uneven and ineffective heating.

Ultrasonics have also been used to treat infected seed, but the heat generated does more damage to seed than is acceptable. Soybean seeds have been freed of soybean mosaic virus by exposure to microwaves but when seed moisture was above 10–12%, germination suffered. Nematodes have been eradicated from rose plants without affecting vigor of plants at 2450 MHz (microwaves) but the results were no better than hot water treatment at 48°C for 30 minutes.

In the 1970s a field applicator of microwaves was developed to eliminate weeds, nematodes, and possibly other organisms from soil. One commercial apparatus was named the "Zapper," a name derived, no doubt, from the colloquial expression of "zapping" test animals with radiations. Results of these applications were inconclusive and the large amount of energy required to power the machine as well as its slow progress across a field virtually eliminated it as a practical means of plant disease control.

REFERENCES

Baker, K. F. (ed.), *The U.C. System for Producing Healthy Container-Grown Plants.* Univ. Calif. Agric. Expt. Sta. Extension Service, 1957.

*Bunt, A. C., *Modern Potting Composts.* Pennsylvania State University Press, University Park, 1976.

*DeVay, J. E., J. J. Stapleton, and C. L. Elmore (eds.), *Soil Sterilization.* FAO Plant Production and Protection Paper 109, Rome, 1991.

*Hardison, J. R., Fire and flame for plant disease control. *Annu. Rev. Phytopathol.* 14:355–379 (1976).

*Johnson, J., Soil steaming for disease control. *Soil Sci.* No. 61:83-91 (1946).

*Katan, J., Solar heating (solarization) of soil for control of soilborne pests. *Annu. Rev. Phytopathol.* 19:211-236 (1981).

*Lawrence, W. J. C., *Soil Sterilization.* George Allen & Unwin, London, 1956.

*Mohsenin, N. N., *Electromagnetic Radiation Properties of Foods and Agricultural Products.* Gordon and Breach, New York, 1984.

*Newhall, A. G., The theory and practice of soil sterilization. *Agric. Eng.* 16:65–70 (1935).

*Newhall, A. G., Experiments with new electric devices for pasteurizing soils. Cornell Univ. Agr. Expt. Sta. Bull. 731 (1940).

*Senner, A. H., Application of steam in sterilization of soils. USDA Tech. Bull. 443 (1934).

* Publications marked with an asterisk (*) are general references not cited in the text.

12 Fungicide Development and Use

Fungicidal protection is a primary strategy in the control of many plant diseases. Fungicides are not the only way to protect plants from disease, but they are the best recognized by the general public. Unfortunately, the image of plant pathologists as "squirt gun botanists" has resulted from the common, but erroneous perception that most plant diseases are controlled by chemicals. Ironically, more plant diseases are controlled, to whatever extent, by nonchemical means than with fungicides. Nevertheless, fungicides, including bactericides, constitute important weapons in the war against plant diseases. They frequently are the only effective control measures and usually the only immediate response to a plant disease crisis.

Fungicides by definition are chemicals that kill fungi. However, in practice many chemicals are fungistatic, not fungicidal. A *fungistat* is a chemical that arrests fungal development, thus preventing spore germination and mycelial expansion. In the broad sense, the term fungicide also includes fungistats and bactericides.

Fungicides are placed in two categories based on their spectrum of biological activity. General fungicides affect a wide range of organisms and include copper, mercury, carbamate, and other groups of fungicides. These kill or inhibit many fungi including water molds (phycomycetes), many ascomycetous leaf and stem fungi, and rust and smut fungi. Fungicides such as sulfur, carboxin, and metalaxyl have narrow ranges of activity and are effective largely against powdery mildews, smut fungi, and phycomycetes, respectively. Some antibiotics such as streptomycin and terramycin are used against certain bacteria but have no effect on fungi.

Fungicides can also be grouped on their general mode of action. Most are *protectants* and must be applied to plants before infection takes place in order to establish a protective barrier between host plant and causal agent of disease. These are effective only on the surfaces to which they are applied. A few are eradicative (see Chapter 9) but can serve as protectants as well. Certain disease control chemicals are, or are claimed to be, systemics. A *systemic* pesticide (i.e., a fungicide, bactericide, nematicide, or insecticide) is one taken up by the plant and distributed to some extent in the plant system, doing its work inside the plant rather than outside. Some workers prefer the term *xenobiotic* to systemic since the word describes a chemical that is foreign to the plant rather than a natural one like catechuic acid. A few fungicides act as vapors rather than solids or liquids. Action of a chemical in the gaseous phase is considered to be *fumigative*. Sulfur and a few other chemicals are heated to produce vapors in greenhouses or other enclosed areas to control powdery mildew and other foliar diseases.

Several general categories of materials are used as fungicides and these are separated or classified in various ways. Elemental sulfur and antibiotics are the only naturally occurring chemicals used for disease control. Other forms of sulfur, inorganic lime-sulfur and organic such as carbamates, are widely used to control many plant diseases. Other than sulfur, there are few, if any, natural compounds that control plant diseases the same way that natural plant products such as nicotine, rotenone, pyrethrum, and sabadilla control insects. Some insecticides derived from plants are still in use. Materials such as urine, manure, wine, and vinegar, used to treat plant diseases in premodern times, have little or no value compared to modern fungicides.

Metallic salts were among the earliest effective fungicides; these usually were chlorides or carbonates of mercury, arsenic, or copper. Of these, only the copper fungicides are used today. Mercuries were used both for seed treatment (see Chapter 9) and foliar sprays, but the high toxicity of mercury limits their use in developed countries. Other metals including barium, selenium, iron, silver, zinc, osmium, cerium, cadmium, and thallium have been tested for fungicidal action and many have proved to be as fungitoxic as mercury or copper. However, some are too expensive, phytotoxic, difficult to formulate, or simply less effective.

Organic compounds constitute the majority of fungicides in use today and these are characterized further in the discussions on names and uses of modern fungicides. Antibiotics are metabolites produced during growth of microorganisms, particularly actinomycetes, bacteria, and fungi that are toxic to some other microorganisms. Some antibiotics, such as streptomycin and cycloheximide, are produced by the same organism (in this case *Streptomyces griseus*), but affect completely different groups of microorganisms; streptomycin kills certain bacteria and cycloheximide kills certain fungi. Relatively few antibiotics are used for plant disease control because of cost, phytotoxicity, difficulties in formulating or applying, and ineffectiveness. Another category of materials that is treated as pesticides for legal purposes is biological agents (i.e., living organisms). These are discussed in Chapter 16.

HISTORY OF DEVELOPMENT OF FUNGICIDES

Sulfur, in elemental form, appears to have been the first fungicide applied to foliage and was considered to be specific for powdery mildews as early as 1824 (or 1821, the date varies with different authors). And it is still used to control powdery mildews. Lime-sulfur, prepared in a number of ways, was the first "manufactured" fungicide. It was recommended for grape powdery mildew in 1833, and a year later for peach leaf curl. Lime-sulfur was mentioned for use on fruit trees in 1803, but this recommendation may have been for insect control.

Copper sulfate was used, or suggested, for control of plant diseases from 1779 until Millardet formulated a safe, effective compound known since 1882 as Bordeaux mixture. This was a mixture of copper sulfate and calcium hydroxide (hydrated lime) that yields copper hydroxide, a relatively water-insoluble product. Copper sulfate by itself is very water soluble and can cause severe plant injury. Copper hydroxide releases enough copper into solution to be fungitoxic (i.e., fungicidal) but not enough to be phytotoxic. Bordeaux mixture is prepared by mixing the ingredients in different proportions such as 12 pounds of copper sulfate and 12 pounds of hydrated lime in 100 gallons of water. This mixture is called 12–12–100 Bordeaux mixture. The proportions of ingredients can be as low as 1/2–1/2–100. The important consideration is that sufficient, usually an excess, of lime be added to neutralize (or fix) all of the copper ions. The end product is referred to as a neutral or fixed copper because of its relative insolubility in water. Bordeaux mixtures of 12–12–100 or 10–10–100 are generally applied only to dormant plants, while more dilute solutions are used on growing plants.

Discovery of Bordeaux mixture opened the era of modern fungicides. Bordeaux mixture, or Bordeaux substitutes such as basic copper sulfate, are still used today. Bordeaux mixture is a good, safe fungicide and bactericide and its discovery stimulated development of other inorganic fungicides, the first generation. Several Bordeaux substitutes emerged, some had improved properties from neutralizing the copper with alkaline solutions other than lime, and frequently were given provincial names, probably to distinguish them from Bordeaux mixture. These included combinations with ammonium hydroxide ("Eau Celeste"), sodium carbonate ("Burgundy Mixture"), and ammonium carbonate ("Cheshunt Compound").

Chlorophenolmercury, marketed in 1913, was the first organic mercury fungicide and it started a second generation of fungicides, the organics. Until 1913 all fungicides were inorganic compounds such as hydroxides, chlorides, carbonates, and similar chemicals. In addition to the organic mercury fungicides used for seed treatment described in Chapter 9, various alkyl- and phenylmercury compounds were used as protectant and eradicant foliar fungicides.

The fungicidal properties of thiram, tetramethylthiuram disulfide (TMTD), a product used to vulcanize rubber, emerged in 1931. Today it is used only for seed treatment. Thiram was the first of several organic sulfur fungicides still used to control plant diseases. The carbamate fungicides started with thiram and metallic salts of dithiocarbamates. The first of these, nabam (sodium salt), ferbam (iron salt), and ziram (zinc salt), were developed in the early 1940s. In 1950, manganese and zinc salts of ethylene bisdithiocarbamate were marketed as maneb and zineb. In 1962 a coordination product of zinc and manganous salts of ethylene bisdithiocarbamate was marketed as mancozeb.

Captan, a phthalimide fungicide (see Table 12.1 for chemical name), was introduced in 1952 and became an important foliar and seed protectant. Even though it contains sulfur it is classed as a phthalimide fungicide (or sometimes carboximide) rather than a sulfur fungicide. Ten years later captan was succeeded as a foliar protectant by captafol, a chemically similar material. However, captan is still used for seed treatment.

The last important fungicide in the second generation was dodine, which became the main fungicide for control of apple scab because it has eradicative as well as protective properties.

Griseofulvin was isolated from cultures of *Penicillium griseofulvum* in 1939. It was thought to have systemic activity based on detection in guttation drops from treated plants. While effective against several fungal pathogens and of low phytotoxicity, it found little practical application.

TABLE 12.1. Common Names of Major Fungicides by Type or Group and the Chemical Name of a Typical Fungicide in Each Group

Group (or Type)	Common Name	Chemical Name
Acylalanines	Metalaxyl	*N*-(2,6-Dimethylphenyl)-*N*-(methoxyacetyl)-alaninemethyl ester
	Benalaxyl	
	Furalaxyl	
Anilides	Carboxin	5,6-Dihydro-2-methyl-*N*-phenyl-1,4-oxathiin-3-carboxamide
	Chlozolinate	
	Fluoroimide	
	Fenfuram	
	Flusulfamide	
	Flulolanil	
	Iprodione	
	Mepronil	
	Methfuroxam	
	Methsulfovax	
	Oxycarboxin	
	Procymidione	
	Vinclozolin	
Benzimidazoles	Benomyl	Methyl-1-(butylcarbamoyl)-2-benzimidazolecarbamate
	Carbendazim	
	Fuberidazole	
	Thiabendazole	
	Thiophanate-methyl	
Carbamates	Maneb	Manganese ethylene bisdithiocarbamate
	Dithofencarb	
	Ferbam	
	Nabam	
	Propamocarb	
	Propineb	
	Thiram	
	Zineb	
	Ziram	
Conazoles (and triazoles)	Propiconazole	1-[2-(2,4-dichlorophenyl)4-propyl-1,3-dioxolan-2-yl methyl]-1*H*-1,2,4-triazole

TABLE 12.1. Continued.

Group (or Type)	Common Name	Chemical Name
	Aconazole	
	Bitertanol	
	Etaconazole	
	Diclobutrazol	
	Difenoconazole	
	Diniconazole	
	Flusilazole	
	Flutriazol	
	Furconazole	
	Hexaconazole	
	Imazalil	
	Imibenconazole	
	Myclobutanil	
	Penconazole	
	Prochloraz	
	Tebuconazole	
	Tetraconazole	
	Triadimefon	
	Triadimenol	
	Tricyclazole	
	Trifumizole	
Guanidines	Dodine	*n*-Dodecylquanidine acetate
	Guazatine	
	Iminoctadine	
Morpholines	Dodemorph	4-Cyclododecyl-2,6-dimethyl morpholinium acetate
	Dimethomorph	
	Fenpropimorph	
	Tridemorph	
Organophosphorus	Edifenphos	*O*-Ethyl *S,S*-diphenyl phosphodithioate
	IBP	
	Iprobenfos	
	Pyrazophos	
	Tolclofos-methyl	
Phthalimides	Captan	*cis* -*N*-Trichloromethylthio-4-cyclohexene-1,2-dicarboximide

TABLE 12.1. Continued.

Group (or Type)	Common Name	Chemical Name
	Captafol	
	Folpet	
Piperizines	Triforine	*N,N'*-1,4-Piperazinediylbis (2,2,2-trichloroethylidene) -bis[formamide]
	Piperalin	
Pyrimidines	Bupirimate	5-Butyl-2-ethylamino-6-methylpyrimidin-4-yl dimethylsulfamate
	Dimethirimol	
	Ethirimol	
	Fenarimol	
	Mepanipyrim	
	Nuarimol	
	Triarimol	
Quinones	Chloranil	2,3,5,6-Tetrachloro-1,4-benzoquinone
	Dichlone	
	Dithianon	
	Oxine	
	Oxalinic acid	
	Pyroquilon	
Substituted benzenes	Chloroneb	1,4-Dichloro-2,5-dimethoxy-benzene
	Chlorothalonil	
	Dicloran	
	Dinocap	
	HCB	
	Quintozene (PCNB)	

Systemic fungicides comprise the third generation. It started when streptomycin was shown to be taken up by seedlings. However, the antibiotic was phytotoxic at bactericidal concentrations. Streptomycin is an important protective chemical used to control fire blight of apples and pears. Emergence of bacterial strains resistant to the antibiotic has reduced its value and has required use of tetracycline antibiotics to which fire blight bacteria remain sensitive. At first these chemicals were more likely to be called chemotherapeutants than systemics. They were among the first disease control chemicals that appeared to cure diseases of plants and the term "systemic," as in systemic insecticide, seemed to belong exclusively in the lexicon of entomology.

Cycloheximide, another antifungal antibiotic like griseofulvin, is produced by *Streptomyces griseus*, the same actinomycete that produces streptomycin. This antibiotic was first used to control leaf spot of cherries and later for diseases of turfgrasses and roses. It appears to have some systemic (or chemotherapeutic) activity and for a time was thought to be the cure for white pine blister rust. Unfortunately its use became embroiled in controversy over its effectiveness (see Chapter 8). Its antifungal properties notwithstanding, cycloheximide's high phytotoxicity and mammalian toxicity ultimately resulted in disuse for plant disease control. It is used, however, as a fungal inhibitor in selective culture media.

The identification of mycoplasma-like organisms (MLOs) as causal agents of diseases such as lethal yellowing of coconut palm, pear decline, X-disease of peaches, and peach yellow leaf roll led to the use of tetracycline antibiotics as therapeutants. Despite the many antibiotics that have been produced very few are used in plant disease control. Reasons for this include their limited supply, high cost, and a reluctance to apply chemicals used in human medicine.

In addition to those mentioned above, several antibiotics are used outside the United States, especially in Japan. In 1959, blasticidin, produced by *Streptomyces griseochromogenes*, was the first antibiotic used in Japan to control rice blast. Now kasugamycin (Kasumin™) has become a very effective control for rice blast and is used to control several fungal and bacterial diseases of fruits and vegetables. Other antibiotics including polyoxin, validamycin, and mildiomycin are used on apples, pears, citrus, grapes, coffee, cacao, cucumbers, carrots, potatoes, small grains, and ornamental plants.

The first systemic fungicides, the benzimidazoles (benomyl, thiabendazole, carbendazim, and thiophanate, which converts to carbendazim), were introduced in the late 1960s and early 1970s. Shortly after benomyl came into use an anilide systemic, carboxin, was introduced primarily as a seed treatment for control of smut fungi in cereals. A related compound, oxycarboxin, is used almost exclusively for the control of rust fungi. This is an excellent example of how slight changes in chemical structure of a compound can affect biological activity. The relationship between structure and activity is discussed in Chapter 13.

The first systemic fungicides and a majority of those developed since move only upward in plants, a property that restricts their use as true systemics. Systemic fungicides move in one of three ways. Acropetal fungicides move only upward in plants. These fungicides do not enter the protoplasm (symplast) and move in the apoplast, especially xylem, cuticle, cell walls, and intercellular spaces. Unless they enter roots this passive movement of systemics limits the distance that they can move in a plant to inhibit pathogen activity. Downward (basipetal) movement such as occurs with fosetyl-Al is rare among systemic fungicides.

Long distance movement in plants is *symplastic,* which involves movement in living protoplasm of cells, primarily the phloem. Movement from cell to cell occurs through plasmodesmata, the fine strands of protoplasm connecting cytoplasm of adjacent cells. Very few fungicides are *ambimobile* (i.e., move in both directions via xylem and phloem).

The majority of systemic fungicides move acropetally. Root epidermis, lacking a cuticle, allows greater uptake of compounds than leaf epidermis. Therefore, an acropetal systemic must be applied to roots in order for appreciable absorption and movement in plants. Most "systemic" fungicides probably function by local diffusion or simply as protectants.

The fourth generation of fungicides originated in the 1970s and includes ambimobile fungicides such as fosetyl-Al and sterol biosynthesis-inhibiting (SBI) compounds [sometimes called ergosterol biosynthesis-inhibiting (EBI) fungicides]. The largest group is the triazoles or conazoles (tridemefon, triademenol, biteranol, propiconazole); it also includes the pyrimidines (triarimol, fenarimol, nuarimol), piperazines (triforine), and others. The sterol inhibiting fungicides are prone to development of fungicidal resistance in fungi because of their narrow, specific mode of action. However, resistant fungi often have reduced fitness and do not persist in the absence of fungicidal selection pressure.

REQUIREMENTS OF A GOOD FUNGICIDE

Edgington (1981), who summarized the structural requirements for systemic fungicides, considered selective toxicity the most important. With the exception of systemics such as fosetyl-Al, that alter host physiology and are not directly toxic to pathogens, systemic fungicides must be nontoxic to the host but toxic against the pathogen. Examples of selective toxicity are carbendazim, carboxin, and metalaxyl, which have specific modes of action against ascomycetes, basidiomyces, and phycomycetes, respectively. Other desirable characteristics are stability within plants, ability to penetrate cuticle, membrane permeability to facilitate entry into protoplasts, and water solubility to aid in application to plant surfaces.

Although many chemicals have fungicidal activity, few are developed commercially or acceptable. Paramount requisites for a good fungicide are marked with an asterisk (*). Other desirable characteristis are included that are less important.

*1. Low lethal dosage to targeted organisms. Cost of chemical is often the greatest expense in pesticide applications, especially if the material is applied in mixtures with other chemicals so that application costs are minimized.

*2. Noninjurious to the host. Some fungicides, especially eradicants, are somewhat phytotoxic, but injury can be minimized by adjusting timing and/or rate of application. If damage is excessive the fungicide is unlikely to be registered.

3. Nontoxic to man and animals. Many of the early metallic fungicides, especially arsenicals and mercurials, were unacceptable or their use severely limited because of high mammalian toxicity. Few modern fungicides have high mammalian toxicity; however, carcinogenic, teratogenic, or mutagenic associations have resulted in more stringent tests and requirements for pesticide registration.

4. Impart no undesirable flavors, odors, or poisonous properties, or detract from aesthetic values. When Bordeaux mixture was first developed grape growers were

concerned that copper would taint the wine. Fortunately it did not. In fact, wells in the region contained as much copper as could be detected on treated grapes. Any reduction in quality of a crop from applied chemicals will sharply detract from their use. Ferbam was an effective fungicide for control of many foliar pathogens, but its use on ornamentals was curtailed because a black deposit remained on foliage.

5. Compatibility with other chemicals with which it is mixed, such as spreaders, stickers, or other adjuvants, insecticides, or other fungicides. Fungicide mixtures have become more common with the advent of specific compounds such as systemics. Mixtures broaden spectrum of activity, stabilize performance, and retard fungicidal resistance. Some fungicides such as the carbamates, dyrene, chloronil, and captan decompose in alkaline solutions and therefore should not be mixed with lime-sulfur, basic coppers, or other alkaline materials. Incompatible mixtures can either reduce effectiveness or increase plant injury. For example, dodine combined with several insecticides can damage certain varieties of fruit trees and captan mixed with sulfur can injure apple foliage. Captan applied with dormant oil can be phytotoxic because of increased penetration.

*6. Adhere tenaciously to plant surfaces and resist weathering. One of the inherent advantages of Bordeaux mixture was that once it dried it bound tightly to plant surfaces. Most fungicides contain a sticker that helps retain them on plants. New plant growth or expansion of small leaves will be unprotected and repeated application of chemical is needed to protect new surfaces.

7. Ease of preparation and application. Most fungicides and other chemicals are applied as aqueous sprays and hence must be dispersible in water. Most foliar fungicides are relatively insoluble in water. Various formulations have been developed to aid uniform mixing of chemicals with water. Fungicides are generally finely ground powders treated so that they do not carry surface charges that cause flocculation or clumping. Since fine particles (even colloids) will settle out of suspension with time, most application equipment includes some means of agitating or stirring to keep ingredients in suspension.

8. Noncorrosive to containers, machinery, or other objects. One problem with mercury fungicides was corrosiveness to application equipment. The same is true of lime-sulfur, but its greatest problem in this respect was a chemical reaction with lead-based paints to produce lead sulfide and impart a dark color. Although newer nonlead paints are not affected, corrosion still occurs in gutters, flashing, and other metal items.

*9. Low cost. Disease control is an expense added to normal costs of crop production and growers are naturally reluctant to spend any more than is absolutely necessary. However, there may be excessive or unnecessary use of chemical, or nonchemical, treatments if cost is low. Such applications are considered low-cost insurance. When mercury seed treatments were available it cost about 10 cents to treat enough seed to sow an acre of wheat so most farmers routinely planted treated seed as a precautionary measure. However, after mercuries were discontinued new treatments cost about a dollar to treat an acre of seed and farmers are sometimes tempted to forego seed treatment even in cases where needed.

NAMES OF FUNGICIDES

Inorganic or organic fungicides are often assigned to a group based on chemical structure or active ingredient. Before modern organic fungicides were developed, fungicides were often considered coppers, mercuries, or sulfurs. Starting with the carbamates, fungicidal groups were given names of the chemical base to which a fungicide was allied, such as carbamates, benzimidazoles, carboxamides, triazoles, pyrimidines, and piperazines. Different group names may be used in different publications just as some compounds can have more than one chemical name depending on the nomenclatural system used. For example, captan is referred to as a carboximide by some writers and a phthalimide by others; and carboxin is sometimes a carboxamide, sometimes an anilide, and occasionally an oxathiin. These names are useful only for grouping similar fungicides and are rarely used in practice. Fungicides have three names, all of which are used in commerce and in practice. The first is the chemical name, which can be long, complex, and meaningless to most fungicide users. However, chemical names are recognized worldwide.

The second name is the common (coined) name and is assigned to a specific chemical to avoid the technical complexities of pronouncing and recognizing the chemical name. Common names are also largely standardized, although different common names can be used in different countries. For example, the fungicide known as carboxin in the United States is called carbathiin in Canada.

Table 12.1 lists the main chemical groups of organic fungicides, the chemical name of a typical fungicide of that group, and common names of the main fungicides. Many of these compounds have been discontinued by manufacturers but are included for historical and comparative purposes.

The third name is the trade or proprietary name of a fungicide and is registered by the manufacturer or formulator of that particular fungicide. There may be multiple trade names for the same fungicide, some of which provide clues to the active ingredient or common name, whereas others have no relationship to the name of the fungicide in a product. For example, ferbam is the common name given to the chemical ferric dimethyldithiocarbamate, which was, or is, marketed under many trade names as dissimilar as Fermate™ and Sup'r-Flo™. The life span of a trade name is often quite short.

The official label for a registered fungicide bears all three names of a fungicide as well as percentages of active ingredient and inert materials. Figure 12.1 is a typical fungicide label showing key components of a label, such as trade name, common name, chemical name, amount of active and inert ingredients, patent and EPA registration numbers, directions for use, and precautionary information.

Many compilations are available that give detailed information about fungicides. These are usually revised annually so that information is current. However, information on older, discontinued products is omitted. Obtaining information on discontinued fungicides for historical or other purposes is then made difficult. Three such compilations (*Farm Chemicals Handbook*, *Agricultural Chemicals—Fungicides*, and *The Pesticide Manual*) are included in the references for this chapter. These sources give

Common Name Trade Name Chemical Name

DU PONT

REG. U. S. PAT. OFF.

PARZATE® C

ZINEB FUNGICIDE

GUARANTEE: Effective for Purposes Claimed

ACTIVE INGREDIENT:

Zineb (Zinc ethylenebisdithiocarbamate) 75.0%

(Total Zinc as Metallic .... 17.7%)

INERT INGREDIENTS 25.0%

SPECIMEN LABEL

U. S. Pat. 2,545,948
USDA Reg. No. 352-231
Canadian Reg. No. 7243 P.C.P. Act

CAUTION! MAY IRRITATE EYES, NOSE, THROAT, AND SKIN. MAY BE HARMFUL IF INHALED OR SWALLOWED.

Avoid breathing dust or spray mist. Avoid contact with skin, eyes, and clothing. Keep away from fire or sparks.

Keep from children

In case of contact, flush skin or eyes with plenty of water; for eyes, get medical attention.

IMPORTANT: Never allow "Parzate" C to become wet during storage. This may lead to certain chemical changes which will reduce the effectiveness of "Parzate" C as a fungicide and create vapors which may be flammable. Keep container closed when not in use.

Distributed in Canada by
DU PONT OF CANADA LIMITED
BIOCHEMICALS – TORONTO, ONTARIO

NET 5 LBS.

E. I. DU PONT DE NEMOURS & COMPANY (INC.)
INDUSTRIAL AND BIOCHEMICALS DEPT. WILMINGTON, DELAWARE

DIRECTIONS

Du Pont "Parzate" C Zineb Fungicide should be used only in accordance with recommendations on this label, or in separate Du Pont Agricultural Bulletins available through local dealers.

Use sufficient spray or dust to provide thorough coverage. Do not apply more than 3 lbs. actual zineb per acre per application to vegetable crops, grapes and hops (4.8 lbs. for cabbage, cauliflower, broccoli).

SPRAYS: Except as noted, use 1½ to 2 lbs. per 100 gals. of water. Start filling the tank with water. Sift in the required amount of "Parzate" C when the agitator paddles are covered or the jet or recirculating pipe from the pump is covered. Fill the tank with water when the "Parzate" C is thoroughly mixed.

DUSTS (Ground or Air Equipment): Except as noted, "Parzate" C should be used at a concentration of 10% (7.5% active) by diluting and thoroughly mixing with talc, pyrophyllite or other suitable non-alkaline dust diluents. Commonly used insecticides may dissolve an equivalent amount of diluent in preparing the dust mixtures.

"Parzate" Du Pont Reg. Trademark in Canada.

Dilution schedule for small quantities: Eight teaspoonfuls of "Parzate" C per 1 gal. of water is equivalent to 2 lbs. per 100 gals.

NOTICE TO CANADIAN USERS

Water volumes refer to U. S. measures. To compute imperial measure, multiply figures given by 5/6.

CONTINUED ON BACK PANEL

DIRECTIONS CONTINUED FROM BACK PANEL

APPLES: Summer control of scab, sooty blotch, fly speck, Brooks spot, black rot—use 1 to 1½ lbs. in cover sprays.

MUSHROOMS: Verticillium (brown spot, dry bubble), Dactylium (mildew, soft decay, cobweb), Mycogone (bubbles), and Trichoderma (green mold)—use 1 lb. of 19% (11.25% active) dust or 10 gals. of spray (1 lb. per 100 gals.) per 4000 sq. ft. of bed. Apply at 3- to 7-day intervals until buttons begin to form; also between "breaks" as required.

HOPS (Pacific States): Downy mildew—use spray or 8% (6% active) dust soon after training vines and repeat at 7- to 10-day intervals as long as needed. Do not apply later than 2 weeks before harvest.

WHEAT: Rust—apply 1½ lbs. per acre in 25 to 40 gals. of water. Make first application when the plants are in the boot or early heading stage or when rust is first reported in the area. Repeat at 7- to 10-day intervals until the milk stage.

TOBACCO: Blue mold—apply spray or dust directly to the young plants, or dust through the plant bed cloth. For each 100 sq. yds. of plant bed, apply 3 to 6 gals. as a spray or 2 to 3 lbs. of 10% (7.5% active) dust depending upon size of plants. Start applications when leaves are about the size of a dime, or when blue mold first appears in the vicinity. Apply at 3- to 4-day intervals until transplanting time.

COMPATIBILITY: "Parzate" C is compatible with common insecticides such as methoxychlor, DDT, organophosphorus compounds, nicotine sprays, arsenicals and rotenone.

Made in U.S.A., Printed in U.S.A.

Figure 12.1. A typical fungicide label. (With permission of E.I. DuPont de Nemours & Company.)

common names, trade names, fungicide type or group (carboxamide, imidazole, triazole, etc.), chemical and sometimes structural formulas, as well as origin (company and year) of a compound. Also included are formulation (wettable powder, dust, etc., and sometimes concentration of active ingredient), toxicity (LD_{50}) and safety guidelines, phytotoxicity, uses (crops and regions or countries if not generally used), and important diseases controlled. There is also limited information on application procedures, when to start and stop treatment, frequency of repeated applications, and dosage rates (e.g., amount of active ingredient per unit of seed or per acre or hectare of crop). Some precautions may be included, such as toxicity to fish, birds, and invertebrates, compatibility with other chemicals, reentry times, and restrictions or unavailability in certain countries where sale is not permitted (e.g., not for sale in United States).

FORMULATIONS OF FUNGICIDES

Fungicides are rarely marketed as pure active ingredient, but are diluted with inert material and additives to make products easier to handle, mix, and apply, more stable in suspension, and more effective. Fungicides are marketed as one or more of the following formulations.

Wettable Powders (WP)

Most fungicides are formulated as wettable powders consisting of finely ground or precipitated particles (<10 μm) that remain in suspension but are not dissolved. These are sometimes marketed as water dispersable powders. In addition to active ingredient, wettable powders also contain a nonionic or anionic wetting agent, usually a long chain alcohol, petroleum sulfonate, fatty acid ester, or other chemical. Nonionic agents are preferred since anionic materials tend to precipitate out of solution in hard water. The formulation may also contain a thickening agent such as alginate or carboxymethyl cellulose, a hydrophilic agent such as bentonite clay, and a suspension agent such as methyl cellulose, polyvinyl acetate, or aluminum silicate. Most wettable powders contain 25 to 50% active ingredient.

Dry Flowables (DF)

These formulations are similar to wettable powders, but the fungicide powder consists of tiny spheres that facilitate flow from product containers.

Dusts (D)

Dusts usually contain only a small proportion of active ingredient, generally less than 10%. Relatively few fungicides are applied as dusts; however, sulfur, the most common dust fungicide, is usually formulated as almost pure sulfur with about 5% talc or similar diluent to enhance flowability. However, fine sulfur dust has the nasty tendency to explode at high temperatures.

Flowables (F)

Flowables are preparations of finely ground powders in slurries that readily mix with water. These formulations consist of wettable powders suspended in oil or liquid bases and usually are less phytotoxic than emulsifiable concentrates.

Emulsifiable Concentrates (EC)

Only recently have emulsifiable concentrates been used for fungicides. The fungicide is dissolved in a water-immiscible solvent plus a surfactant that disperses it in water to form a stable emulsion.

Liquid Concentrates (L or LC)

Because of general insolubility of fungicides in water, liquid formulations are seldom used. One of the earliest was phenyl mercury acetate, which had sufficient water solubility to be applied as solutions. Many of the antibiotics are formulated as liquids.

Granules (G)

Granules are formed by mixing dissolved fungicide with some carrier such as clay (analogous to pelleting of seed described in Chapter 9) to form particles 30 to 60 mesh size. Granules are mostly used for soil applications.

FUNGICIDE USES

Fungicides are applied to control diseases in one of three ways. Greatest use involves spraying or dusting fungicide onto foliage, fruit, stems, or other plant parts to protect surfaces from invasion by plant pathogens. In some cases fungicides act as eradicants as well as protectants. Some chemicals such as fenaminosulf (Dexon™) are broken down rapidly by sunlight and are not suitable for foliar application but can function effectively in soil applications.

Another broad application, albeit with considerably smaller amounts of fungicides, is seed treatment (discussed in Chapter 9).

Several fungicides including PCNB, dicloran, metalaxyl, and chloroneb are applied to soil. The materials are usually sprayed or dusted into planting furrows or applied as granules. Risk of fungicidal resistance is high with soil applications in comparison with foliar fungicides because of prolonged exposure of fungi to a chemical and additional screening opportunities.

METHODS OF APPLICATION

Fungicides are applied by various methods. The one generally selected is determined by (1) application equipment available, (2) formulation, and (3) the kinds of crops, that is, tree fruits, row crops, greenhouse crops, ornamentals, forest trees, or other kinds.

After Bordeaux mixture was developed, a way to apply it was needed. Originally the mixture was splashed onto grapevines with large brushes. This evolved to using hand-operated rotary brushes and eventually small hand-pump sprayers. Today fungicide spraying is done by a variety of pressurized sprayers attached to tractors (Figure 12.2), airplanes, and helicopters.

Sprays are sometimes characterized by amount of spray solution applied to an acre of crop, but these categories vary in different regions and with different crops. In orchards of western United States high volume (HV) sprays apply fungicides in 350 to 800 gallons per acre (g/A), medium volume (MV) sprays apply 100 to 350 g/A, and low volume (LV) (concentrate) sprays apply only 10 to 100 g/A. Ultra-low volume (ULV) sprays apply almost pure chemical, usually less than 2 quarts per acre. ULV is used to apply some antibiotics but rarely other fungicides. In Europe high volume refers to more than 250 liters per hectare (liters/ha), medium volume is 50 to 250 liters/ha, low volume 5 to 50 liters/ha and ultra low volume 0.5 to 5 liters/ha.

Dusts are applied by ground dusters or by air. Airplane application of some dusts such as sulfur can be hazardous at high temperatures because of the explosive nature of sulfur dust. Granules can be applied through fertilizer application equipment such as Gandy™ or other applicators.

Chemigation is a general term for application of chemicals by irrigation. Sprinkler irrigation systems can be used to apply a few water-soluble fungicides and fumigants. Metam-sodium (Vapam™), a methyl isothiocyanate (MIT) releasing soil fumigant that can be applied through sprinkler irrigation systems, has given good control of pod

Figure 12.2. Air blast sprayer applying fungicide for apple scab control. Note penetration of foliage by the spray.

rot and Verticillium wilt of peanuts and is used to control some other soilborne pathogens. Depth of soil penetration by chemicals is regulated by amount of water applied. Application of captafol and chlorothalonil for control of potato early blight is almost as effective by chemigation as aerial application of the same fungicides. However, use of large amounts of water is disadvantageous because it dilutes the fungicide and washes it off the foliage. Flowable formulations of captafol, chlorothalonil, and mancozeb have been applied through sprinklers for control of foliar diseases and fruit rots of cucumbers. Fenamiphos (Nemacur™), a nematicide sprinkled into soil, controlled root-knot nematodes in vegetables and increased yields without injuring plants.

A few fungicides are effective when applied as fumes. One of the earliest applications involved painting sulfur on steam pipes in greenhouses where it would vaporize. This system is very effective for controlling powdery mildews and is still common, although sulfur is now placed in small pans over heating elements. Other fungicides, imazalil and vinclozolin for example, are formulated in smoke generators for the same purpose.

FACTORS IMPORTANT TO EFFECTIVE DISEASE CONTROL

Simply applying fungicides or other disease control chemicals does not guarantee success. Several factors must be considered if maximum benefits are to be realized in applying chemicals for disease control.

One of the first considerations is relative effectiveness (i.e., the efficacy) of a fungicide. There often are more than one fungicide available for control of a specific pathogen, but usually all are not equally effective. Of those available the most effective may not be the best selection because of cost, toxicity, difficulty of application, residues, or other feature. Frequently a fungicide used in commercial agriculture may be either unavailable or unsuitable for home gardeners.

Good coverage, or placement, is probably more critical for fungicides than insecticides or herbicides. Insects are mobile and in their movement over a leaf are likely to come in contact with a toxicant. Herbicides are absorbed by plant surfaces and adequate concentrations can be applied without complete coverage. However, plant pathogens are so small and their entry into the plant so localized that unprotected spaces on plant surfaces treated with a fungicide may leave gaps where infection can occur. For this reason, sprays and dusts must consist of very fine particles that must cover the entire plant, including lower surfaces of leaves.

Coverage also relates to particle size of a fungicide. If particles (dust or spray droplets) are too large it is difficult to obtain complete and uniform coverage without using excessive amounts of fungicide. As a rule, the smaller the particle size the better the coverage since fine sprays act as a mist to envelope most plant surfaces. However, particles of small size have a tendency to drift. Fortunately, this is less serious with fungicides than with insecticides (bee poisoning, etc.) or herbicides (plant damage).

Adhesion of the fungicide to the plant surface is essential for effective control and hence most fungicide formulations include a spreader and sticker to enhance coverage and adhesion. Bordeaux mixture has a self-contained sticker and once dry it adheres tenaciously to plant surfaces. Even the most tenacious fungicides will erode with rain or irrigation. Also, some deteriorate in sunlight. Furthermore, as plants grow, new unprotected surfaces appear and more fungicide must be applied.

Target exposure also affects coverage. Dense plant canopies of row crops and orchards, and wide rows of vine or cane plants (tomatoes, grapes, blueberries, raspberries, etc.), interfere with penetration of sprays so that lower leaves or inner parts of rows are not covered. This can be compensated to some extent by using high pressure or air blast sprayers (see Figure 12.2).

Perhaps the most important consideration for effective fungicide application is proper timing of sprays. It probably matters not so much what is splashed on, among the effective fungicides, but when it is splashed on. Fungicide applications must coincide with vulnerable points in a disease cycle, which is usually the period after inoculation but before infection, this despite new fungicides—both eradicants and systemics—that allow some leeway. Although a variety of forecasting systems have been developed as described in Chapter 5, fungicide applications are usually made on a calendar basis. That is, they are applied at some growth stage of a plant and repeated at certain intervals, usually 10 to 14 days, or as long as environmental conditions favor disease development.

REGISTRATION OF FUNGICIDES

Early concern about safety of pesticides was directed largely at the known poisons, arsenic and mercury in particular. Although acceptable levels of chemicals (tolerances) were established for insecticides in the late 1800s and early 1900s, there was little regulation of fungicides in the United States until passage of the Federal Insecticide, Fungicide, and Rodenticide Act (FIFRA) in 1947. The first pesticide regulation in the United States was the Insecticide Act of 1910, which was concerned primarily with heavy metal pesticides (lead, arsenic, and mercury). It was intended to protect the buyer or user from fraudulent advertising and prevent sale of adulterated or misbranded products.

FIFRA initially was concerned with product registration and labeling to ensure that a product did what was claimed. The act also required that directions for proper use would ensure expected results without injury to animals, humans, plants, or nontarget invertebrates (honeybees, worms, crayfish, clams, etc.). The burden of proof at this time was on the government to show that the product constituted a health hazard. The act was administered by the Department of Agriculture (USDA).

The 1954 Miller Pesticide Amendment to the Food, Drug and Cosmetic Act compelled pesticide manufacturers to prove that a product was safe to use. This amendment also made the Food and Drug Administration (FDA) responsible for establishing tolerances, set procedures for doing so, and decreed that no pesticide could be registered until a tolerance was granted.

The Environmental Protection Agency (EPA) was created in 1970 and took over many of the functions of the USDA and FDA for registration and setting tolerances. The FDA still enforces tolerances. It samples foods sold in the marketplace for residue levels. In 1972 the Federal Environmental Pesticide Control Act (FEPCA) revised FIFRA to expand registration, cancellation, and suspension criteria to include factors beyond the content of the label. The EPA could now consider adverse effects on the environment and weigh risks and benefits of a particular pesticide. It also specified that a pesticide could not be sold in the United States unless registered with the EPA. A provision was also made for certification of pesticide applicators, although training and certification of applicators are conducted by the Cooperative Extension Services and Departments of Agriculture of the various states.

The primary or perhaps only toxicity test at that time was acute toxicity. In 1975, however, FIFRA was amended to broaden the requirements for testing to include subacute toxicity (three dosage levels) and short term feeding tests. Chronic toxicity is determined by long-term (up to 2 years) feeding trials as well as tests for teratogenic, mutagenic, and carcinogenic effects. There must also be tests for toxicity of fungicide breakdown products.

Sometimes breakdown products cause more concern than the parent compound. For example, ethylene thiourea (ETU) is a product of ethylene bisdithiocarbamate fungicides (maneb, zineb). ETU is a carcinogen and this resulted in the bisdithiocarbamate fungicides being put on the EPA RPAR (rebuttable presumption against registration) list. This presumption can be rebutted by the owner (patent holder) of a pesticide, or by user groups. Some of these fungicides have since been reregistered.

Tolerances are set using the most sensitive test animal and a "no effect" dosage times a safety factor as a basis for the final decision. If the no effect dosage is 10 ppm a safety factor of 100 means that the allowable tolerance is 0.1 ppm. Also taken into account is the acceptable daily intake (ADI), which considers the amount that might be accumulated by an individual as part of normal eating habits. This means that the tolerance will be weighted somewhat depending on crop usage. A fungicide would likely have a much lower tolerance on grapes or lettuce than on pineapples or bananas, which are peeled, or on cotton, which is not ingested. Environmental concerns include effects on wildlife habitat, survival, and reproduction. In addition, advantages in overall performance of a chemical in terms of effectiveness in controlling disease, improved yield, higher quality crops, reduced phytotoxicity, and so on, are also considered.

States can register fungicides for a use not indicated on the label (e.g., different diseases) if a Federal tolerance has been set for that crop (e.g., corn). This provision is sometimes called a "section 24c," in reference to the portion of FIFRA that applies, or a "special local needs" (SLN), and usually is requested by the manufacturer of the fungicide.

There is also a provision under section 18 for temporary use of an unregistered fungicide to meet emergency (crisis) needs. This does not require a Federal tolerance, but there must be *some* residue and toxicological data. The request for an emergency label is generally made by a user group (wheat growers, citrus growers, etc.) usually in cooperation with researchers who have information on potential disease losses and efficacy of the fungicide. Typically, there is a limit to the total quantity of product that can be applied as well as a limit on acreage treated. There may also be a fixed time frame for this emergency use and some monitoring of application and crop residues.

There is another provision that allows unregistered fungicide to be used. An Experimental Use Permit may be issued to developers allowing them to test fungicides in commercial fields. The main provision is that the crop cannot be marketed and must be destroyed after yield, quality, residue, and other pertinent data are recorded.

REFERENCES

*Anon., *Farm Chemicals Handbook*. Meister, Willoughby, OH, 1991.

*Eckert, J. W., and J. M. Ogawa, The chemical control of postharvest diseases: Deciduous fruits, berries, vegetables and root/tuber crops. *Annu. Rev. Phytopathol.* 26:433–469 (1988).

Edgington, L. V., Structural requirements of systemic fungicides. *Annu. Rev. Phytopathol.* 19:107–124 (1981).

*Horsfall, J. G., *Principles of Fungicidal Action*. Chronica Botanica, Waltham, MA, 1956.

*Juergensmeyer, J. C., and J. B. Wadley, *Agricultural Law*, Vol. II, Chapter 27. Little, Brown, Boston, 1982.

*Matthews, G. A., *Pesticide Application Methods*. Longman, London, 1979.

*Roberts, J. W., Recent developments in fungicides: Spray materials. *Bot. Rev.* 12:586–600 (1936).

*Siegel, M. R., and H. D. Sisler (eds.), *Antifungal Compounds*, Vol. 1, *Discovery, Development and Uses*. Marcel Dekker, New York, 1977.

*Somers, E., Formulation. In Torgeson, D. C. (ed.), *Fungicides: An Advanced Treatise*, Vol. 1. Academic Press, New York, 1967. Pp. 153–193.

*Thompson, W. T., *Agricultural Chemicals*. Book IV: *Fungicides*. Thompson, Fresno, CA, 1991.

*Van Valkenburg, W. (ed.), *Pesticide Formulations*. Marcel Dekker, New York, 1973.

*Waller, J. M., Plant disease control. In Haskel, P. T. (ed.), *Pesticide Application: Principles and Practice*. Clarendon Press, Oxford, 1985. Pp. 427–455.

*Worthing, C.R. (ed.), *The Pesticide Manual*. The British Crop Protection Council, UK, 1991.

* Publications marked with an asterisk (*) are general references not cited in the text.

13 Fungicide Characteristics

The most important quality of a fungicide is its efficacy. There must be a simple, quantitative, and reproducible way to compare fungitoxicity of the many compounds under study. Eventually, promising chemicals must be tested on plants for specific diseases both in small, highly controlled environments and in the field. Such tests usually establish optimum dosages, time of application, and application methods.

Mode of action may be determined for new, site-specific fungicides. This information delineates potential applications of fungicides and may foretell possibility of pathogens developing resistance to these compounds. Fungicidal resistance has replaced toxicity and environmental damage as a main concern in development of new fungicides. Marketing of new fungicides may involve strategies for mitigating or circumventing fungicidal resistance as much as it does direct control of specific diseases.

FUNGICIDE TESTING

Spore germination tests, which detect both fungitoxic and fungistatic activity, are the simplest and fastest means of determining fungitoxicity and comparing chemicals. Tests involve mixing fungal spores with known dilutions of test compounds and determining percentage of spores that germinate. In 1807 Benedict Prévost was first to use this technique to test effects of several chemicals on germination of wheat smut teliospores. Steeping wheat seed in milk of lime in copper containers controlled smut while steeping seed in wooden containers did not. Prévost tested copper in various forms and other metallic salts such as arsenic and mercury, but considered the latter metals too dangerous (poisonous) for general use. Eighty-some years later similar spore germination tests were conducted with rust spores. In the early 1900s the spore germination method was somewhat standardized and became the established procedure for screening chemicals for fungicidal potential. Spores must be fresh and produced on some uniform substrate since spore age and nutrition affect germinability.

Spores must germinate uniformly and consistently or comparison of one test with another is invalid and requires adjustment. Spores should be large enough to be easily seen with moderate magnification, but not so large that the required number in a sample cannot fit in the field of view. Too many or too few spores per field is unsatisfactory. The direct method deposits a standarized spore suspension in a cavity on a microscope slide with a measured amount of chemical and after a prescribed incubation period the percentage of spores that germinate is determined.

What constitutes a germinated spore is sometimes subjective, but as a rule the germ tube must be twice the length of the spore to be considered one that has germinated. The tendency has been to keep the counts conservative. At times orange juice or other supplement has been used to enhance spore germination, but these additives may neutralize or otherwise compromise toxic effects of a test compound and not give a true indication of fungicidal activity.

Botrytis cinerea has been a common test fungus for several reasons. For example, (1) it produces large numbers of spores in culture, (2) the spores are hyaline and of a moderate size, being neither too small nor too large, and (3) the spores germinate quickly in high numbers.

There is an indirect method of measuring the effect of chemicals on spore germination. This involves mixing measured amounts of chemical with agar media and seeding the agar with spore suspensions of the assay fungus. After an appropriate incubation period, the number of fungal colonies is counted. Results are then compared with chemical-free checks or with other concentrations of the test compound. Bacterial cells are substituted for fungal spores when determining bactericidal properties. This method requires some expendible materials, namely culture media and culture dishes. Space for incubating cultures is also needed but the method makes counting easy. However, because spore clumps appear as single colonies or if development of spores buried in the substrate is delayed, there may be more variability than with direct counts.

A range of fungicide concentrations is tested to determine lower (zero effect) and upper (100% effect) limits of toxicity and a dose–response curve is prepared. Since the lowest and highest rates involve relatively few spores, a point between the two extremes is selected. This often is the dosage required to kill or inhibit 50% of the spores. This value is called the ED_{50} (effective dose to kill 50% of the spores) of the fungicide.

To test compounds on fungi that do not produce spores, such as *Rhizoctonia solani*, mycelial plugs of the fungus are placed on agar containing various concentrations of a test chemical. Growth measurements are made over time to determine the degree that growth is inhibited. If there is a marked reduction of mycelial growth, the plugs can be removed and placed on chemical-free media to determine if the effect is fungicidal or fungistatic.

With the techniques described, large numbers of candidate chemicals can be screened for activity on a wide variety of pathogens, particularly fungi; dosage levels can also be determined. After *in vitro* tests show fungicidal potential, chemicals are tested on living plant tissue. These tests may start with leaf disks, excised leaves, or other plant parts in the laboratory, followed by intact plants in greenhouses or growth chambers. To ensure adequate and controlled application of a fungicide, there must be a suitable and uniform means of inoculation. Only then can results be considered comparable from test to test. Such procedures also provide a means of quantification.

In vivo tests conducted under controlled conditions further define dosage levels, application techniques, and timing of fungicide applications. The tests also help to determine possible ways to control pathogens such as obligate parasites not amenable to laboratory testing. The artificial environment of laboratory tests is better suited to establish effects of a chemical on pathogen development than on the disease. Commercial companies now begin primary screening of chemicals on selected groups of diseases in greenhouses. The diseases selected vary with region, company, and time depending on which diseases and crops are important. The diseases might include rust and powdery mildew of wheat, rice blast, rice sheath blight, grape downy mildew, and potato late blight. The most promising materials from initial screenings are further tested on other crops.

As parameters for use are better delineated, chemicals are tested under field conditions. This entails small experimental plots where either natural or applied inoculum establishes disease. Plots are replicated to reduce variations in soil, climate, and other uncontrollable factors. At the end of an experiment fungicidal effects on disease control, increased (or depressed) yield, phytotoxicity, dosage response, and application frequency are quantified.

When dosage and timing have been reasonably defined, chemicals are tested on a larger scale in commercial fields, often with farmers making the applications. This is to reduce unintentional, but nevertheless preferential attention to detail that can exceed normal procedures and thus introduce bias. Field trials usually include tests at several locations to minimize variation in soil type, farming practices, and climate. Moreover, tests are conducted for several years to avoid confounding factors such as disease escape or unfavorable conditions for disease development.

Until a fungicide is approved and registered by the appropriate authority (EPA in the United States or its counterpart in other countries) the crops produced in these tests may not be used and must be destroyed. During this phase of testing fungicide residues on crops are determined and residue tolerances established. Provisions for field testing under experimental use permits or other exemptions to extend use to other crops are discussed under fungicide registration.

TOXICITY TESTS

Fungicides generally pose less hazard to humans and other animals than insecticides, perhaps because fungi are physiologically closer to plants than animals. Also, many of the biological processes are not shared by animals, or are not at risk from fungicides. Historically, insecticides such as nicotine sulfate, lead arsenate, and nerve poisons of post-World War II have caused much more alarm than fungicides. Of the fungicides only the mercuries are considered to pose an overt hazard to humans. These concerns were based entirely on acute toxicity, which is a single dose sufficient to cause rapid death. The dose is usually expressed as the *lethal dose* required to kill a certain percentage of a test population, 50% being the common standard. A compound with an LD_{50} of 30 means that 30 mg of chemical per kg of body weight per test animal is lethal to 50% of the animals.

White male rats are used in standard tests for acute toxicity, although when mice are used it is so specified. Ingestion of the chemical may be required on more than one animal species. There may also be dermal application, usually on rabbits, inhalation by rats, and applications to eyes of rabbits. Table 13.1 gives LD_{50} values for a range of fungicides. These are generally ranked in three toxicity groups, namely, (1) **high,** chemicals with an LD_{50} of <50, (2) **medium,** LD_{50} values of 50 to 500, and (3) **low,** LD_{50} > than 500. The majority of fungicides are in the low toxicity group.

Ever-rising concerns of chronic toxicity, carcinogenic, teratogenic, and mutagenic effects of chemicals require additional tests for recently developed fungicides, especially those classed as systemics. The additional tests required can be short term, conducted primarily on rats and rabbits, or long term, 2-year feeding studies with rats and dogs, as well as several generations of rats or mice for mutagenic, teratogenic, or carcinogenic effects.

TABLE 13.1. Mammalian Toxicity (LD_{50}) of Commonly Used Fungicides

TOXICITY	
High (LD_{50} <50)	
Cycloheximide	2
Inorganic Mercury	30–100
Organic Mercury	30
Medium (LD_{50} 50–500)	
Blasticidin	56
Dexon	60
Fentin hydroxide	156–345
IBP	490
Low (LD_{50} >500)	
Organic cadmium	600
Thiram	800
Vapam	820
Dinocap	980
Dodine	1000
Dichlone	1300
Propiconazole	1517
Iprodione	2000
Fenarimol	2500
Dyrene	2710
Thiabendazole	3100
Carboxin	3820
Carbamates	1400–6750
Captafol	4600

TABLE 13.1. Continued.

TOXICITY	
Captan	9000
Streptomycin	9000
Benomyl	9590
Folpet	10000
Dicloran	10000
Triforine	16000
Kasugamycin	22000

The Ames test is a rapid screening test for mutagenic activity developed by a group working with Dr. Bruce Ames at the University of California, Berkeley. In this test a histidine-deficient mutant of the bacterium *Salmonella typhimurium* is seeded on a histidine-free agar medium. The test chemical, absorbed on a paper disk, is placed in the center of the plate. Since the medium does not contain histidine only bacterial cells that revert (mutate) to the wild type (i.e., histidine-sufficient) will grow. Colonies growing proximal to the paper disk indicate mutagenicity and colony numbers reflect mutation rate.

In addition to such standard tests, others also may be required on possible nontarget organisms. These tests are often volunteered by manufacturers to demonstrate relative safety of a product. For example, the fungicide diniconazole was tested for nonmammalian toxicity (sometimes called fish and wildlife toxicity) and the following values were obtained.

Acute oral LD_{50}	Bobwhite quail	1490 mg/kg
	Mallard ducks	>2000 mg/kg
48-Hour EC_{50} (effective concentration)	*Daphnia magna* (water flea)	7.4 mg/liter
96-Hour LC_{50} (lethal (concentration)	Killifish	6.84 mg/liter
	Rainbow trout	1.58 mg/liter
	Carp	4.0 mg/liter
8-Day LC_{50}	Mallard ducks	5075 ppm
Acute contact LD_{50}	Honeybee	>20 mg/bee

MODES OF ACTION OF FUNGICIDES

The way a fungicide kills or inhibits is its *mode of action*. However, its activity can be considered from several aspects, for example, the relationship between structure of a fungicide and its biological activity. It is noteworthy that fungicides in the same chemical group such as triazoles, benzimidazoles, and others often have similar biological activities; conversely, there are many examples of similar compounds that have completely dissimilar biological activity.

Chemical Similarities and Biological Activity

Thiram (tetramethyl thiuram disulfide) is one of the first organic fungicides marketed. The similar compound, tetraethyl thiuram disulfide (marketed as Antibuse™) is used to treat alcoholism by making those who consume alcohol violently ill. Thiram can cause the same unpleasant reaction. This has caused problems for workers who consume alcohol after handling thiram-treated seed.

Some chlorinated benzenes used in pest control have completely different ranges of activity and mammalian toxicity. The structural and chemical formulas of four examples, their uses, and LD_{50}s are given in Figure 13.1.

Benzenehexachloride (BHC) is an excellent insecticide. It has little or no fungicidal activity but is highly toxic to mammals. Hexachlorobenzene (HCB) differs from BHC chemically only in being an unsaturated benzene ring and so lacks six hydrogen atoms of BHC. HCB has very selective fungicidal activity, and is used specifically to control common smut of wheat and a few other smut fungi. Unlike BHC, it has low mammalian toxicity.

Structure	Formula	Name	Use	Toxicity
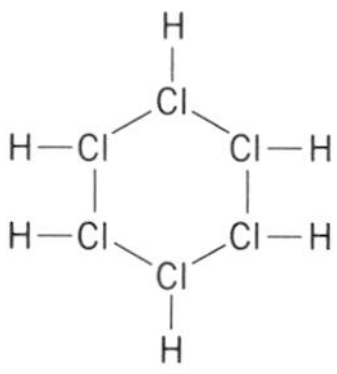 (ring: Cl–H at each of six positions)	$C_6H_6Cl_6$	Benzenehexachloride (BHC)	Insecticide	100
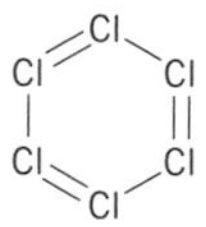 (ring: Cl ×6)	C_6Cl_6	Hexachlorobenzene (HCB)	Smut Fungicide	3500
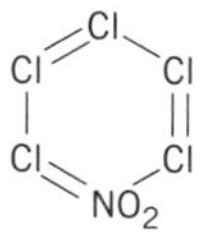 (ring: Cl ×5, NO_2)	$C_6Cl_5NO_2$	Pentachloronitrobenzene (PCNB)	Soil Fungicide	15000
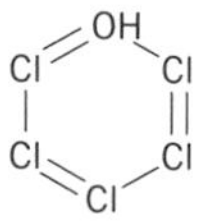 (ring: Cl ×5, OH)	C_6Cl_5OH	Pentachlorophenol (PCP)	Wood Preservative	50

Figure 13.1. Structure, use, and toxicity of four chlorinated benzene compounds.

Pentachloronitrobenzene (PCNB), known also as Terraclor™, differs from HCB in that the benzene ring has five chlorine atoms and one nitro group. This compound has a wide spectrum of activity, is toxic to many fungi, and has low mammalian toxicity. Pentachlorophenol (PCP) differs from PCNB in that a hydroxyl group replaces the nitro group. PCP is too phytotoxic to use on living plants but was widely use as a wood preservative. Of the four chemically similar compounds discussed, PCP has the highest mammalian toxicity.

Captan is widely used as a seed and foliar protectant. Captafol (Difolatan™) is almost identical in structure, but has an additional –CClH; it also has a wide spectrum of activity and is more effective than captan against some diseases such as peach leaf curl. Captan is considered to be quite safe, but captafol produces an allergic reaction in certain human populations, particularly Asians. Before it was discontinued in the United States, captafol was used for peach leaf curl in eastern states but not in the west where a relatively high proportion of farm workers are of Asian ancestry. Captafol also was one of the main fungicides used in single application techniques (SAT) for control of tree fruit diseases.

Diniconazole is a relatively new fungicide with the molecular formula $C_{15}H_{17}OCl_2N_3$. It is a sterol biosythesis-inhibiting fungicide with systemic acivity and is proposed for use on a wide range of crops. It differs chemically from a similar compound, uniconazole, which has the molecular formula $C_{15}H_{18}OClN_3$. The former has one less hydrogen and one more chlorine than the latter compound. Uniconazole, however, is being developed as a plant growth regulator and has the effect of reducing plant height.

Mode of Action in Relation to Disease Development

Fungicidal mode of action is broader at the agricultural (or epidemiological) level than at the cellular level. Figure 13.2 illustrates the limits of effectiveness of fungicides with different types of activity through early stages of disease development. Protective

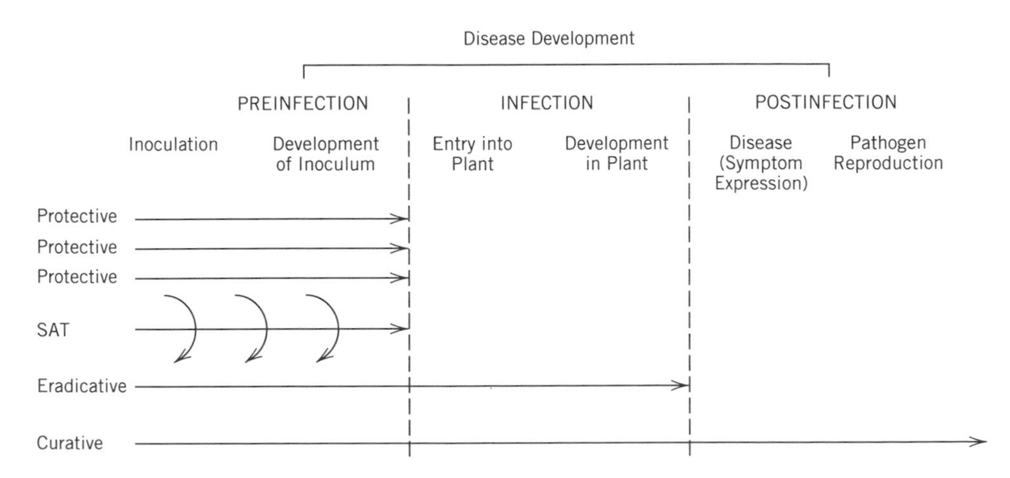

Figure 13.2. Extent of fungicidal protectant (including SAT), eradicant, and curative action in the early stages of disease development.

fungicides are effective only through the preinfection portion of a disease cycle, rarely into the early part of the infection process. Because protectants have a short period of activity, such fungicides must be applied repeatedly to prevent diseases from developing (Table 13.2).

The SAT is a method whereby a single massive dose of fungicide is applied to a plant surface. Over time the fungicide is redistributed on plant surfaces by rain, dew, or irrigation so that the foliage remains protected much longer than with standard applications of fungicides. This technique increases the probability of developing fungicidal resistance, especially if the fungicide is site-specific. However, most SAT fungicides have broad spectrums of activity and resistance has not developed thus far.

Fungicides such as dodine, which are eradicants, kill or inhibit fungi through the infection process. This appears to "cure" plants if disease is not well established. This is sometimes referred to as "burning out" infections and is chemical excision of diseased tissue that prevents further development in treated infection sites.

TABLE 13.2. Control of Apple Scab and Black Rot of Grapes with Postinfection Application of Fungicides

Fungicide	Protectant Program (14–15 Sprays) (%)	Curative + Protective (%) (5 sprays) Hours after Infection Period 72	Curative + Protective (%) (5–7 sprays) Hours after Infection Period 96
Primary Apple Scab on cv Golden Delicious			
None	96%	—	—
Captan	2.0	41	—
Dikar	1.0	—	—
Bitertanol	0.3	1.3	2.8
Triforine	—	1.0	15.8
Etaconazole	—	1.0	3.8
Fenarimol	—	0.5	3.0
Black Rot Infection on Grape Leaves			
None	91%	—	—
Ferbam		42	52
(full season spray)	0.2	—	—
Fenarimol	—	19	44
Tridimefon	—	0.4	0.4
Penconazole	—	0.2	0.1

Source: Condensed from Ellis, M.A., L.V. Madden, and L.L. Wilson. *Plant Dis.* 68:1055–1057 (1984) and *Plant Dis.* 70:938–940 (1986).

Curative fungicides include some systemics that are applied after infection. Such fungicides provide disease control well into the disease development process. The antibiotic oxytetracycline is infused into trees with diseases caused by mycoplasma-like organisms such as X-disease of peaches or pear decline; in so doing the antibiotic brings about a cure of sorts. Actually there is only a retardation of symptoms and annual treatments are usually required. Benzimidazole fungicides injected into elm trees to treat Dutch elm disease also must be applied at repeated intervals.

The term "kick-back" originated from greenhouse studies on postinfection applications of fungicides to fruit trees infected with several foliar diseases including apple scab, cedar-apple rust, and cherry leaf spot. The study involved inoculating test trees with pathogens and applying fungicides 24–96 hours after the start of an infection period, which is well into the postinfection period (Figure 13.2). The eradicant fungicide, phenylmercuric acetate, controlled all but a trace of scab, whereas all unsprayed checks were diseased. Several systemic fungicides, including fenarimol and triforine, gave excellent control of all three diseases. After-infection activity of fungicides has been referred to variously as "curative" and "eradicative" as well as "kick-back." Now the preferred term is "postinfection activity" but kick-back lives on in the form of Reach-Back™, a trademark for the fungicide fenarimol.

Postinfection application of systemic fungicides can be very effective (Table 13.2). Some systemics gave excellent control when applied 4 days after the start of the infection period. Not only is there more flexibility in timing spray applications, but fewer applications may be needed. To be effective, however, the beginning of infection periods must be precisely determined. The disease prediction programs described in Chapter 5 can be used for this purpose.

At the Cellular or Biochemical Level

Many books and articles discuss how fungicides work. The general views on this subject are summarized here. Effects of fungicides on metabolism of pathogens have been demonstrated, but exact mechanisms often are not well understood. Much of the professed knowledge is extrapolated or inferred from other systems, and some is purely speculative.

Metabolic pathways might by likened to a staircase in which each step represents one transformation in the pathway (Figure 13.3A). A substrate (symbolized by the tread, 1, 2, 3, etc.) is transformed to the next compound by an enzyme (symbolized by the riser, a, b, c, etc.). If one enzyme or substrate is inactivated by a fungicide, that step is eliminated or blocked. The gap created by a missing step (Figure 13.3B) can be bypassed or stepped over with relative ease because the inhibition is rarely complete. If, however, many steps are removed the gap is much larger (Figure 13.3C) and much more difficult to bypass. This is a simplistic analogy of how fungicides are thought to interfere with enzyme systems and metabolic pathways of microorganisms. It also suggests how with relative ease or difficulty pathogens develop resistance to fungicides.

There are two general categories of fungicides that have been shown to impair metabolic processes. One category includes fungicides that interfere with energy production. Such fungicides are both multisite and nonspecific inhibitors. They are

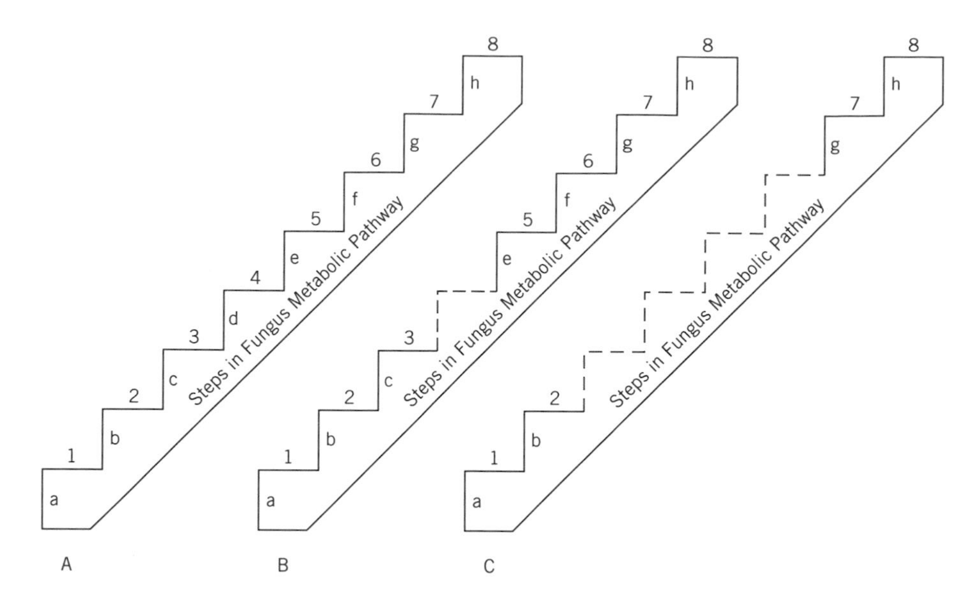

Figure 13.3. Stepwise representation of a metabolic pathway of a fungus. (A) Without interruption by a fungicide. (B) Interruption by a single-site fungicide. (C) Interruption by a multi-site fungicide.

broad-spectrum, protective fungicides, like ethylenebisdithiocarbamates, phthalimides, chlorophenyls, and chloroquinones. These, as well as most inorganic fungicides such as sulfurs, coppers, and mercuries, react with sulfhydryl (–SH) enzymes and thereby interfere with many enzyme systems. Some nonspecific inhibitors such as carboxin, dimethyldithiocarbamates, and triphenyltin interfere with metabolic pathways such as the succinic dehydrogenase enzyme system or glycolytic/citrate cycle.

The other general group includes those that interfere with various physiological processes such as respiration, nuclear activity, and cell membrane permeability. Processes in this category can be further divided and associated with certain fungicides or groups of fungicides. Cycloheximide, blasticidin, kasugamycin, and streptomycin inhibit protein synthesis; ethirimol, fenarimol, metalaxyl, furalaxyl, and hymexazol inhibit nucleic acid synthetase; benzimidazoles, aromatic hydrocarbons, and anilides interfere with the nuclear process; polyoxin inhibits cell wall sythesis (chitin formation); many sterol biosynthesis-inhibiting fungicides including piperazines, morpholines, pyridines, pyrimidines, imidazoles, and triazoles inhibit lipid biosynthesis; and dodine interferes with cell structure and increases cell membrane permeability resulting in leakage of cell contents.

Of these different mechanisms the best understood are those involving the nuclear process and lipid biosynthesis. The action of benzimidazoles (benomyl, thiophanate, and thiabendazole); anilides (carboxin, oxycarboxin, methsulfovax, mepronil, methfuroxam, flutolanil, vinclozolin), and aromatic hydrocarbons (chloroneb, chloronitrobenzene, hexachlorobenzene, pentachloronitrobenzene) on fungal nuclear division is similar to the effect of colchicine in higher plants.

Tubulin is the protein in the microtubules that make up the spindle apparatus. Fungicides that bind with tubulin prevent microtubule assemblage and prevent spindle formation, thus interfering with nuclear division.

Ergosterol is the main sterol, or one of the principal ones, in fungi. It comprises part of fungal lipid membranes and is comparable to cholesterol in mammalian systems. Based on this comparison there are several steps in the ergosterol pathway that are inhibited, but the most important one appears to be the splitting-off of methyl groups (demethylation) at specific sites, especially the C_{14} and C_4 positions, of 24-methylene-dihydrolanosterol, an ergosterol precursor. Thus, many of the so-called sterol biosynthesis-inhibiting fungicides are also referred to as sterol (or steroid) demethylation inhibitors.

Fungicides that affect the processes discussed above generally bind to single sites on molecules and, therefore, are extremely specific in mode of action. This specificity is also the "Achilles' heel," since fungitoxicity is bypassed in some fungi by relatively minor alterations in metabolic pathways, which in some instances result in resistance to a particular fungicide.

FUNGICIDAL RESISTANCE

Failure of fungicides to control diseases against which they were effective and the potential for pathogens to develop resistance cast a pall over fungicide development and use. Fungicidal resistance is a relatively recent phenomenon and is not a concern with older fungicides that are broad-spectrum and non-site-specific.

Fungicidal resistance is defined as a stable, inheritable adjustment by a fungus to a fungicide, resulting in less than normal sensitivity to that fungicide. The term tolerance is not considered an equivalent term and generally is not accepted. Fungicidal tolerance is considered to be acquired or adaptive resistance, which is not genetically transferable, and develops from exposure to increasing concentrations of a fungicide. Fungicide-tolerant organisms revert back to being sensitive when no longer exposed to the fungicide. Adaptive resistance (i.e., tolerance) is temporary while genetic resistance is permanent.

Since 1970 fungicidal resistance has increased dramatically. Delp (1980) illustrates number of genera of organisms resistant to various types of fungicides. The mercuries, coppers, dithiocarbamates, and other conventional broad-spectrum fungicides have encountered very few resistant genera, whereas aromatic hydrocarbons, dodine, and pyrimidines have slightly more. These numbers have remained constant for 10–20 years. However, starting with the benzimidazoles and continuing with antibiotics, carboximides, ergosterol biosynthesis inhibitors—all site-specific fungicides—there has been a marked increase in the number of pathogens resistant to these compounds. This pattern has serious practical consequences, namely failure of some fungicides to control disease. An additional consequence is the phenomenon of cross-resistance, where resistance to a given fungicide applies as well to compounds with similar site-specific modes of action. Multiple resistance is not common but occurs when pathogens develop resistance to two or more fungicides that have different modes of action.

Fungicidal resistance compromises disease control and restricts use of site-specific fungicides. Resistance to dodine by the apple scab fungus, to benzimidazoles by the wheat strawbreaker footrot fungus, and to streptomycin by pear fire blight bacteria has required changes in control practices and development of strategies to overcome fungicidal resistance.

Fungicidal resistance generally does not arise *de novo* but can exist at low levels in pathogen populations (i.e., the occasional fungus spore or bacterial cell). A fungicide acts essentially as a screen (Figure 13.4) through which only resistant cells can pass, thereby having a competitive advantage, albeit a temporary one. The more often a fungicide is applied the greater the opportunity for resistant cells to pass through the screen.

Several mechanisms are postulated to explain development of resistance to fungicides. These include reduced permeability or exclusion of toxicants from cells, detoxification by chemical breakdown or binding to nontarget sites, decreased conversion into toxicant, or modification of sensitive sites. Some of these changes are controlled by single genes and others by multiple alleles.

Polyoxin D interferes with chitin synthesis in fungi. However, resistance is not related to changes in the chitin synthetase target but rather to changes in the fungal membrane that result in reduced uptake of fungicide. Similarly, the protein- synthesizing system of a blasticidin-resistant strain of *Pyricularia oryzae* remained sensitive to blasticidin. It is assumed that less toxicant penetrates cytoplasmic membranes of resistant cells.

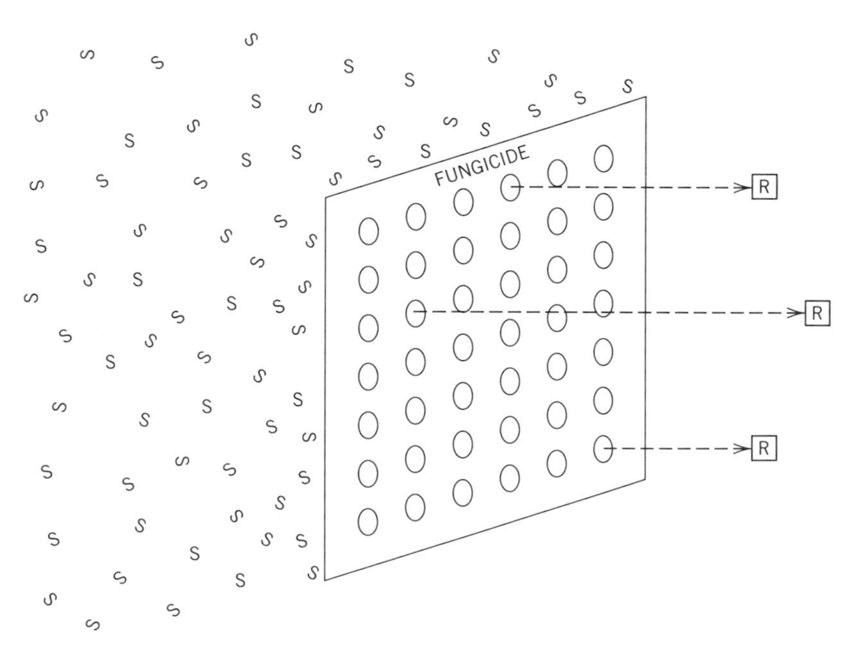

Figure 13.4. Screening effect of a fungicide on sensitive and resistant strains of a fungus. S, sensitive: R, resistant.

Detoxification has been ascribed to certain fungi that possibly take up toxicant and excrete less-toxic metabolites. Decreased toxicity of a compound might also result from the toxicant being tied up at some nontarget site in a cell and therefore unable to interfere with a particular physiological process. It has been postulated that fungal pigments in *Pyrenophora avenae* bind organomercury fungicides.

Some fungicides are not fungitoxic until converted to a toxic metabolite in the pathogen. For example, the organophosphorus fungicide pyrazophos is toxic only when converted by the fungus to a pyrimidine compound. *Pythium debaryanum* cannot effect this conversion and is therefore resistant to pyrazophos.

Modification or changes at the site of fungicidal action is probably the main mechanism by which pathogens become resistant to fungicides. Slight shifts in location of the binding site of a toxicant apparently are mediated by mono- or polygenetic variations. The exact mechanism is not completely understood except for a few fungicides such as the benzimidazoles.

Several factors are important in development of resistance. The principal ones are rate of pathogen sporulation and dispersal. As a rule, pathogens that produce large numbers of spores that are widely dispersed, such as powdery mildew fungi or *Botrytis cinerea*, are more likely to develop fungicidal resistance than nonsporulating root and stem disease fungi such as *Rhizoctonia solani,* or fungi such as *Pseudocercosporella herpotrichoides*, the spores of which are not dispersed long distances. Benomyl is used to control both apple scab and cedar-apple rust and resistance has developed to scab, which has a repeating cycle, but not to rust that does not.

Ability of resistant strains to compete (i.e., fitness) with sensitive strains (wild types) is probably the most frequently mentioned requisite for success of fungicide-resistant strains. If resistant strains cannot compete favorably, either as pathogens or as saprophytes, resistant populations cannot sustain themselves. Fitness is a complex of qualities involving probabilities of infection, rate of colonization of host tissues, and amount of sporulation (or rate of multiplication of bacteria). Aggressive, heavily sporulating fungi (or rapidly dividing bacteria) are apt to survive in competition with fungicide-sensitive pathogens. Reduced fitness of resistant strains may be the result of biochemical or physiological alterations such as substitution of ergosterol precursors for ergosterol or changes in osmotic properties of membranes. Resistance to some fungicides such as the benzimidazoles and acylalanines does not appear to be linked to reduced fitness and such strains favor a shift toward a population of efficient and resistant pathogens.

The threshold of infections needed for disease to develop is yet another factor that contributes to development of resistance. If a single fungicide-resistant spore or bacterial cell can establish an infection that results in disease, chances of resistant strains becoming entrenched are greater than when multiple infections are necessary for disease to develop. The chances that resistant populations of powdery mildew fungi or *Botrytis* will emerge, where a single conidium can establish infection and lead to rapid buildup of disease, are greater than is the case with wheat strawbreaker foot rot where many infections on the same plant are necessary to cause severe disease. This may help explain why disease control fails so suddenly in powdery mildews but takes longer for strawbreaker foot rot to fail.

Nuclear condition of a pathogen (haploidy, diploidy, or polyploidy) and nuclear status (monokaryosis, dikaryosis, or heterokaryosis) have been investigated very little with respect to fungicidal resistance. Limited genetic studies suggest that diploids are more stable than haploids and homokaryons more stable than heterokaryons, unless the latter have some advantage such as fungicidal resistance. Theoretically, sexual reproduction is more likely to yield fungicide-resistant segregants but asexual reproduction will perpetuate resistance more effectively. In fungi that reproduced by conidia, fungicidal resistance has a good chance of increasing because propagation is basically vegetative cloning and little genetic segregation is involved. If a conidium carries a gene for resistance the trait will be passed to most progeny of the spore.

STRATEGIES TO AVOID OR MINIMIZE RESISTANCE

Even though resistance to antibiotics was recognized in the early 1960s it was not until resistance to benzimidazoles emerged about 10 years later that much concern was expressed about this phenomenon. Since then fungicidal resistance and strategies to reduce its impact have become paramount issues for pesticide manufacturers and users.

Resistance to streptomycin by fire blight bacteria commonly occurs in pear orchards where streptomycin is used routinely. The primary method of avoiding streptomycin resistance is to use another bactericide. However, only two alternatives for fire blight control are available. Copper fungicides are effective but sometimes cause russetting, which reduces quality on some pear varieties such as Anjou. Tetracycline antibiotics are also effective, oxytetracycline being registered for blight control. However, companies that market baby food do not use antibiotic-treated fruit in their products. As a matter of general policy, antibiotics or other chemicals that have value for human or veterinary medicine are little used in agriculture because of the potential for organisms to develop resistance. Some segments of agriculture exlude certain fungicides for similar reasons. For example, benomyl is effective against many tree fruit diseases. Currently, it is one of the most widely used effective postharvest fungicides. Ironically, its use in orchards is discouraged lest postharvest pathogens become resistant to benomyl.

Use of alternative chemicals with different modes of action is not always possible, feasible, or desirable. Chemicals equally efficacious may not be available or registered for use. If the alternative is also site-specific in mode of action using it may just trade one type of resistance for another even though such resistance is not apt to develop immediately. The ideal strategy is to use the least specific fungicide that is available and effective. Another option uses mixtures of specific and broad-spectrum (contact or general) fungicides. An alternative to mixed fungicides is mixed application schedules where application of a site-specific fungicide is alternated with a general fungicide.

Mixed application schedules are considered to carry the lowest risk of resistance developing because resistance occurs only when the specific or high-risk chemical is used. When a mixture of fungicides is used there is still opportunity for continual selection pressure. Possibly the best option is to use a mixture containing a specific and a nonrisk chemical alternated with the nonrisk chemical alone. In short, specific fungicides should never be used by themselves.

Fungicides should be applied only when needed and at recommended dosages. This is sound advice not only to avoid fungicidal resistance but also for economic and environmental reasons. When possible, continual treatment of large plantings with the same fungicide should be avoided. This reduces the possibility of resistant strains moving in from adjacent areas. Although resistance appears to be a qualitative characteristic it is inadvisable to use low rates of systemics or other fungicides despite the temptation to cut costs. Doing so may only exacerbate marginal levels of resistance.

Catastrophic effects of resistance can also be avoided by integrating nonchemical disease control measures with chemical applications. Strawbreaker foot rot of wheat can be controlled doing a combination of things, namely, rotating crops, modifying seeding dates, using resistant varieties, and applying fungicides. While nonchemical practices can be routine, application of fungicides should be based strictly on need. With strawbreaker foot rot the need can be determined by establishing disease thresholds (Chapter 5).

Prediction of resistance may be established with test fungi (not necessarily plant pathogens) before fungicides are released for commercial use. Such predictions are useful in developing countermeasures. When resistance is likely to occur, it is advisable to periodically monitor fields for resistance. Reliability of monitoring depends largely on sensitivity of the test and how the results are interpreted. Alleged failures of fungicides to control diseases can lead to rumors of resistance where none exists. In one instance, onion growers reported that dicloran failed to control white rot. Moreover, they blamed that failure on inferior fungicide and on fungicidal resistance. Many tests were conducted and no evidence of resistance was found. Instead, circumstantial evidence suggested that growers had been reducing the amount of dicloran applied to soil to cut costs.

DEGRADATION OF FUNGICIDES

Most organic fungicides have short half-lives, but breakdown products can persist for a relatively long time. For example, benomyl and thiophanate-methyl are soon converted to carbendazim in soil, but the latter persists in soil for several months and slowly converts to 2-aminobenzimidazole. The aromatic hydrocarbons HCB and PCNB can persist in soil for several years but metallic fungicides such as copper, arsenic, and mercury can persist for decades.

Reduced effectiveness of iprodione for control of onion white rot has been linked to repeated use, which suggests enhanced degradation of a microbial nature. However, there is little evidence that fungicide degradation in soil results from enriched populations of specific microorganisms. In addition to microbial or other biological degradation, fungicides can be degraded by chemical or photolytic action or rendered inactive by adsorption on soil colloids.

REFERENCES

*Baldwin, B. C., and W. G. Rathmell, Evolution of concepts for chemical control of plant disease. *Annu. Rev. Phytopathol.* 26:265–283 (1988).

*Cohen, Y., and M. D. Coffey, Systemic fungicides and the control of oomycetes. *Annu. Rev. Phytopathol.* 24:311–338 (1986).

*Davidse, L. C., Benzimidazole fungicides: Mechanism of action and biological impact. *Annu. Rev. Phytopathol.* 24:43–65 (1986).

*Dekker, J., and S. G. Georgopoulos (eds.), *Fungicide Resistance in Crop Protection.* Centre for Agricultural Publishing and Documentation, Wageningen, 1982.

Delp, C. J., Coping with resistance to plant disease control agents. *Plant Dis.* 64:652–657 (1980).

*Erwin, D. C., Systemic fungicides: Disease control, translocation, and mode of action. *Annu. Rev. Phytopathol.* 11:389–422 (1973).

*Green, M. B., and D. A. Spilker (eds.), *Fungicide Chemistry: Advances and Practical Applications.* American Chemical Society, Washington, D.C., 1986.

*Köller, W. (ed.), *Target Sites of Fungicide Action.* CRC Press, Boca Raton, FL, 1992.

*Lyr, H. (ed.), *Modern Selective Fungicides: Properties, Applications, Mechanisms of Action.* John Wiley, New York, 1987.

*Marsh, R. W. (ed.), *Systemic Fungicides.* 2nd ed. Longman, London, 1977.

*Matsumura, F., and C. R. K. Murti, *Biodegradation of Pesticides.* Plenum Press, New York, 1982.

*Moller, W. J., M. N. Schroth, and S. V. Thomson, The scenario of fire blight and streptomycin resistance. *Plant Dis.* 65:563–568 (1981).

*Owens, R. G., Chemistry and physiology of fungicidal action. *Annu. Rev. Phytopathol.* 1:77–100 (1963).

*Rucke, K. D., and J. R. Coats, *Enhanced Biodegradation of Pesticides in the Environment.* American Chemical Society, Washington, D.C., 1990.

*Siegel, M. R., and H. D. Sisler (eds.), *Antifungal Compounds.* Vol. 2. *Interactions in Biological and Ecological Systems.* Marcel Dekker, New York, 1977.

*Staub, T., Fungicide resistance: Practical experience with antiresistance strategies and the role of integrated use. *Annu. Rev. Phytopathol.* 29:421–442 (1991).

*Szkolnick, M., Techniques involved in greenhouse evaluation of deciduous fruit tree fungicides. *Annu. Rev. Phytopathol.* 16:103–129 (1978).

*Vyas, S. C., *Nontarget Effects of Agricultural Fungicides.* CRC Press, Boca Raton, FL, 1988.

*Publications marked with an asterisk (*) are general references not cited in the text.

14 Modifying the Environment

Protection is usually perceived as application of fungicides to plant surfaces, but many plant diseases are controlled by establishing environmental barriers between host and pathogen. Such barriers are created by modifying or manipulating environmental factors that influence disease development and are accomplished by changes in routine or additional farming practices. Many of the tactics evolved over decades or centuries of agriculture. Initially their application was accidental or based on growers observations. During the past 80–100 years research has modified old cultural practices, or developed new ones, that reduce disease severity, eliminate inoculum, impede infection, or promote plant vigor. Pathogens may be affected directly by physical barriers or indirectly by a modified environment.

Mechanisms by which cultural practices reduce disease development and/or severity may be disease escape, biological control, eradication, or exclusion. However, the basic strategy is to protect the host from infection and tactics are directed at plants rather than pathogens as occurs with exclusion or eradication.

Pathogen activities, host phenology, and physiology are influenced by environmental factors. The present vogue of plant health management emphasizes the importance of maintaining plant vigor as a facet of plant disease control (or management). Even though plant vigor is difficult to define, healthy, vigorous plants generally are able to repel or tolerate attack by some pathogens. Some pathogens such as rust and powdery mildew fungi, and bacteria are more severe on young succulent tissue than older, mature tissues. Others, usually facultative parasites such as *Sclerotinia sclerotiorum*, *Alternaria solani*, *Microdochium nivale*, and *Armillaria mellea* thrive only on weak or senescent tissues. The direction and degree to which one or more environmental factors must be modified to influence disease severity depend largely on pathogen and host response to those changes. Modification of environment by cultural practices can injure hosts. Some practices retard certain pathogens but favor others. For example, changing from conventional to no-till or minimum tillage in wheat may reduce losses from strawbreaker foot rot but result in greater losses from Rhizoctonia bare patch. Excessive nitrogen fertilization of apple trees may reduce losses from apple scab but increase incidence of fire blight. Sprinkler irrigation can reduce losses from crown and root diseases of beets or peanuts but increase losses from foliar diseases.

PROTECTION BY ESTABLISHING ENVIRONMENTAL BARRIERS

Modification of Moisture

Moisture may be the most important environmental requisite for disease development; it occurs in three forms: (1) free moisture on plant surfaces, (2) relative humidity, and (3) soil moisture. Moisture levels can be altered by various cultural practices such as irrigation, plant density, and drainage. The nature of the disease to be controlled must be considered. Diseases favored by dry soils include dryland root rot of wheat, common scab of potato, sweetpotato pox, and charcoal rot of sorghum and cotton. Conversely, diseases such as Phymatotrichum root rot, take-all of wheat, southern blight, white mold, Cephalosporium stripe of wheat, and black root rot of tobacco are favored by wet soils.

Irrigation Free moisture, be it from rainfall or sprinklers, promotes development of many foliar diseases. Rainfall cannot be controlled, but irrigation methods can be altered and schedules and rates regulated to reduce moisture placed on plant surfaces or the time it remains. Changing from sprinkler to furrow (rill) irrigation eliminates moisture on foliage, but such a change may not be feasible because of topography or established irrigation systems. Applying more water at less frequent intervals reduces the time that foliage remains wet. Irrigating early in the day rather than late afternoon allows foliage to dry before nightfall when drying is slower. Similarly, avoiding irrigation during cloudy weather also reduces prolonged periods of wet foliage.

Several diseases of wheat, including black point and Fusarium head blight, are more severe under sprinklers than in fields irrigated by furrows. Misting in greenhouses, propagation chambers, and transplant nurseries is more likely to intensify foliar diseases than surface irrigation. Diseases such as take-all of wheat and stem rot of peanuts are uncommon in arid climates unless crops are irrigated.

Sprinkler irrigation, and misting to a lesser extent, spreads inoculum by splashing spores, bacteria, or nematodes from plants or soil onto susceptible hosts. This is a common way that pathogens such as *Phytophthora* spread in greenhouse and nursery plantings, particularly between plants grown in containers where soil may be wetter than is desirable. Splashed spores and bacteria also intensify foliar diseases such as Cercospora leaf spot of beets and angular leaf spot of cucumbers.

Drainage Diseases caused by phycomycetes are often most severe in poorly drained soils. A variety of methods are used to drain water from agricultural fields, the most common being drainage ditches. Drain tiles or perforated plastic pipes buried in soil also effectively remove excess subsurface water. Lowering water content of soils helps control diseases such as Phytophthora root rots and Pythium damping-off and root rot.

Another way to reduce soil moisture near the surface is to grow plants on raised beds, ridges, or mounds. Growing pineapples on raised beds greatly reduces Phytophthora diseases by improving drainage.

Flooding Flooding has several effects on plant diseases. Primarily it carries pathogens to uninfested fields and creates conditions favorable for certain pathogens, especially water molds. In these instances flooding obviously should be avoided. Ironically, flooding also is a disease control practice, best known from work with Fusarium wilt of banana. Before wilt resistant Cavendish varieties were grown, susceptible Gros Michel was the predominant cultivar in Central America and flooding was attempted as a means of control, but the long fallow of 6 months or more and short protection period made this approach impractical. Oxygen starvation kills pathogens in flooded soils. Sclerotia produced by pathogens such as *Sclerotinia sclerotiorum* and *Verticillium dahliae* are predisposed to attack by other microorganisms in flooded fields. However, flooding also has a negative effect by destroying beneficial organisms in soil; in fact, recolonization by pathogens can be more damaging after flooding than before.

Spacing of Crops Wide spacing of rows and plants within rows reduces both relative humidity and free moisture on foliage. Wheat plants along vehicle tracks in fields frequently have noticeably less powdery mildew because of better aeration and drier atmospheres around widely spaced plants. Gray mold is more severe on raspberries where rows are close together and plants in the rows are dense. Dense plantings reduce air movement required to dry foliage and fruit. Similar effects contribute to Botrytis bunch rot of grapes and brown rot of stone fruits.

Modification of Temperature

Temperature probably is second only to moisture in importance as an influence on disease development. It can be moderated by a number of cultural practices.

Time of Planting An example of the influence that temperature has on plant diseases is seen in the relationship of soil temperature to Gibberella seedling blights of wheat and corn. The fungus grows in culture from 3 to 32°C, maximum seedling blight of wheat occurs from 12 to 28°C, and of corn from 8 to 20°C. There is no blighting of wheat below 12°C and none on corn above 20°C. Wheat grows at temperatures below that at which maximum seedling blight occurs and corn grows well above temperatures that the disease is severe. Therefore, damage from seedling blight on wheat and corn can be reduced by planting wheat earlier in spring or later in fall when soil temperatures are lower than optimum for disease and planting corn later in spring when soil temperatures are higher than optimum. By alterating planting dates, seedlings or sprouts develop at temperatures that preclude or mitigate disease and favor plant growth. Shifts in seeding or planting dates are used to control other diseases (Table 14.1). While temperature is the primary environmental factor involved, the mechanisms of control are completely different and discussed later in this chapter in the section on indirect barriers.

Shading Temperatures of soil and plants can be reduced by providing shade. However, this is usually accompanied by increases in moisture, especially elevated relative humidities. Many cultural practices that modify temperature or moisture concurrently influence the other factor.

TABLE 14.1. Influence of Seeding Date on Severity of Four Wheat Diseases

Cephalosporium Stripe[a,b]

Date of seeding	Percentage Infection	Yield (bu/A)
Kansas		
Sept. 27	94	
Oct. 12	90	
Oct. 26	77	
Nov. 4	64	
Washington		
Sept. 6	99	19
Sept. 16	97	20
Sept. 23	66	44

Strawbreaker Foot Rot[c]

Selection 101 (High Susceptibility)		Nugaines (Moderate Susceptibility)	
Seeding Date	Yield (bu/A)	Seeding Date	Yield (bu/A)
Sept. 9	37	Sept. 8	52
Sept. 16	46	Sept. 15	54
Sept. 23	62	Sept. 23	60
		Oct. 2	62

Barley Yellow Dwarf[d]

	Disease Incidence (%)	
Seeding date	1982	1984
Before Sept. 15	27	39
After Sept. 15	9	16

Snow Mold[e]

Seeding date	Relative recovery
Aug. 28	39
Sept. 7	27
Sept. 18	15
Sept. 28	8
Oct. 9	0

Source: Extracted and adapted from the following sources:
[a] Raymond, P. J., and W. W. Bockus, *Plant Dis.* 68:665–667 (1984). [b] Bruehl, G. W., *Plant Dis. Rept.* 52:590–594 (1968).
[c] Bruehl, G. W., et al., *Plant Dis. Rept.* 58:554–558 (1974).
[d] Wyatt, S. D., et al., *Plant Dis.* 72:110–113 (1988)
[e] Bruehl, G. W., and B. Cunfer, *Phytopathology* 61:792–799 (1971).

Powdery mildews typically are more severe in areas shaded by trees, buildings, fences, and weeds. Eliminating shade quickly controls mildew by increasing temperatures, reducing humidity, and altering host structure and physiology.

Phytophthora cinnamomi causes root rot in many woody plants, but the fungus fares poorly at soil temperatures above 33°C. Mulches have been suggested to maintain elevated temperatures to control *P. cinnamomi*, but the technique probably is not feasible because of shading under tree crops such as avacado.

Ventilation Air moving through fields, greenhouses, storage facilities, or other areas, lowers air temperature and humidity. The positive influence of ventilation on disease control is readily seen in storage rots of fruits and vegetables, foliar diseases such as black spot of roses, and needle diseases of forest trees. Disease is less severe in open grown, well-ventilated plants than in clumps or dense plantings where air movement is restricted.

Burning Fire is an extreme method of modifying temperature to influence disease development. Burning has several effects, the most direct being destruction of inoculum. Perhaps the best known example of this is brown spot needle disease of longleaf pine described in Chapter 11. Other examples of diseases controlled, at least in part, by burning are discussed in Chapter 15. Fire also affects soil temperatures by blackening soil surfaces and increasing heat absorption. Alternative hosts that harbor pathogens or disseminating agents are often destroyed by burning.

The jarrah (*Eucalyptus marginata*) forests of Australia are affected by a dieback disease caused by *Phytophthora cinnamomi*, in which fire plays an important role. If fire intensity is insufficient, disease-susceptible trees and shrubs become established and maintain the pathogen. However, fires of greater intensity stimulate germination of resistant and desirable species such as *Acacia pulchella*, which in time reduces the amount of inoculum persisting in the soil.

Burning crop residues is an undesirable practice to be used only in extreme circumstances and then with caution.

Irrigation Application of water to soil or plant surfaces has a cooling effect. Cotton grown on ridges has less Verticillium wilt compared with rill- or flood-irrigated plants because drained soil is warmer. Rill irrigation cools soils with less danger of exacerbating foliar diseases than sprinkler irrigation. Sprinkler irrigation or misting cools plant surfaces and reduces injuries from physiological conditions such as sunscald. Drip irrigation probably has little effect on soil temperatures. However, water temperature can be a factor. Sugar beets irrigated with warm water from wells intensified rot in fields.

Modification of Aeration

Increasing or decreasing air movement around plants indirectly affects plant diseases by altering temperature and moisture. Aeration of soil, however, has a more direct effect on root development, and survival and activities of pathogens and beneficial microorganisms. With few exceptions, all plant pathogens are aerobes and require oxygen. Decay of untreated fenceposts and utility poles is rapid and intensive just below the soil surface and is much slower deeper in soil. The majority of soil organisms are in the upper 10–12 inches of soil because of greater availability of oxygen. The following methods are used to increase aeration around plants and in soil.

Pruning and Thinning Pruning removes leaves, shoots, blossom clusters, and branches, usually of woody plants; thinning, however, removes whole plants. Pruning and thinning are routine horticultural practices to improve plants or plantings for commercial or aesthetic purposes. Plants produce more fruit, apples develop better color, border plants are better spaced, rose bushes have better form, and so on. Pruning and thinning are also important methods of removing diseased organs and/or plants from otherwise healthy populations (i.e., eradication).

Removing foliage by pruning or thinning increases air movement through plant canopies. This in turn alters temperature and moisture. Temperatures are raised by greater exposure to sunlight or decreased by cooling effects of improved ventilation. However, the greatest benefits are reduction of moisture on plant surfaces and lowering relative humidity. Fruit rots, leaf blights and spots, and powdery mildews occur more commonly on inner or lower parts of plants where there is shading and little air movement. Opening plant canopies to better aeration generally reduces severity of these diseases.

Row Orientation In many regions, prevailing winds come from more or less specific directions depending on the season. Plant rows of orchard trees, vine and cane crops, and row crops can be oriented so that winds blow parallel to the rows and thus dry and cool them. If the prevailing wind blows across the planting the first row or rows act as windbreaks, which result in dead air spaces within plantings. This favors disease development by increasing relative humidity and settling of airborne pathogens.

Tillage Tilling prepares ground for planting, breaks up compacted clods, creates air channels, and improves soil aeration. Deep plowing and subsoil chiseling aerate deeper soil layers and permit roots to penetrate deeper in soil. Subsoiling creates conditions favorable to beneficial soil organisms and enhances better root growth, often allowing roots to grow at depths where there are no parasitic fungi or nematodes.

Drainage Water occupies air channels in poorly drained soils, interferes with healthy root development, and favors development of soilborne pathogens such as nematodes and phycomycetes. Draining soil improves soil aeration and corrects these undesirable effects.

Modification of Soil Reaction (pH)

Soil pH influences several important plant diseases. Alkaline soils promote take-all of cereals and grasses, common scab of potato, oat cyst nematode, and Phymatotrichum root rot. Acid soils favor club root of crucifers, Cephalosporium stripe of wheat, most plant parasitic nematodes, southern blight, Fusarium wilt of banana, and Rosellinia root rot of many tropical plants. Most plants can grow in slightly acid or slightly alkaline soils. Since many diseases are most severe in acid soils it is possible to control some soilborne diseases by adjusting soil pH to slightly above neutral (above pH 7). Soil reaction can be modified by incorporating several types of materials such as inorganic fertilizers and organic amendments.

Inorganic Compounds Two classical examples of diseases controlled by altering soil pH are common scab of potato, favored by alkaline conditions, and club root of crucifers, favored by acid soils. Early treatment for scab involved adding elemental sulfur to soil to lower pH and for club root control, lime was added to raise pH.

High rates of chlorides (sodium, potassium, magnesium, ammonium, etc.) reduce take-all of wheat in some areas. The presumed mechanism is inhibition of nitrification, accumulation of ammonia, and lowered pH. N-Serve™, 2-chloro-6(trichlorophenyl) pyridine, is a nitrogen stabilizer that inhibits nitrification and has a similar effect. High rates of sodium are reported to reduce numbers of *Phymatotrichum omnivorum* propagules in soil, but it is difficult to see how this treatment would do anything but intensify an alkaline soil reaction that favors this fungus.

Fertilization Nitrogen fertilizers can alter soil pH. Ammonium forms lower pH and nitrate forms raise it. Repeated applications of ammonium fertilizers in the Pacific Northwest have resulted in a general lowering of soil pH. This is believed to be one reason for increasing severity of Cephalosporium stripe and strawbreaker foot rot, two wheat diseases increased by ammonium fertilization. In western Washington, where soils tend to be acid, nitrate fertilizers elevate soil pH enough to increase severity of take-all in cereals and grasses. Palti (1981) lists more than 30 diseases of crops including cereals, vegetables, and fruits that are favored by either ammonium or nitrate fertilizers. It is noteworthy that the number favored by each is almost evenly divided.

Other fertilizers such as calcium sulfate (gypsum) and ferric ammonium sulfate can be used to alter soil pH when additional nitrogen is not needed.

Organic Amendments Various organic materials are added to soil to improve nutrition, soil structure, and pH. Animal manures provide nitrogen; green manures improve soil structure and add nitrogen if the plants are legumes. Composts provide nutrients, improve soil structure, and promote an increase of beneficial soil microorganisms. Straw, wood chips, sawdust, ground bark, bagasse, or just about any other plant residue is applied as surface mulches to retain soil moisture and moderate temperatures. If materials other than manures are incorporated into soil additional nitrogen is needed to facilitate breakdown because they have high carbon:nitrogen ratios. Peatmoss (sphagnum) lowers pH and is often added to soil mixes to promote growth of acid loving plants such as rhododendrons and orchids, and improve water-holding capacity. All of the materials mentioned alter pH to some extent, usually by lowering it.

Modification of Light

Light influences vegetative growth and sporulation of pathogens. Effects of light have been studied extensively *in vitro,* but the greatest influence on disease development probably is indirect effects via moisture and temperature. Certain phycomycetes such as downy mildew fungi sporulate primarily in the dark, but this type of activity is not readily interrupted by cultural practices. Certain wavelengths, particularly ultra-violet, increase cuticle thickness. This protective barrier over plant surfaces is important in preventing infection by many fungal and bacterial pathogens. A thinner cuticle may be one reason that powdery mildews are more severe in greenhouses and shaded areas where UV is less intense. Light also has a direct effect on opening and closing of stomata through which some pathogens such as certain rust fungi and bacteria enter plants. Thus, modifying light exposure can reduce the period a plant is vulnerable to infection.

UV light also stimulates sporulation in some fungi such as *Botrytis squamosa* and *Alternaria* spp. Diseases of tomatoes and cucumbers caused by these fungi have been controlled by using UV absorbing vinyl films in greenhouses in Japan and Cyprus. Infrared-absorbing films have also been used but they appear to raise and maintain higher greenhouse temperatures. Early blight, leaf mold, and gray mold of tomato were less severe, probably because of improved plant vigor resulting from warmer temperatures.

A possible direct influence of sunlight is its effect on powdery mildew and leaf rust on bluegrass lawns. Lawns of rust and mildew susceptible bluegrass variety Merion have more severe mildew in shaded areas, particularly on north and east sides of houses, while rust is prevalent in sun-exposed, south and west sides. The line of demarcation at edges of shaded areas can be distinct. Perhaps these diseases are more severe in these situations because of thinner cuticle in shade and open stomata in exposed areas.

Pruning and Thinning Opening up plants by pruning, or rows of plants by thinning, probably has its greatest effect in altering temperature and moisture. However, some pathogens and resultant diseases are affected directly by exposure to light. Dwarf mistletoes propel their seeds horizontally for distances up to 40 feet. Thinning a stand of infected trees opens lanes, which increases the distance mistletoe seeds can travel before striking a tree. In addition to greater inoculation distances dwarf mistletoes produce greater numbers of seeds as a result of increased light.

Grape growers often remove foliage from tops of plants ("hedging") to allow sunlight to penetrate and hasten maturity of grapes. Along with this horticultural benefit, bunch rot is reduced.

Shading Effects of shading on temperature and moisture were described earlier but some diseases respond to direct effects of light. Shade is reported to reduce Cercospora leaf spot of coffee and Sigatoka leaf spot of banana. Providing shade is a common practice in these and other tropical crops. On the other hand, shade increases incidence or severity of Verticillium and Fusarium wilts of tomato, Sclerotinia stem rot of sunflower, corn anthracnose, and late blight of potato. Symptom expression of some viral diseases such as raspberry mosaic is enhanced by cloudy weather.

An effect of shade on virus transmission is reluctance of some vectors to enter shaded areas to feed. The leafhopper vector of beet curly top virus avoids shaded plants. Unfortunately, many important suscepts of curly top virus such as tomato require high light intensities for normal growth.

Snow Removal by Blackening Agents Snow molds of cereals are caused by fungi that attack plants under heavy snow cover where temperatures are just above freezing. Prolonged snow cover weakens plants because they metabolize and draw nutrients from reserves in crowns and roots. Nutrient depletion renders plants less able to recover in spring. In extreme situations a blackening agent such as coal dust is applied to the snow cover. Blackening snow hastens melt so that plants receive light and photosynthesis can begin. Exposure also is lethal to fungal hyphae on plant surfaces. This tactic is feasible on north slopes and drifts where snow is deeper and lasts longer. A major limitation, other than cost, is the possibility that subsequent snowfalls cover the blackening agent and negate its action.

Reflective Materials Some aphids are repelled by reflective surfaces. Placing reflective materials such as aluminum foil, plastic sheets, or white mats under plants protects them from infection by certain viruses. This technique may be practical for home gardens and small "organic" truck farms, but probably is not feasible for most commercial operations.

PROTECTION BY PHYSICAL, TEMPORAL, AND SPACIAL BARRIERS

Interfering with Inoculation

Physical barriers placed between host and pathogen can prevent inoculation and thereby protect hosts from infection. These can be direct barriers through or across which pathogens cannot move. Others are indirect barriers consisting of breaks in time or space that eliminate inoculum or render it ineffective.

Direct Barriers Many fruits such as apples, peaches, oranges, lemons, pomegranates, and tomatoes are individually wrapped both for appearance and to protect fruit during shipment. Many tropical fruits are bagged in polyethylene films for shipping to reduce moisture loss and protect against injuries. Wraps are sometimes impregnated with protective chemicals such as waxes and mild fungicides, or wraps are preceded by fungicidal sprays or dips. Protection from some decay fungi is simply mechanical exclusion of spores. Wrapping hands or bunches of bananas in polyethylene bags or wraps is the main method of preventing diseases such as cigar-end rot, pitting, and speckle.

Trenching prevents spread of pathogens from infected plants to adjacent uninfected ones through soil or root contact. The Dutch elm disease fungus spreads from diseased trees to adjacent healthy elms through root grafts. Digging narrow trenches to a depth of several feet around diseased trees isolates them and protects neighboring trees. This prevents root transmission but not infection via bark beetles, which disseminate the pathogen.

Trenching has also been used in Canada to protect Douglas fir plantings from infection by the laminated root rot fungus in adjacent tree roots and stumps (Morrison, et al., 1988). Trenching also has controlled fairy ring in cranberry bogs and root rots in tea plantations.

Barriers of sheet metal, concrete, or heavy oil buried in trenches 3 feet deep prevent spread of Phymatotrichum root rot from affected fruit trees to healthy trees in adjacent rows.

Root-free barrier zones are created in some ponderosa pine forests of western United States by bulldozing out Armillaria infected trees. Effectiveness requires that infection centers be well defined. Zone width depends on size of residual trees. Larger trees require wider zones, varying from 15 feet for trees of 5 inch mean dbh (diameter at breast height) to 25 feet for trees of 11 inch mean dbh.

Buffer zones have been used for many years to restrict spread of burrowing nematodes in citrus groves. Such buffer zones are established by first determining distribution of nematodes in groves by sampling and then removing a row of trees surrounding the infested area. Soil in this zone is fumigated twice a year with ethylene dibromide, or other fumigant to kill germinated citrus seedlings, thus providing root-free bare zones through which nematodes do not move. Trenching is now proposed to replace ethylene dibromide, which has been discontinued.

Screenhouses protect plants from insect-transmitted viruses. Interregional programs for maintaining virus-free fruit trees described in Chapter 18 keep virus-free mother trees in screenhouses to exclude vectors of various viruses. Where viruses pose threats in production of high value crops, screenhouse production may be as economically feasible as is greenhouse production of tomatoes, cucumbers, and ornamentals.

In England clear polyethylene sheets used to protect peach trees trained against buildings have also reduced incidence of peach leaf curl to half that of exposed trees. Peach trees are covered from late January until mid-May, but in mild climates the covering can be left throughout the year to give an added benefit of protecting blossoms from late frosts.

Windbreaks protect crops from wind damage and can also affect pathogen dispersal and deposition. Reduction of windblown inoculum by windbreaks has been demonstrated for southern corn leaf blight, fire blight, citrus canker, and other bacterial diseases and suggested for *Verticillium*. Diseases such as rusts, leaf spots, and barley yellow dwarf of cereals are frequently more prevalent on the immediate leeward side of natural or artificial windbreaks. Quiet air in this area allows pathogens, or their vectors, to settle out much as a snow fence results in accumulation of snow proximal to the fence. Beyond the dead air space is a zone relatively free of pathogens or their vectors, just as there is a snow-free zone beyond the snow fence. However, these zones are narrow and unless windbreaks are repeated at appropriate distances across windswept areas the zone protected from pathogens is relatively small. Disease control benefits are probably coincidental with other benefits of windbreaks and their use only for disease control is unrealistic. However, growers can take advantage of natural windbreaks such as ridges, trees or other natural features when establishing plantings.

Cover crops are commonly used in orchards, vineyards, plantations, parks, and other permanent plantings to control dust and weeds. They also serve as physical barriers to movement of soilborne pathogens by minimizing spread of contaminated soil on machinery, animals, and man. One benefit is elimination of weeds that can harbor pathogens such as root-knot and root-lesion nematodes and *Verticillium dahliae*, which can infect commercial crops.

Interplanting a variety of crops is standard farming practice in many parts of the world, especially on subsistence farms in the tropics. Annual crops are sometimes planted between rows of newly planted orchard trees or vine crops to generate income until the main crop comes into production. In irrigated land of western United States tomatoes or potatoes were often interplanted in young orchards. Unfortunately, these hosts can intensify *Verticillium* microsclerotia in soil and the practice has been discontinued. Young cherry orchards were especially vulnerable because they are highly susceptible to Verticillium wilt.

Interplanting several rows of corn, sorghum, or other tall crop plants between soybeans and other legumes is recommended as a barrier to protect soybeans from infection by the bean pod mottle virus. A 1 m band of corn protects peppers from mosaic virus. Four to 12 rows of grain sorghum free of weeds is a barrier preventing spread of Phymatotrichum root rot to susceptible field crops such as cotton, but is not effective in orchards where lateral roots grow beyond the barrier zone.

Strip cropping, where relatively wide bands of different crops are grown in the same field, is an extension of interplanting. It is generally a soil conservation measure but by breaking up solid plantings also serves as barriers to spread of pathogens.

Mulches are physical barriers placed on soil surfaces, often to conserve moisture, increase soil temperature, or control weeds. Mulches consist of plastic films, layers of straw, bamboo mats, and strips of burlap, cloth, or other fabrics. Mulches placed along rows of strawberries, melons, cucumbers, and other fruits produced at the soil surface prevent fruit from contacting soil, thereby protecting them from decay.

Indirect Barriers Distance is an effective barrier between pathogen and host. Figure 7.3 shows that pathogen numbers drop sharply as distance increases from their source. This fact provides the basis of protective zones required for most seed certification programs. In the white pine blister rust control program a *Ribes*-free zone of 1000 feet was considered adequate to protect adjacent pine stands from infection. Healthy cacao plants are protected from swollen shoot virus in Ghana and Nigeria by a 30–50 m protective zone that impedes mealybug vectors on nearby infected trees. Carrot fields more than 2 miles from volunteer carrots have less carrot thin leaf virus than fields closer to sources of inoculum and isolation zones of 3 miles essentially eliminates spread of this viral disease.

Timing agricultural operations can minimize contact between pathogen and host when the latter is most susceptible. For example, delaying or advancing seeding dates may produce small plants and hence smaller targets for fungal spores. This allows plants to reach nonsusceptible stages, thereby bypassing periods of pathogen activity. In the Pacific Northwest at least four wheat diseases can be controlled to some extent by advancing or delaying seeding dates. Table 14.1 shows disease severity or intensity resulting from altered seeding dates, thereby minimizing losses from strawbreaker foot rot, Cephalosporium stripe, barley yellow dwarf, and snow mold.

Although the tactics for the above diseases are similar, the reasons for altering seeding dates are completely different. Early seeding of winter wheat produces larger targets for rain-splashed spores of the strawbreaker foot rot pathogen. Early seeding also allows more time after infection in fall for the fungus to penetrate several layers of leaf sheaths and colonize stems where it interferes with water movement, thereby reducing grain yields. Delaying seeding a few weeks beyond normal seeding dates produces yields comparable to chemical control.

Cephalosporium stripe is restricted to fall-seeded wheat, presumably because the fungus enters plants through roots injured by frost heaving of soil. Delayed seeding produces smaller plants going into the winter. Smaller root systems move as intact units whereas larger root systems tend to break during soil movement.

Barley yellow dwarf virus is vectored by a number of aphids, most important being the bird cherry-oat aphid. Grain yields are reduced in direct proportion to stage of growth when infection occurs. Yield reductions are not severe when infection is delayed beyond the seedling stage. Wheat and barley, the main economic hosts, can develop at temperatures too low for aphid activity. Therefore, delaying seeding in fall or advancing it in spring allows plants to develop beyond the most vulnerable stages when aphids are not active.

Snow mold fungi colonize wheat plants under prolonged snow covers and deplete food reserves needed for regrowth in spring. Damage is reduced by planting very early in fall so that large plants go into winter. This gives them a better chance of surviving than smaller plants with less food reserves. At the other extreme, very small plants resulting from very late seedings deny snow mold fungi a substantial food base and plants essentially escape infection. However, delayed seeding in all cases creates a danger of plants too small to survive extreme winters. Therefore, a balance must be struck between delayed seeding to produce acceptable disease control and plants large enough to survive until spring.

Location of fields, orchards, vineyards, or other plantings can protect plants from certain diseases. Plantings upwind from inoculum sources such as alternative host plants, other plantings of the same crops, volunteer plants, or crop residues reduce opportunities for infection. Topography can also predispose plants to disease. Frost damage is associated with infection by *Pseudomonas syringae*, but there is a question as to whether this connection is causative, that is, do ice nucleating properties of bacteria result in frost damage, or does infection take place in frost damaged tissues. Nonetheless, locating orchards where there is good air drainage reduces frost damage and Pseudomonas blight.

Assaying soils for nematodes such as *Xiphenema, Trichodorus,* and *Longidorus*, all of which transmit some viruses, aids in selecting or preparing suitable planting sites to protect against infection by these viruses. The same holds for fungal vectors of plant virues if assay methods are available.

Disrupting the Pathogen–Suscept Interaction

Removing or neutralizing pathogens also protects plants and can involve some forms of eradication and exclusion. However, in many situations pathogens are still present in an area and have been neither excluded nor eradicated.

Early or Late Planting This tactic was discussed in the previous section and barley yellow dwarf is a good example of altering seeding dates to prevent convergence of host and pathogen. Another example is white rot of onions as it develops in parts of the Pacific Northwest. The white rot fungus develops best at low or moderate temperatures. In the Walla Walla area of Washington state a sweet Spanish onion famed for its mild flavor is planted in fall and harvested in early summer. Because the plants grow much of the time at low temperatures during winter the fungus has extensive opportunity to infect and colonize them. If these same onions are planted in spring when they grow at higher temperatures white rot is virtually absent, even though the fungus is still abundant. By this shift in planting dates onion plants are protected from infection and normal yields are obtained. Unfortunately, this also means that these onions are marketed at the same time when other areas are shipping onions. Biologically spring planting is an effective disease control measure but is economically unsatisfactory in this case.

Elimination of Infection Courts Removal or retardation of tissues or organs through which pathogens enter can protect plants from disease. Pear trees are subject to fire blight infection through blossoms when temperatures are 16–27°C. Pear blossoms produced early when temperatures are not high enough for bacterial activity escape infection. However, pears also produce late blooms that provide important infection courts. Orchardists mechanically remove these second (or late) blossom clusters to protect trees.

Fire blight occurs in the Pacific Northwest with some regularity but may occur on different hosts in different years. For example, one year pears are severely blighted but another year it may be apple, mountain ash, or hawthorn. The reason for this variable occurrence may depend on which host is in bloom when temperatures favor bacterial activity. If blossoming could be advanced or retarded by growth regulators or cultural practices, perhaps trees could be protected from infection.

Branch pruning also eliminates potential infection courts. Removing lower branches of white pine trees to 10 feet above ground reduces chances of lethal blister rust cankers by 98% (Lehrer, 1982). The high control potential, at least in some regions, makes pruning feasible for forest plantings as well as ornamental trees.

Grapevines require 2 prunings in southern India and the second pruning is done in September about the time of the monsoon. Bacterial canker increases as a result of bacteria being spread to fresh wounds by splashing and blowing rain. By delaying pruning for one month until after the monsoon bacterial canker is greatly reduced although there is a lowering of grape yield and quality.

Elimination of Alternate Hosts Removal of alternate hosts, required for completion of life cycles of many rusts, is commonly considered to be eradication. This is when the host to be eliminated is the perennating host of the pathogen. However, in the case of white pine blister rust and some other conifer stem rusts, economic hosts are the perennating hosts. If noneconomic hosts are removed the causal fungi still exist in an area and, even though they cannot reinfect primary hosts, pathogen eradication has not been accomplished. Removal of *Ribes* species for white pine blister rust control simply protects white pines from infection within a certain distance from alternate hosts.

Elimination of Alternative Hosts and Vector Reservoirs Many pathogens, both obligate and facultative parasites, infect large numbers of hosts, many of which have no direct economic value (e.g., weeds and noncrop plants). For example, beet curly top virus affects not only many crops such as beets, tomatoes, peppers, and cucurbits, but also many ornamental and noncommercial plants, including noxious weeds. Many wild perennials (or winter annuals) serve both as sources of virus and maintenance and breeding hosts for beet leafhoppers, vectors of the virus. Weed control is an important means of eliminating these sources of inoculum. Destruction of weeds along irrigation canals and ditches interrupts movement of inoculum to new crop plants in spring.

Many viral diseases can be controlled by removing weeds or other virus reservoirs in the vicinity of plantings. This is especially effective for nonpersistent viruses transmitted by vectors that do not move long distances. A good example is tomato spotted wilt, caused by a virus transmitted by thrips and which is serious in greenhouse tomatoes and other plants. Sources of both virus and vectors frequently are herbaceous ornamentals and weeds in the vicinity of crop plants. Many of these reservoir plants are perennials, biennials, or winter annuals that carry viruses over winter to next seasons crops. Some nematode-transmitted viruses such as raspberry ringspot and tomato black ring infect many common weeds. Raspberry is a poor host for the nematode vectors, *Longidorus* species, and populations decline around raspberry plants, but increase when weeds are present. Weed control effectively controls these viral diseases in raspberry plantings.

Another important function of alternative hosts from an economic, but not biologic, aspect is maintenance of inoculum for short periods until susceptible economic hosts are available. Take-all of wheat can be controlled by 1-year rotations out of cereals to nonhost crops (e.g., peas, potatoes, and alfalfa) but the rotation crop must be free of grass weeds or the fungus simply carries over on these. A similar bridging phenomenon occurs with wheat streak mosaic virus. This virus is vectored by wheat leaf curl mites, which require green hosts to survive. If there is a break of 2–3 weeks between maturation of one wheat crop and the emergence of a new crop, mites cannot survive. However, if successive crops overlap such as may occur with late maturing wheat varieties and an early seeding in an adjacent field then vectors simply move, virus and all, to new green seedlings where they feed and transmit wheat streak mosaic virus. Volunteer wheat plants and susceptible grass weeds also provide green bridges for survival of vectors and virus.

Elimination of alternative hosts is critical to successful disease control through crop rotation. Root-knot nematodes lower quality of potato tubers for processing and cause severe losses to potato growers. Prior to 1970 recommended rotation crops included corn and wheat, which are not hosts for northern root-knot nematodes. However, rotations with these cereals lost effectiveness and the Columbia root-knot nematode was discovered to be responsible. This nematode can invade and reproduce in many of the plants formerly used in potato rotations (O'Bannon, et al, 1982). Now identity of root-knot nematodes decides whether a rotation will succeed and what plants to include.

Growing Early Maturing Varieties Some diseases do greatest damage late in plant development and disease losses can be reduced by growing early maturing varieties.

Potato late blight fungus infects tubers by spores washing from diseased foliage into soil where they infect tubers. Moisture and temperature conditions are less favorable for infection before fall rains and cool temperatures develop. Tubers of varieties that mature before these conditions occur escape infection. Similarly, late maturing varieties of wheat are more severly damaged by stem rust because this rust requires higher temperatures than do other cereal rusts for disease to develop. Early maturing wheat varieties ripen before extensive rust cycles occur.

Manipulation of Seeding and Planting

Altering seeding dates to allow plants to escape infection was discussed earlier. Other modifications of seeding or planting processes also can protect subsequent crops from disease.

Seeding Rate Increasing numbers of individual plants in plantings minimizes drastic losses by providing a surplus of plants. It is anticipated that only a portion will be affected and survivors will constitute an adequate population for profitable production. Direct seeding or double row planting tomatoes for curly top control illustrates this principle. However, overplanting creates problems other than cost of additional seed. Most diseases occur in patches rather than in uniform patterns and surviving plants are concentrated in groups leaving gaps where disease occurs. This results in lower yield per plant because of competition in high density areas and no yield where plants are missing. There is a similar problem when disease does not occur as expected in that thinning is required to create plant populations suited to optimum production.

Seeding or Planting Depth Some pathogens can infect seedlings only in early stages of development. Thus, deeper placement of seeds takes longer for seedlings to emerge and prolongs periods of susceptibility. Some cereal smuts such as flag smut of wheat and head smut of corn infect coleoptiles soon after germination. Shallow seeding hastens emergence of seedlings and hence reduces opportunity for infection. However, some diseases such as dwarf smut and ergot increase in shallow seedings.

Shallow planting also reduces *Rhizoctonia* infection of young vegetable plants (celery, potatoes, beans) by minimizing amount of susceptible crown and stem tissue in soil where it is exposed to the fungus. Tissues become less susceptible as they mature.

Deep planting also protects against infection. Ergot sclerotia sometimes contaminate cereal seed and farmers are concerned that the sclerotia, which are about the same size as seed, constitute a threat to planted grain. Sclerotia germinate in soil to produce perithecial stromata on stalks 1–2 inches long. By planting seed, and contaminating sclerotia, 3 inches deep, cereal seedlings emerge but perithecial stalks are not long enough to reach the soil surface. Thus, no inoculum is produced and plants escape infection.

Vegetative versus Seed Propagation Some diseases such as onion smut usually infect plants only in the seedling stage. Starting plants from transplants or sets significantly reduces incidence of smut. White rot, however, is more likely to be carried into plantings on onion transplants than on sets or seeds. Necrotic ringspot of bluegrass turf appears more frequently in lawns established with sod than in lawns established from seed even though the disease can occur in seeded turf. A degree of protection is accomplished by starting lawns with seed instead of planting sod.

Tomatoes are usually started in fields from transplants but direct seeding provides more plants, some of which are often eliminated by curly top. Direct seeding is probably more feasible for home gardens or small commercial plantings than for large-scale operations. Lettuce transplants suffer less disease losses from bacterial corky root than direct seedings when disease severity is high but not when it is low.

Removal of Toxic Excretions

Many nonparasitic diseases result from toxic excretions from plants and pollutants emitted from industrial sources. Occasionally toxic excretions cause diseases of a quasiparasitic nature. Protection against toxic materials usually requires elimination or inactivation of toxicants.

Replant Problems Roots of many plants such as peach, asparagus, and chrysanthemum release compounds toxic to members of their own kind. This prevents replanting the same type of plant immediately after removal of old plantings. The usual procedure is to remove as many roots as possible or accelerate their decomposition by cultivation, fertilization, and irrigation. Some replant problems, such as the apple replant problem in Washington state, are corrected, in part, by soil fumigation, which indicates that microorganisms are involved in some way.

Allelopathy Some plants excrete chemicals toxic to other kinds of plants. A classical example is juglone released by roots of black walnut. This compound is toxic to other plants, especially tomatoes. Allelopathic relationships must be considered when establishing plantings and selecting rotation or cover crops.

Toxic Gases Ethylene is generated as fruits ripen (apples, tomatoes, etc.) and senescent tissue. It produces abnormal growth such as epinasty, especially on plants in confined atmospheres. Corrective measures remove sources of ethylene or improve ventilation to remove or dilute the gas.

Air pollutants from industrial sources cause various undesirable effects such as split suture in peaches from fluoride, and chlorotic flecking of tobacco from ozone. Susceptible plants can be protected from atmospheric pollutants by locating plantings upwind from sources, or maintaining a distance between sources of pollution and plantings that allows dilution of pollutants. Fluoride injury can be reduced by spraying plants with lime. Calcium hydroxide reacts with fluorine to produce calcium fluoride, which is innocuous. This measure is probably not feasible except in extreme circumstances for use on very valuable plants. Moreover, it depends on ability to forecast periods of high fluoride emission.

Toxic Metabolites A few parasitic or quasiparasitic diseases result from toxic metabolites that are produced or function outside of plant tissues. Best established is production of HCN (hydrocyanic acid) and other nitrogenous compounds by fairy ring fungi. At less than toxic concentrations, these compounds result in the extra-green zone of stimulated grass on the outer edge of the fairy ring. Toxic levels produce an inner ring of dead grass. These toxic materials are removed or diluted by soaking affected areas with water over a prolonged period, usually 1 month. Although other mechanisms are also involved the toxic materials are leached out of root zones.

Slime flux of elm and other deciduous trees is caused by bacteria that are not parasitic or directly pathogenic. In fermenting xylem fluids, the bacteria produce substances toxic to inner bark and cambium. Bark is protected by inserting drainage tubes into the outer xylem to drain toxic flux away from the trunks.

REFERENCES

*Christenson, P. E., and N. D. Burrows, Fire: An old tool with a new use. In Groves, P. N., and J. J. Burdon (eds.), *Ecology of Biological Invasions*. Cambridge University Press, Cambridge, 1986. Pp. 97–105.

*Cook, R. J., Management of the environment for the control of pathogens. *Phil. Trans. R. Soc. London Ser. B* 318:171–182 (1988).

*Fargette, D., R. M. Lister, and E. L. Hood, Grasses as reservoirs of barley yellow dwarf virus in Indiana. *Plant Dis.* 66:1041–1045 (1982).

Lehrer, G. F., Pathological pruning: A useful tool in white pine blister rust control. *Plant Dis.* 66:1138–1139 (1982).

Morrison, D. J., G. W. Wallis, and L. C. Weir, *Control of Armillaria and Phellinus Root Disease: 20-Year Results from the Skimikin Stump Removal Experiment*. Candian Forestry Service, Pacific Forestry Centre. BC-X-302, 1988.

O'Bannon, J. H., G. S. Santo, and A. P. Nyczepir, Host range of the Columbia root knot nematode. *Plant Dis.* 66:1045–1048 (1982).

Palti, J., *Cultural Practices and Infectious Crop Diseases*. Springer-Verlag, New York, 1981.

*Stevens, R. B., Cultural practices in disease control. In Horsfall, J. G., and A. E. Dimond (eds.), *Plant Pathology: An Advanced Treatise*, Vol. 3. Academic Press, New York, 1960. Pp. 357–429.

*Stover, R. H., Disease management strategies and the survival of the banana industry. *Annu. Rev. Phytopathol.* 24:83–91 (1986).

*Thurston, H. D., Plant disease management practices of traditional farmers. *Plant Dis.* 74:96–102 (1990).

*Zitter, T. A., and J. N. Simms, Management of viruses by alteration of vector efficiency and by cultural practices. *Annu. Rev. Phytopathol.* 18:289–310 (1980).

* Publication marked with an asterisk (*) are general references not cited in the text.

15 Altering Cultural Practices

Modifying the environment to protect plants is usually accomplished by altering cultural practices. Farmers alter practices expecting diseases losses will be reduced but improper, excessive, or untimely usage might increase losses.

Cultural practices used to control or manage plant diseases include application, manipulation, or modification of any practice involved in crop production. Some practices function directly or indirectly to encourage biological control (as broadly defined in Chapter 16). These practices minimize contact between pathogen and host (i.e., protect) by three general strategies.

1. Suppress inoculum development or destroy existing inoculum by tillage, crop rotation, pruning, solarization, flooding, burning, sanitation, and weed control.
2. Assist crops to escape potential attack from pathogens by site selection, resistant varieties, seeding dates, planting layout, pruning and thinning, drainage, cultivation, and control of vectors or disseminating agents.
3. Regulate plant growth to minimize susceptibility by avoiding weak or lush growth, and maintain plant vigor by proper seedbed preparation, rate of seeding or planting, cultivation, fertilization, and irrigation.

Some practices function more than one way. For example, cultivation buries inoculum (no. 1 above) and improves plant growth (no. 3). Modified irrigation helps plants escape infection by eliminating disease-favoring environmental conditions (no. 2) and promotes thrifty plant growth (no. 3). The following cultural practices are presented in the same general order that they are considered and applied by farmers.

SITE SELECTION

Selecting a site (e.g., location) suited to a particular crop is the first step a grower takes to produce healthy plants. Planting sites include several characteristics that affect disease development.

Exposure refers to general openness of an area. An area may be shaded by trees, mountains, or other features. A protected site has less air movement, higher relative humidity, and milder temperatures than exposed areas.

Aspect, the direction a site faces, affects the amount of sunlight falling on an area. North aspects receive less direct light than south facing sites in the northern hemisphere and vice versa south of the equator. Amount of sunlight reaching a site influences both temperature and moisture, the two most important environmental factors that affect diseases.

Air drainage is important in a planting site to protect frost- sensitive plants. Orchard trees are more likely to be damaged by Pseudomonas blight when trees are located in frost pockets. White pine blister rust and many conifer needle diseases are more severe in valleys where cool moist air favors infection. Many foliar and stem diseases prevail in low areas where air movement is restricted. Red leaf of cranberries is most abundant where inadequate air drainage results in excessive moisture.

Soil drainage prevents many soilborne diseases. Diseases caused by water molds such as Pythium root rots of peas, wheat, and other crops and Phytophthora root rots of apple, alfalfa, rhododendron, raspberry, and many other plants are most severe in poorly drained soils. Poor drainage may relate to soil type. Heavy clay soils hold more moisture than lighter loam or sandy soils. Cephalosporium stripe of wheat is severest in heavy soils because they tend to frost heave because of higher moisture content. However, head smut of corn is more common in soils with low moisture-holding capacities, and is more severe in sandy loams than silty loam soils.

Elevation influences disease as a result of lower temperatures and more rainfall and snow. Snow mold and dwarf bunt of wheat are more severe at higher elevations because of prolonged snow cover required for infection and development. The same is true of snow blights of conifers, which occur at high elevations where snows are deep and longlasting.

TILLAGE

Once a site is selected it must be prepared for planting. This may require initial removal of established vegetation and then a series of tillages to prepare soil for planting. Good seedbed preparation loosens soil and improves drainage and aeration, promoting vigorous seed germination, uniform seedling emergence, and root development. Although the terms tillage and cultivation are often used interchangeably and synonymously, tillage here refers to practices before planting and cultivation to those after planting.

Conventional or traditional tillage is deep tillage usually with moldboard plows that invert soil to depths of 8–12 inches. This is followed by one or more secondary tillages employing disk or tooth harrows. These operations bury plant residues and break up soil clumps to create fine textured seedbeds.

Deep plowing turns inoculum under soil surfaces and controls several plant diseases. Plowing reduces incidence of Fusarium wilt–nematode complex in cotton when done during the dry season and after harvest reduces numbers of *Phymatotrichum* sclerotia in soil. Deep plowing places sorghum downy mildew oospores below planting depth of seeds and buries sclerotia of *Sclerotium oryzae* and *Sclerotinia sclerotiorum*, thus rendering them ineffective. Shallow plowing buries teliospores of the mint rust fungus so that basidiospores are not produced to initiate new infections.

However, consecutive cycles of deep plowing may simply return active inoculum to the soil surface. Repeated moldboard plowing can create compacted plow-pans. This compacted zone results in more moisture and less oxygen in the root zone and impedes deep root penetration.

Disking and harrowing are considered conventional tillage practices. Although they leave crop residues on the soil surface, they mix residues with soil better than moldboard plowing. Residues on soil surfaces affect plant disease by providing accessible inoculum, influencing seed germination, seedling emergence, plant development, and vigor, thereby decreasing or increasing disease severity. These tillage methods uniformly distribute infested plant residues, or pathogen structures such as oospores and sclerotia, through soil. Plant residues decompose faster when mixed with soil but slowly on the surface or when deeply buried as compacted layers.

Conventional tillage reduces incidence of diseases such as Cephalosporium stripe, take-all, and strawbreaker foot rot of wheat, Stemphylium purple spot of asparagus, stalk rot of corn, and other diseases where pathogens cannot compete with other soil organisms. Decomposition of infested plant material eliminates any competitive advantage of pathogens.

Conventional tillage also brings infected plant parts to the soil surface where they are less likely to survive. The potato late blight fungus overwinters in infected tubers in soil. Inoculum develops the following spring on sprouts from surviving tubers. Disking or harrowing after harvest brings many tubers to the surface where they freeze during winter, simultaneously killing the tubers and the fungus.

Conservation tillage is any tillage practice that reduces loss of soil or water. These practices leave more infested residues on soil surfaces, increase soil compaction and soil moisture, and decrease soil temperature, nutrient availability, and root growth. Conservation tillage can be either minimum till or no-till. Minimum tillage prepares soil for seeding, germination, and plant growth. No-till has no seedbed preparation other than soil disturbance by the seeding drill. Previous crop residues remain intact. Weeds and other live plants such as grass sods are killed with herbicides. Seeds are planted directly into ground overlaid with plant residues.

In general, minimum tillage increases plant diseases because it enhances survival and even intensification of pathogens in plant residues. Bare patch of wheat is not a problem until no-till is practiced. This apparently results from colonization of plant residues by *Rhizoctonia*, subsequent invasion of wheat plants, and survival of the fungus in undisturbed soil. Stalk rot, leaf blights, and Goss's bacterial wilt of corn, bacterial blight of soybean, halo blight of bean, head blight, tan spot, and Septoria leaf blight of wheat survive in plant debris on soil surfaces and increase with reduced or no tillage. Minimum tillage also allows development of volunteer plants that serve as bridging hosts for pathogens from a previous crop.

Strawbreaker foot rot of wheat is often less severe in fields planted with minimum tillage because of retarded plant development. Seed germination and seedling emergence are slower in minimum-tilled soils and have the same beneficial result as delayed seeding. Pythium root rot of wheat is more severe in no-till (i.e., direct seeded) wheat because soils remain cooler and wetter.

Deep chiseling (subsoiling) is tantamount to tilling soil below the surface without inverting the top horizons. It is accomplished with subsurface blades or long teeth or chisels that break up plow pans, hard pans, or other compacted layers. This allows deep penetration of roots into soil. Bean roots concentrated on compacted layers of soil can be severely affected by Fusarium root rot. When compacted soil is broken by deep chiseling, bean roots penetrate deeper in soil than the usual plowing depth. Pathogen populations generally are scarce below the 1-foot level and hence deep roots escape infection. This permits greater development of root systems and plants tolerate greater amounts of disease. Barley roots penetrate compacted layers leaving channels through which roots of subsequent bean crops develop deep in soil and escape root rot.

SELECTION OF VARIETIES

One of the earliest decisions that growers apply to control diseases in a cropping cycle is selecting plant varieties (cultivars), strains, biotypes, and seed provinences that are resistant to specific diseases and adapted to given sites. For many crops and diseases there are no suitable resistant varieties, but site adaptation is always important. Sometimes economic or regulatory conditions restrict use of certain plant types even though they are resistant to disease. For example, California's one-variety law requires planting a certain type of cotton (having to do with fiber length). Even possessing seed of other types is discouraged. This precludes planting some resistant varieties.

Plant growth habit can reduce losses from some diseases. Boll rots are less severe in cotton cultivars with certain shaped leaves and bracts that promote drying within and beneath plant canopies because of better light penetration and air movement. Lack of extrafloral nectaries in other cotton varieties denies some pathogens points of entry. Peas with tendril-like leaves modify microclimates of plants (more light, better air movement, drier atmosphere) and reduce incidence and severity of some diseases. Beans grown on wire trellises have less white mold and increased seed yields because of improved air circulation and light penetration. This suggests that engineering bean plants with stiff, sturdy, upright growth has value for disease control just as it produced lodging resistance in cereals.

Resistance to specific diseases is probably the most cost- effective disease control measure for both commercial farmers and home gardeners. This involves using disease-resistant cultivars or varieties. Botanically and horticulturally these terms have different meanings, but the term "variety" is used throughout this book because it is more familiar than "cultivar". Many varieties of plants, especially annuals and biennials, have been developed for disease and pest resistance. These should be grown where they meet grower needs. Unfortunately, resistant varieties are not always comparable to susceptible lines in yield or quality.

For many years wheat varieties have been developed to include resistance to cereal rusts, smuts, and foliar diseases. Alfalfa has been selected and improved to include resistance to insects and many diseases including Verticillium wilt, stem nematode, bacterial wilt, Phytophthora root rot, and anthracnose. Tomatoes are available that are resistant or tolerant to Fusarium and Verticillium wilts, root-knot nematodes, tobacco mosaic, and curly top. The list of varieties of crops resistant to specific diseases is long.

Resistance to some diseases is less than perfect for a number of reasons; lower yields, poorer quality, grower preference, and limited adaptibility. However, resistant varieties are advantageous in reducing other control costs. For example, wheat varieties Hyak and Madsen were developed in the Pacific Northwest for resistance to strawbreaker foot rot. An 8-year comparison of these varieties with foot rot susceptible ones such as Nugaines and Daws revealed that yields of the two resistant cultivars would have benefited from fungicide applications only 2 of 8 years whereas fungicides would have significantly increased yields of susceptible varieties every year. Similarly, development of scab-resistant apple varieties reduces the number of fungicidal sprays needed through a growing season. Reduction in number of fungicidal sprays needed to control hop downy mildew on resistant varieties has been demonstrated in Bavaria. Fungicides were applied 14 times a season on a calendar basis. This number was reduced to 8 when based on a disease forecasting system. Only 4 sprays were needed on resistant varieties.

Adaptation to a site or region is important to general plant health. Many alfalfa varieties resistant to Phytophthora root rot are non- or semidormant. These varieties are not adapted to colder climates and therefore of little value to farmers in such regions. Correlation of adaptation to site and susceptibility to specific diseases is common in forest trees. Douglas fir is a major forest conifer in the western United States and is planted in many parts of the world. There are two recognized biotypes of Douglas fir, coastal and Rocky Mountain. When one biotype is grown in regions or sites suited to the other, diseases such as Rhabdocline needle cast can be more severe than on indigenous biotypes. This variation may be related to pathogenic races or strains of the fungus or to phenological differences, such as time of bud break, that render different types more susceptible on sites where not adapted. Similarly, Monterey pine grown in New Zealand or Africa has proven to be much more susceptible to Dothistroma needle blight than it is in its native habitat. Swiss needle cast fungus of Douglas fir, although native to the northwestern United States, was not described until this tree species was planted in Europe. In its native habitat Swiss needle cast did not pose a problem to forest trees but has since caused damage in Douglas fir plantations. Ponderosa pine from the Black Hills region of South Dakota was widely planted in the western Rocky Mountains to replace white pine killed by blister rust. Seed from these sources were not adapted and pine stands established with this seed have stagnated (i.e., they failed to grow and develop) resulting in heavy mortality from Armillaria root rot, bark beetles, and other pests that attack trees of low vigor..

Rootstock–scion combinations can contribute to survival of many woody plants. European grapevines were grafted onto American grape rootstocks during the 1800s to reduce losses from grape phylloxera. European grapes are very susceptible while American grapes are tolerant.

A method of reducing losses from collar rot is to graft apple trees onto resistant or tolerant rootstocks, the first of which came from the EM (East Malling) and MM (Merton Malling) collection of growth-regulating (i.e., dwarfing) rootstocks. MM104, MM106, and EM7 are very susceptible and should be avoided where there is potential threat of collar rot. EM4 and EM9 are the most collar rot resistant of the series and should be used where the disease is certain to strike. Additional resistant rootstocks are being discovered and will be available at some future time. Mahaleb cherry roots are less severely affected by Verticillium wilt than are Mazzard roots, which are rapidly killed. Although probably too costly to have practical application, grafting tomatoes onto more resistant eggplant or other *Solanum* species reduces losses from bacterial wilt.

Grafts can also be selected to produce foliage that is resistant to disease. *Hevea brasiliensis* is the best *Hevea* latex producer but it is also very susceptible to South American leaf blight. *H. guianensis* is resistant but is a poor latex producer. Top grafting (crown budding) *H. guianensis* or other resistant scions onto 8- or 10-foot stems of *H. brasiliensis* produces trees that have resistant tops with good latex-producing trunks.

CROPPING SEQUENCE

Cycling of crops, which involves both occupation of land by one or more crops and frequency that specific crops are grown, has a great influence on intensity of plant diseases. In nature there appears to be stability akin to monoculture but ecological successions and plant mixtures rarely allow diseases to develop to epidemic proportions. As agriculture evolved there has been a tendency to grow fewer types (genera, species, or varieties) of plants on increasingly larger areas. This creates conditions that permit diseases to develop to epidemic proportions.

Multiple cropping or intercropping occurs when two or more crops are grown at the same time in the same field. This is common in tropical and subtropical climates and where limited arable land dictates intensive use, but it also is practiced in temperate areas. Interplanting young fruit trees with potatoes or tomatoes to generate income also increased Verticillium in sweet cherries. However, in some mixtures of crops, diseases are reduced by physical barriers formed by plants or plant excretions toxic to pathogens. Intercropping with *Brassica* or *Tagetes* reduces Diplodia and Fusarium diseases on chrysanthemum; corn, beans, or cowpeas reduce incidence of bacterial wilt in potatoes; and tomatoes interplanted with sugarcane have less fungal wilts.

Sequential cropping produces two or more successive crops in the same field per year. In China, short season vegetables such as cabbage, radish, or turnip may precede a rice or wheat crop. As part of the wheat/pea rotation in the Pacific Northwest, spring planted peas are harvested in midsummer followed by fall-planted wheat. In southcentral Washington, spinach planted after onion harvest in early summer is cut for processing in early fall and onions planted in October. Unfortunately, spinach harvesting equipment also spreads the onion white rot fungus to uninfested fields.

Monocropping grows a single crop in a field per year. Different crops can be grown in successive years as *crop rotation* or the same crop can be grown as *annual cropping*. Monocropping itself has little influence on disease severity, but if there is diverse genetic material in the single crop it can be protective. This is demonstrated by use of wheat multilines to reduce losses from stripe and leaf rusts. In central America mixtures of beans, sometimes consisting of as many as 3 species and 17 varieties in a single planting, reduce losses from diseases.

Repeated or annual cropping poses the greatest risk of disease development because it allows, even encourages, extensive buildup of pathogen populations. However, a few diseases eventually decline with repeated growing of the same crop. Take-all decline of wheat is perhaps the best known example of this phenomenon, which is believed to result from increased populations of organisms antagonistic to the pathogen in soil and on roots. A similar decline has been reported in necrotic ringspot of turfgrasses. Some rotation crops interrupt take-all decline. Legumes appear to have an unfavorable effect on antagonistic microflora and counteract decline.

Monoculture generally refers to growing the same crop such as wheat, corn, rice, and potatoes over large areas. This is particularly dangerous with aerially dispersed pathogens. When the same genotype is grown severe disease epidemics can result. Monocultures have resulted in epidemics of potato late blight in Ireland, rice brown spot in Bengal, and cereal rusts and southern corn leaf blight in the United States.

Crop rotations reduce pathogens, balance nutrients, reduce weed populations, decrease toxic residues, improve soil structure, and probably have other beneficial effects. Rotation is considered to be oldest of disease control practices, going back 2000 years or more and alluded to in the Bible. While it likely is the most widely practiced disease control measure much of its application probably has been unintentional, coincidental, or accidental. Discussion of crop rotation in agricultural literature in the mid-1800s centered mainly on its value in preventing exhaustion of plant nutrients. Today it is probably the most prevalent disease control recommendation for soilborne pathogens, especially fungi and nematodes.

Most practical rotations are 3 years or less. In most commercial farming operations a specific crop must be grown every other year, or every third year at most, for economic and logistical reasons. Often the crop rotated out to reduce disease is the most important economically, so a grower incurs losses during the time the crop is not grown. Some crops require special equipment and if it is idle another type of loss occurs. Markets are also jeopardized if farmers cannot supply commodities on reliable schedules or risk selling on open markets, which often bring lower prices.

Take-all of cereals is controlled effectively by 1-year rotations, but Cephalosporium stripe requires 2 years out of winter cereals or grasses to have a beneficial effect in soils where disease is severe. Some pathogens survive in soil for a long time. For example, sclerotia of the onion white rot fungus remain viable for more than 15 years without a suitable host. No rotation can realistically be this long, but virtually any rotation helps because decline in pathogen populations usually follows. And no rotation, no matter how long, will completely eliminate a pathogen from soil. So it is a matter of obtaining an effective compromise, rotations long enough to reduce pathogen populations to acceptable (or tolerable) levels and short enough to minimize economic liabilities.

Trap and decoy crops are nonhost plants that stimulate development of pathogens but do not permit completion of their life cycles. Some writers use the terms synonymously, but in practice trap crops allow invasion by parasites and prevent reproduction whereas decoy crops stimulate parasites to hatch or germinate and start the infection process but do not penetrate the plants. The terms antagonistic, hostile, and catch crops are also used to describe these plants. Sometimes a trap crop refers to partially resistant or early maturing hosts that are removed (harvested) before the parasites mature. Use of trap and decoy crops has been directed mostly at plant nematodes, but they also are effective against several soilborne fungi and parasitic seed plants.

Julius Kühn tried unsuccessfully to control beet cyst nematodes with a "trap crop" of rape. Recent workers have planted wild beets with sugar beets and found that nematode eggs hatched but larvae did not mature. Many reports that allude to trap or catch crops are difficult to evaluate. Some are undocumented and based on single or limited personal observations. Some simply report pathogen decline, which could be a nonhost interaction such as any rotation crop and not stimulatory or entrapment mechanisms.

Marigold, chrysanthemum, and castorbean have been used as trap crops for several nematodes, especially root-knot nematodes. Larvae enter roots but die within a few days. *Crotalaria* is suggested as a cover crop in citrus groves to control burrowing nematodes since they cannot reproduce and survive in this plant. *Crotalaria* has also been reported to control root-lesion and spiral nematodes.

Some trap or decoy crops are very effective. A common weed in rice fields in Asia, *Sphenoclea zeylanica*, gives almost complete control of rice root nematodes within 2 months due to nematicidal root exudates. This action is similar to the effect sudangrass and rape have on root-knot nematodes and Verticillium wilt of potato. Some plants that are not hosts of broomrape and witchweed, including sunflower, alfalfa, cowpea, chickpea, safflower, cotton, and peanut, stimulate seeds of these parasitic plants to germinate, but the germlings cannot form haustoria and die for lack of attachment to hosts. However, practical application is hampered by the long time that seeds are viable in soil and exacting requirements for germination.

Several decoy plants including ryegrass, corn poppy, and mignonette stimulate resting spores of cabbage club root fungus to germinate, but zoospores are unable to infect these nonhosts. Potato powdery scab fungus can infect jimsonweed, but does not produce galls or resting spores. Marigold has been reported to reduce populations of *Verticillium albo-atrum* in olive groves but neither the mechanism nor significance of this has been explained. Many trap or decoy plants have little commercial value but some such as corn, peanut, and asparagus have considerable value and should be used where practical.

Aside from using living plants to deplete pathogen populations, some chemicals cause pathogens to begin development in the absence of hosts. Nematode hatching factors were discussed earlier and are an example of this strategy. Control of onion white rot is being attempted with this approach. The fungus produces sclerotia that survive in soil for many years and germinate when *Allium* spp. (i.e., onions, garlic, shallots, chives, etc.) are grown. Sclerotia also germinate in response to sulfhydryl-containing organic compounds that simulate chemicals in onions and garlic. Synthetic onion or garlic oil stimulates sclerotia to germinate, but they decay and die without a host. White rot control using these synthetic oils is being tested in various parts of the world but two problems have emerged. One is that germinated sclerotia produce secondary sclerotia. These are smaller but may survive long enough to initiate disease. Second, treatment requires 2 years for best results and many onion growers are not prepared or willing to leave their land out of onions that long.

Use of trap and decoy crops or stimulants is not likely to become a major method of disease control. It may serve a role in subsistence or sustainable agricultural programs but probably will have to be supplemented or integrated with other control measures.

SEEDING AND PLANTING

Most crops, especially annual crops, are started from seed. Modifications of seeding or planting processes can reduce losses from certain diseases. The strategies behind these modifications were discussed in Chapter 14 and include altering seeding dates, rates, and depths.

Planting dates can be advanced (early) or delayed (late) depending on specific disease and condition the modified practice is intended to mitigate and vary from region to region. In addition to those mentioned earlier, some diseases reduced by early planting include downy mildew of corn, common smut of wheat, fungus root rots of wheat, leaf blight of corn, powdery mildew of pea, and blackleg of rape. Some diseases reduced by late planting are black rot of peanut, barley yellow dwarf in rice, stalk rot of corn, downy mildew of sunflower, and Fusarium root rots of peas and beans. Late (November 30) planting of tomatoes in Florida and delayed planting of potatoes in Japan and India reduced bacterial wilt because of lower temperatures.

Seeding or planting rates determine stand densities, which can reduce or increase disease losses. Dense plant populations can create microclimates favorable for disease development. In general, lower plant densities decrease severity of foliar diseases while dense plantings reduce losses from soilborne diseases. This relates to differences in epidemiology of the two groups of diseases. Many foliar diseases are polycyclic, which increase exponentially as a function of number of plants available for infection and subsequent inoculum production (see Chapter 4). The more plants infected in the first cycle the more will be infected proportionately in secondary cycles. Most soilborne diseases are usually monocyclic and intensity depends on initial infections. Since only one infection cycle occurs in a season, plants that escape initial infection remain healthy. If 50% of the plant population is infected 50% remains to produce a crop. Doubling initial plant populations may produce normal yields.

Plant density sometimes creates conditions that discourage infection. Solid stands of alfalfa have less alfalfa witches broom compared to rows because leafhopper vectors are repelled by dense shady stands.

Planting depth was discussed in the previous chapter and, as a general rule, shallow planting is preferable to deep planting. For example, deep planting rice retards growth and delays maturity. This may result in greater disease incidence.

PLANTING LAYOUT

Planting design influences diseases by avoiding or reducing conditions that promote them. *Plant spacing* regulates row width and space between plants in rows, increases air circulation, temperature, and light, and lowers relative humidity around plants. Fewer plants reduce infections and inoculum production. Wide spacing reduces losses from several diseases including Sclerotinia stem rot of beans, sunflower, and other susceptible crops, foliar diseases of rice, and Rhizoctonia web blight of rhododendron. Some diseases such as Phymatotrichum root rot are affected more by spacing of plants within rows than between rows because rhizomorphs develop in the direction of tillage and cultivation, and because of root contact between adjacent plants in rows. Cultivation breaks root contact between plants in adjacent rows.

Spacing is important in controlling coffee rust. Wide tree spacing reduces rust incidence in unshaded coffee plantings in Brazil and Kenya. The onion white rot fungus does not grow through soil, rather it spreads by contact between plants. White rot spreads rapidly down rows and proportion of plants infected is inverse to distance between plants. Plants spaced 2 inches apart had 23% white rot, 3 inches apart 13%, and 5 inches apart 6%.

Row orientation with prevailing winds permits air movement along rows and reduces diseases such as bean rust and fruit rots of strawberries and caneberries.

PRUNING AND THINNING

These practices remove sources of inoculum, improve plant growth, and reduce conditions that favor disease development.

Pruning removes cankers from woody plants and is a primary method for controlling fire blight of apple and pear, cane blights of raspberry, cankers of roses, sycamore anthracnose, gall rust of pine, and many other diseases. Pruning sometimes prolongs lives of valuable but systemically infected trees. Therapeutic pruning is useful in containing vascular wilts in ornamentals and fruit trees.

Verticillium wilt infections reoccur annually and some trees such as sweet cherry acquire greater resistance with age. Removal of wilt-affected branches often carries trees beyond susceptible periods. Therapeutic pruning is also performed on elm trees affected by Dutch elm disease. Pruning affected branches eliminates sources of the pathogen and breeding sites for elm bark beetles, which disseminate the fungus.

Therapeutic pruning succeeds because infections are isolated or localized in restricted bands of xylem. Major roots usually serve distinct sections of a tree. Pruning isolates these diseased sections, prolonging life of trees as long as new infections do not occur. Pruning for Dutch elm disease control must be part of an integrated program that includes preventing beetles from transmitting the fungus, disrupting root grafts between adjacent trees, and systemic fungicide therapy.

Pruning opens plant canopies and modifies microenvironments. It improves air movement and reduces incidence of powdery mildews, rusts, black spot of roses, brown rot of trees and shrubs, and many other foliar diseases. Removing 4–6 leaves from upper shoots on grapes 3–4 weeks after bloom reduces losses from summer bunch rot by increasing wind speed and sunlight in grape canopies. Fruit sunburn can be minimized by leaving several leaves to shade each cluster.

Pruning wounds also serve as entry points for pathogens of diseases such as fire blight of pome fruits, cane blight of raspberry, dying arm of grape, canker diseases of roses, and others. This problem is eliminated or greatly reduced by disinfecting pruning tools to prevent spread of pathogens, pruning during times when pathogens are inactive (usually very cold or very warm periods), and proper pruning to hasten callus formation and rapid healing. Use of wound dressings, protective paints, pastes, and the like has been subject to considerable debate, but can do little or no harm.

Winter pruning is recommended for control of diseases such as pear fire blight and grape dying arm because pathogens are less likely to be active then. However, olive knot bacteria are more likely to be spread by winter pruning because bacteria are released through freeze-induced cracks in bark, hence spring pruning is recommended for olive knot control. Late pruning of hops reduces severity of downy mildew in regions with dry summers such as those in central Washington. Early pruning stimulates new growth when weather is favorable for disease. Pruning several weeks later than usual allows new shoots to develop during dry weather and shortens time for infection.

Thinning removes individual plants or groups of plants to create less dense stands. Thinning is sometimes accomplished by roguing infected plants to leave healthy stands that yield normal crops. Roguing rice plants with tungro and potatoes with leaf roll significantly reduces these diseases. Roguing row crops for disease control is probably feasible only for seed crops. Thinning is done mostly to improve plant growth and quality, like well-formed carrots, larger melons, and well-colored tomatoes. Disease control is probably coincidental with improved quality.

WATER MANAGEMENT

All plants, even xerophytes, need water to grow. No matter how wet a climate might be, supplemental water is often applied at some time in cropping cycles. Sometimes water must be removed to make soil suitable for plant growth. Application of water often plays a major role in control (or increase) of plant diseases. Within limits, growers can regulate amount of water, when, how, and how often it is applied, and reuse.

Proper water management benefits plants and reduces disease in several ways. Good management prevents moisture stress on plants, facilitates nutrient uptake, avoids waterlogging, delays onset of wilt symptoms, stimulates production of adventitious roots, and speeds postirrigation drying of foliage. Water management includes timely application of water for plant growth, flooding soil for irrigation or fallowing, and drainage to improve soil aeration.

Irrigation

Amount of irrigation water applied is normally controlled by farmers and can have a major influence on disease severity. In arid regions take-all of wheat is a problem only in irrigated fields whereas dryland footrot occurs only in dry soils. Less irrigation can reduce the former and additional irrigation can reduce the latter. Irrigation can be regulated to control diseases such as white mold, gray mold, and many foliar diseases caused by bacteria and fungi. Heavy irrigation saturates soil profiles and intensifies soilborne diseases.

Daily soaking of fairy rings in lawns is an effective control. The affected area is perforated every foot with a spading fork or metal rod to a depth of 6–8 inches and the area soaked every day for a month, preferably during mid-summer. The first application should contain a nonionic detergent (i.e., a wetting agent) to facilitate wetting of soil and organic debris. The beneficial effects are presumed to be threefold. First, toxic materials such as hydrocyanic acid excreted by the causal fungus are diluted and leached out of the root zone. Second, the water-impervious barrier produced by the fungus is eliminated. Third, the fungus does not survive in water-saturated soil.

Type of irrigation also has a marked influence on some diseases. Sprinkler irrigation intensifies diseases two ways. First, splashing water spreads pathogens, and second, prolonged wetting creates conditions favorable for infection. Sprinkler irrigation results in serious losses from bacterial diseases such as halo blight of beans and is particularly damaging if the crop is being grown for seed. It also increases peanut diseases caused by *Sclerotinia*, *Pythium,* and *Cercospora*. Sprinkler irrigation combined with contaminated irrigation water is a contributing factor in diseases such as bacterial stalk rot of corn and sprinkler rot of peaches and pears. Irrigation schedules can be modified to decrease time foliage remains wet, the main reason why sprinkler irrigation tends to promote foliar diseases.

However, sprinkler irrigation is preferred over surface irrigation for controlling Fusarium pod rot of peanuts because it better controls the amount of water applied and speeds drainage and drying of topsoil. Infrequent irrigation with adequate amounts of water is better than frequent irrigation with smaller amounts because the topsoil, where the pods form, dries between irrigations.

Overtree sprinkler irrigation increased pear scab in Oregon to twice the level of undertree irrigation. Misting apple trees to delay blossoming as a means of frost protection resulted in nine times more fire blight than in nonmisted trees. Similar increases of fire blight from sprinkler irrigation were also reported in pears. When overtree sprinkler irrigation became prevalent in Washington state orchards concern was expressed about potential increases in apple and pear scab and fire blight. However, observations revealed that powdery mildew was less severe under sprinklers than with surface irrigation. Control of powdery mildew by sprinkling has been known for a long time. It is a common and practical control for powdery mildew on roses. There are two reasons and both render spores ineffective. Conidia are simply washed off of plants and being very hygroscopic they absorb free water until they rupture.

Furrow (rill) irrigation greatly reduces incidence of diseases on aerial parts but can increase severity of soilborne diseases. Sclerotinia rot is often more severe where foliage is in prolonged contact with continuously wet soil. In some ways furrow irrigation is more difficult to regulate than sprinklers. Water collects in low areas and at ends of runs. Debris- or soil-blocked furrows result in pooling of water on one side and dry areas on the other. These extremes favor diseases of one sort or another.

Drip or trickle irrigation probably has the least effect on plant diseases and would be the best type to use in order to avoid disease. However, cost of establishing and maintaining drip irrigation systems probably is too great to justify their use, except for high value crops like tree fruit or vine crops.

Reuse of irrigation water is a common practice and poses many risks for consumers downstream because pathogens are spread in contaminated water. Root-knot, alfalfa stem, and other nematodes, onion white rot fungus, *Verticillium, Phytophthora,* and many other pathogens are spread in contaminated water. *Phytophthora cactorum* attacks more than 60 genera of plants and moves in irrigation water. *Phytophthora lateralis* spreads to its host, Port Orford cedar, via rivers and streams. The onion white rot fungus spread rapidly in the Walla Walla area because of movement of irrigation water from one field to another. The small sclerotia are well suited for dispersal by water. Irrigation water from these onion fields eventually drains into the Columbia River. In one instance a vegetable grower about 100 miles downstream had a sudden unexplained outbreak of white rot in bunching onions. This grower occasionally pumped water from the Columbia River for irrigation and probably picked up the fungus in that way.

The burrowing nematode moves 200 feet downhill in 1 year but only 25 feet uphill. This suggests that nematodes are being carried by percolating water. Greater long-distance spread of root-knot and root-lesion nematodes has been demonstrated through waste irrigation water. Numbers of nematodes decrease markedly beyond pools or reservoirs where they settle out of suspension.

Flooding

Flooding is used to irrigate some crops, especially rice, and is used in California to reduce populations of *Verticillium* in cotton fields rotated with rice. Flooding for 17 weeks during warm spring and summer months eradicates the fungus. Flooding is, or has been, used to control several important plant diseases, particularly Fusarium wilt of banana and many nematode diseases. Periodic flooding is a standard practice in cranberry bogs primarily to facilitate harvest of the berries. Rose bloom is controlled by flooding for 36–40 hours when diseased shoots reach full size and before spores are discharged. Presumably the fungus is killed because of lack of oxygen. A similar explanation is given for control of many soil fungi by flooding, but lengths of flooding periods are much longer, usually weeks or months.

Flooding is the main irrigation method for taro; fields are flooded to a depth of at least 2 inches before planting. Plantings also are flooded to a depth of 1–2 inches while the setts (corms) are rooting. Wet-grown taro remains flooded until harvest. If the water is warm and becomes stagnant Pythium soft rots can be severe. But flooding discourages southern blight, which often starts at the waterline of corms instead of at their base. Flooding taro fields is standard practice to control southern blight in Hawaii.

Drainage

Soil drainage eliminates or reduces conditions that favor some diseases, especially those caused by phycomycetes. Collar rot of apple, Pythium root rot and damping-off of many plants, Aphanomyces root rot of peas, and Phytophthora root rots of ornamentals are just a few diseases that are controlled by good soil drainage. In Japan and China timing drainage of irrigation water is part of management to control rice blast. Paddies may be drained anytime from 1 week to 40 days before heading, but when drained too early there is excessive drying of soil. Nitrogen availability, excessive transpiration, and increased nutrient uptake in dry soil affect rice physiology so that plants are more susceptible to blast. Drainage should be delayed until 7–10 days before harvest when blast is present.

NUTRITION (FERTILIZATION)

Farmers can influence diseases by type and amount of fertilizer they apply. Application of fertilizers for plant nutrition has at least four general effects on plant disease development and severity.

1. Some nonparasitic diseases result from nutrient deficiencies. Iron, zinc, boron, and manganese deficiencies cause specific diseases such as chlorosis of cherry, rosette of apple, blossom blast of pear, and gray speck of oats. Applying the proper balance of nutrients corrects these deficiency diseases.
2. Nitrogen fertilizers can stimulate production of juvenile growth, which favors diseases such as powdery and downy mildews. Lack of nitrogen leads to formation of senescent tissue, which favors diseases such as Alternaria, Botrytis, and Sclerotinia blights.
3. Altering soil pH by some fertilizers was discussed earlier.
4. Certain nutrients, especially phosphorus and potassium, promote tissue maturation, increase plant hardiness, and reduce diseases such as snow mold and take-all of wheat.

Green manures and other organic amendments improve soil structure, build up nutrients, and suppress certain pathogens. However, in some cases heavy applications of green manures or certain crop residues increase diseases, especially when following crops are planted before the green manure crops decompose. Oats cut for hay and left on the ground for any length of time greatly increase numbers of *Fusarium culmorum* chlamydospores in soil. This results in greater losses from foot rot in a wheat crop that follows. If potatoes are planted the same spring that residues of alfalfa, clover, sugar beets, and possibly other crops are incorporated into soil, *Rhizoctonia* may be severe. This probably results from buildup of the fungus on plant residues similar to what occurs in bare patch of wheat.

Mulches of hardwood chips provide a food base for *Armillaria mellea* and can result in severe root rot in trees and shrubs in contact with these mulches. Woody mulches around fruit trees and tea plants increased spread of Rosellinia root rot in Japan, whereas legume cover crops reduced incidence of the disease.

Sudangrass and rape produce hydrocyanic acid and isothiocyanate, which are toxic to fungi and nematodes. Green manures of sudangrass markedly reduce populations of *Verticillium* microsclerotia in soil. Several legumes including peas, beans, soybeans, clover, and sweetclover are used as green manures to control soilborne diseases, including take-all of wheat, Phymatotrichum root rot, potato scab, and *Rhizoctonia*.

A wide range of organic amendments have been applied to soil, often to control plant parasitic nematodes. Many are also effective against fungi. Animal manures, sewage sludge, waste plant materials (straw, sawdust, bark, peelings, etc.), and "cake" remaining after oil is pressed from peanuts, cottonseed, soybeans, and other oil seeds often significantly reduce nematode damage. Some organic amendments can also spread pathogens. Sclerotia and spores of some fungi pass through digestive tracts of animals and remain viable. Baled cereal straw used as mulch for wheat, barley, and rye is believed to have provided inoculum that led to an outbreak of Cephalosporium stripe.

CULTIVATION

Cultivation loosens soil around plants and serves several functions. It improves aeration and water penetration, and controls weeds. It reduces losses from some diseases such as Fusarium root rot of beans by hilling ridges of soil around plant stems. This stimulates production of adventitious roots to replace those lost to disease.

Movement of soil around plants causes "dirting," which consists of loose mounds of soil that function as moist chambers around plant stems creating an ideal environment for development of some fungi. Strawbreaker foot rot of wheat was greatly intensified by spring applications of shanked-in fertilizers around fall seeded plants. The culprit was not the fertilizer but simply mechanical stirring of soil. Cultivation of peanuts also causes dirting and increases losses from Sclerotium stem rot.

Careless cultivation injures roots and stems and provides entry points for pathogens. Crown gall is caused by soilborne bacteria that commonly enter through wounds, and airborne bacteria frequently enter through injuries on leaves, stems, and fruit. Many root, wood decay, canker, and blight fungi enter through wounds.

Frequent cultivation eliminates weed hosts of pathogens and insect vectors and is an effective means of reducing inoculum produced on alternative hosts. It also hastens breakdown of diseased crop residues and hence eliminates or reduces inoculum.

WEED CONTROL

Weeds affect plant health directly and indirectly. Direct effects include competition for nutrients and moisture, and allelopathic influences. Indirect effects involve modifying the physical environment by excluding light and raising relative humidities by restricting air movement. Weeds also serve as alternative hosts for many pathogens and as reservoirs of viruses and vectors.

Many pathogenic fungi, including both facultative and obligate parasites, have wide host ranges that include many common weeds. For example, in the United States *Verticillium dahliae* has been reported on more than 120 genera of plants, and *Rhizoctonia solani* on 500 or more genera. Some foliar pathogens also have wide host ranges. *Erysiphe cichoracearum* occurs on 135 genera. Root-lesion and root-knot nematodes parasitize and reproduce in many plants, and weeds serve to maintain high nematode populations. But the champion by far, considering its restricted geographic range, is probably *Phymatotrichum omnivorum* (a fitting name), which occurs on over 2000 plant species in more than 400 genera. Presence of weeds increased the incidence of *Phytophthora cinnamomi* in container-grown azaleas even when the fungicide metalaxyl was applied to the soil.

Some pathogens have relatively few weed hosts but the few can constitute a considerable threat to susceptible crops. X-disease mycoplasma occurs widely in chokecherry, its principal wild host. By removing chokecherries from the vicinity of peach and cherry orchards losses from X-disease can be significantly reduced. Carrot thin leaf virus occurs in poison hemlock and wild carrots. These weeds may be primary sources of inoculum in some areas and should be destroyed where they constitute a threat to commercial carrot fields. Cabbage black rot bacteria frequently are found in cruciferous weeds such as bird rape, wartcress, and peppergrass. These are alternative hosts and should be controlled in and around crucifer seed production fields.

From a disease standpoint, weeds probably are most important as reservoirs of viruses and their vectors. Viruses such as pea streak and pea enation mosaic also occur in other legumes including weeds such as hairy vetch, black medic, and sweetclover. However, these hosts are not important sources of infection because the aphid vectors prefer to feed on alfalfa. Other viruses such as beet curly top and tomato spotted wilt infect many plants in many families of angiosperms. For example, tomato spotted wilt virus infects 113 species in 35 families under field conditions, and 85% of these hosts also serve as breeding sites for thrips that vector the virus. In some viral diseases such as wheat streak mosaic, which is vectored by the wheat leaf curl mite, host ranges of virus and vector are not necessarily the same, which is to say that some plants that host the virus are not hosts for mites, and vice versa. This virus, moreover, can be transmitted mechanically and vectors are not as essential for transmission as with many viruses.

Controlling weeds that serve as alternative hosts for pathogens or reservoirs for vectors is important in controlling many diseases. Weeds can be eliminated by chemicals, cultivation, and flaming.

CONTROL OF VECTORS AND DISSEMINATING AGENTS

Vectors of plant viruses include four groups of organisms: (1) insects, mainly aphids and leafhoppers but also thrips and whiteflies, (2) eriophyid mites, (3) nematodes of the dorylaimoid group, *Xiphenema, Longidorus,* and *Trichodorus*, and (4) soilborne lower fungi, particularly *Polymyxa* and *Olpidium*.

Disseminating agents of fungi and bacteria are generally beetles. Elm bark beetles disseminate the Dutch elm disease fungus, cucumber beetles carry cucumber wilt bacteria, corn flea beetles disseminate Stewart's wilt bacteria, cerambycid beetles carry pinewood nematodes, pine bark beetles spread blue stain fungi, and Colorado potato beetles spread wilt bacteria to solanaceous plants. These vectors and disseminating agents can be controlled in several ways.

Pesticide applications are the most effective way to control insects that spread plant pathogens. The chemicals used include both systemic and nonsystemic insecticides. Systemic insecticides like aldocarb and disulfoton do not necessarily prevent transmission of persistent viruses to the first feeding host after alighting, but they can prevent vector buildup and secondary spread. The elm beetle spray schedules that are part of most Dutch elm disease control programs are usually delaying actions that slow intensification of the disease.

Soil fumigation is the primary control measure for viral diseases vectored by nematodes and soil fungi. In some circumstances soil fungicides and nematicides also control these vectors but their effectiveness is more contingent on time of application and soil depth than are fumigants.

Pheromones are chemicals that attract specific insects. They are used in traps to collect and monitor insect movements and population thresholds. Their use for disease control is primarily to determine when to start insect spray programs.

Elimination of overwintering hosts of vectors can markedly reduce or delay spread of viruses from sources of infection at the start of a growing season. Infected seed tubers or volunteer potatoes are the source of potato leaf roll virus. Green peach aphids overwinter on peach and apricot trees and in spring acquire virus from these initial sources of inoculum. Since the virus is persistent, once the insects aquire virus they can transmit it to healthy plants as long as the aphids live. Spraying dormant peach and apricot trees in areas of potato production destroys many overwintering aphids, greatly reducing the number that can spread virus the following growing season. It is difficult, however, to pursuade homeowners to spray backyard peach trees to protect some distant potato farmer's crop from leaf roll.

SANITATION

Sanitation includes practices that remove and destroy sources of inoculum from individual plants such as a tree or shrub or from plantings such as orchards or fields. The basic strategy is to eliminate as much primary inoculum as possible thereby reducing infection in subsequent disease cycles. Some physical methods such as plowing, pruning, and burning were discussed earlier. Sanitation is accomplished in several ways.

Burying plant refuse or other sources of inoculum is a practical way to control many diseases. Burying renders inoculum unavailable, dissemination unlikely, and hastens destruction by soil organisms. Potato cull piles are frequent sources of late blight inoculum and should be destroyed before new growth begins in spring. Destruction of unharvested potatoes in fields was discussed under tillage but the tactics are different. Instead of being buried, these tubers are destroyed by freezing. Plowing infested corn residues to prevent stalk rot or wheat head blight, and shallow plowing to bury mint rust inoculum were also described.

Removal of infected plant parts is an important way to control diseases of woody plants. Pruning holdover fire blight cankers from apple and pear trees is generally done in winter. Infected branches are easy to locate in winter because leaves are retained on blighted shoots. Diseased branches should be cut at least 12 inches and preferably 15 to 18 inches below visible margins of cankers. This ensures removal of tissues containing bacteria. Prunings should not be left in orchards because bacteria remain alive and can be spread by rain and insects until branches dry. Excised branches should be destroyed by burying or burning them. Pruning tools should be disinfested between cuts, or at least between trees, by dipping them in dilute solutions of household bleach. This prevents spread of bacteria from tree to tree.

Pear scab survives on infected twigs and in diseased leaves. Partial control is obtained by pruning diseased parts. Canker diseases of ornamental plants are controlled to some extent by pruning infected branches. It is important always to remove and destroy diseased material. Removal of infected branches from trees infected with Dutch elm disease was discussed earlier and it should be added that destruction of this material is critical to successful control of the disease. Many communities have laws or regulations requiring removal and destruction of Dutch elm diseased trees and often specify how the material is to be destroyed. The main reason for removing dead or diseased elm branches is to eliminate breeding sites for elm bark beetles. Owners of large diseased elms often cut the trees for firewood. Unless the bark is removed, even from relatively small branches, beetles can breed in this cut wood. Material removed must be destroyed by burying or burning or, if saved for firewood, logs or blocks must be peeled to remove bark.

Hand picking of plant parts affected by azalea leaf gall, peach leaf curl, and corn smut is sometimes suggested for reducing inoculum. This usually is not practical but may be used where few diseased plants are involved.

Mechanical vine pulling of potato stems is a relatively new and novel attempt to eliminate tuberborne inoculum and reduce *Rhizoctonia* tuber infection. Apparently potato tubers are more susceptible to sclerotial formation at this stage if tops remain attached. The effect may be more on host physiology than destruction of inoculum.

Raking and burning leaves or other infected plant parts is a common recommendation for controlling foliar diseases such as cherry leaf spot, hawthorn leaf spot, poplar leaf blight, apple scab, and dogwood anthracnose. This method of disease control is more practical for home gardeners than for commercial farmers.

Field burning or flaming, despite many environmental problems, is an effective way to eliminate inoculum. Burning grass-seed fields is a standard practice in many areas to control several fungal, bacterial, and nematode diseases. Burning to control brown spot of longleaf pine and Cephalosporium stripe of wheat has been discussed. Burning rice stubble and straw effectively controls Sclerotium stem rot and rice stem nematode. Annual burning of plant residues increases control of ink disease of bulbous iris over that obtained with fungicides alone. Untreated iris yielded only 30% of the crop obtained by both burning and fungicidal sprays. Burning yielded about 40% and spraying 60% that of the combination treatments. In these examples burning relies on natural fuels and the fires are generally fast and of low heat intensity so that any material insulated by soil or not incinerated remains a source of inoculum.

Propane or oil burners are used to control several diseases and are practical as long as fuel costs are not excessive. Burners do not pollute as much as field burning. Burning plant debris beneath cotton plants with tractor mounted or hand-held butane flamers reduces fungal boll rots and increases cotton yields. Flaming potato stems before harvest greatly reduces numbers of *Verticillium* microsclerotia that are returned to the soil. Covered propane flamers are used to burn peppermint and native spearmint (not Scotch spearmint) in fall and spring to control rust. This destroys teliospores in fall and systemically infected (bull) shoots in spring that produce aeciospores. Because rust spores are airborne, flaming is most effective when done over large areas.

Propane or oil flamers are also used to destroy weeds along irrigation canals and ditches during winter when they are dry. Many of these weeds are alternative hosts for pathogens, especially viruses, and serve also as overwintering habitat for vectors.

HARVESTING AND UTILIZATION OF PRODUCTS

Maximum disease losses occur when nothing is salvaged from crops. One way to reduce losses is to harvest and utilize diseased plants in some way. Long before the nature of plant diseases was understood some diseased plants were being utilized.

Indians of Central America used immature corn smut galls as food since recorded time and continue to do so. In the 1800s and continuing into this century tulips infected with viruses were sold at premium prices as "Rembrandt" varieties. In the 1960s cherry growers of the Pacific Northwest were enthused about a new spur-type cherry. This tree would produce fruit in clusters and make harvesting easier, and less costly to growers. However, the pseudospur characteristic proved to be the effect of a virus that reduced internodal growth. This results in bunching of nodes where fruiting spurs develop. Needless to say, this "spur-type" cherry was a disappointment and was dropped.

Salvage of usable products from a diseased crop occurs in many forms and a few are described here. Using products from diseased plants can be beneficial and profitable in many cases. Decayed wood formerly was considered to be worthless, but wood decayed by white rot fungi still contains usable cellulose fibers for pulp and paper products. Wood with pocket rots and blue or other stains give wood unique patterns for decorative panelling. Such panelling often fetches the same or higher prices than sound wood of the same species.

Gray mold causes bunch rot of grapes, but mature grapes decayed by this fungus have modified sugar contents and are used for a special late harvest ("Noble Rot") wine. Apples and other fruit degraded by fungi or blemished are used for juice or animal feed.

Potato tubers of some varieties infected with leaf roll virus develop internal defects known as net necrosis. Net necrosis develops in storage and potatoes intended for processing are rejected because of it. If a farmer grows a net necrosis-prone variety, such as Russett Burbank, and knows that the crop has a significant amount of leaf roll, as determined visually or by serological or other assay method, the potatoes should go for fresh market or immediate processing.

Ergot is a disease of many grasses and cereals. The sclerotia are poisonous to humans and animals. Severely contaminated grain requires extensive cleaning by flotation, screening, or blowing to remove them. The sclerotia have value in the pharmaceutical industry and if available in sufficient quantities they can be sold to defray some of the losses and costs.

Leaf spots of alfalfa and other forage crops cause losses through premature defoliation. By mowing infected forage crops earlier than normal much of this foliage can be harvested. Although yields are less than desired quality is higher than if hay consists mainly of stems.

Harvesting is the final operation in the production cycle of a crop and the operation may damage harvested crops or future crops. Harvesting spreads pathogens. For example, stem nematode is spread in alfalfa hay and *Verticillium* is spread on debris mixed with alfalfa seed. Angular leaf spot bacteria exude from wet infected cucumber leaves and pickers should be kept out of fields during morning hours while vines are still wet. This greatly reduces spread of bacteria by workers.

Harvesting also injures plants. Perennial herbaceous crops such as alfalfa and clover are damaged if cut too close, or cut too late in the season. This causes major wounds in which facultative parasites become established and cause decay of crowns. Late cutting does not permit plants to produce enough food reserves to survive adverse winters. Mechanical harvesting also bruises fruits and vegetables and leads to decay. Cranberries are harvested either dry or wet. Dry harvesting rakes berries from vines with a comb-like device and little damage. These berries are marketed fresh. Wet harvesting involves flooding bogs and beating berries from vines. This bruises berries as well as damages vines. Because of injuries, wet harvested berries tend to decay and generally are frozen for processing. Bruises and cuts on potato tubers during harvest are primary entry points for tuber rots. Use of mechanical harvesting machines for caneberries, grapes, nuts, and similar crops can also damage trees or bushes and lead to invasion by weak parasites that otherwise would be unimportant.

REFERENCES

*Brown, R. H., and B. R. Kerry (eds.), *Principles and Practice of Nematode Control in Crops.* Academic Press, New York, 1987.

*Chaube, H. S., and U. S. Singh, *Plant Disease Management: Principles and Practices.* CRC Press, Boca Raton, FL, 1991.

*Curl, E.A., Control of plant diseases by crop rotation. *Bot. Rev.* 29:413–479 (1963).

*Eckert, J. W., and J. M. Ogawa, The chemical control of postharvest diseases: Subtropical and tropical fruits. *Annu. Rev. Phytopathol.* 23:421–454 (1985).

*Engelhard, A. W. (ed.), *Soilborne Plant Pathogens: Management of Diseases with Macro and Microelements.* APS Press, St. Paul, 1989.

*Hardison, J. R., Role of fire for disease control in grass seed production. *Plant Dis.* 64:641–645 (1980).

*Huber, D. M., and R. D. Watson, Nitrogen form and plant disease. *Annu. Rev. Phytopathol.* 12:139–165 (1974).

*Koepsell, P. A., and J. W. Pscheidt, *Pacific Northwest Plant Disease Control Handbook.* Oregon State University Cooperative Extension, Corvallis, OR, 1991.

*Miller, D. E., and D. W. Burke, Effects of soil physical factors on resistance in beans to Fusarium root rot. *Plant Dis.* 69:324–327 (1985).

*Moore, K. J., and R. J. Cook, Increased take-all of wheat with direct drilling in the Pacific Northwest. *Phytopathology* 74:1044–1049 (1984).

*Palti, J., *Cultural Practices and Infectious Crop Diseases.* Springer-Verlag, New York, 1981.

*Parkinson, V., Y. Efron, L. Bello, and K. Dashiell, Trap crops as a cultural measure in *Striga* control in Africa. *FAO Plant Protection Bull.* 35:51–54 (1987).

*Rotem, J., and J. Palti, Irrigation and plant diseases. *Annu. Rev. Phytopathol.* 7:267–288 (1969).

*Shipton, P. J., Monoculture and soilborne plant diseases. *Annu. Rev. Phytopathol.* 15:387–407 (1977).

*Sugar, D., and P. S. Lombard, Pear scab influenced by sprinkler irrigation above the tree or at ground level. *Plant Dis.* 65:980 (1981).

*Sumner, D. R., B. Doupnik, Jr., and M. G. Boosalis, Effects of reduced tillage and multiple cropping on plant disease. *Annu. Rev. Phytopathol.* 19:167–187 (1981).

*Thresh, J. M., Cropping practices and virus spread. *Annu. Rev. Phytopathol.* 20:193–218 (1982).

* Publications marked with an asterisk (*) are general references not cited in the text.

16 Biological Control

Biological control of plant diseases is an exciting strategy that has generated considerable interest, enthusiasm, optimism, and publications for 25 or more years. Since 1964 the *Annual Review of Phytopathology* has carried at least 22 articles with biological control as title or subject and now includes this category as a major heading. This review journal also distinguishes biological control from cultural control. There have been many papers, symposia, conferences, compilations, and books, especially with multiple authors, on this topic.

Biological control has attracted great interest because of increasing regulation and restriction of pesticides or unsuccessful control attempts by other means. It is especially attractive for soilborne diseases because these pathogens are difficult to reach with specific pesticides. Excessive use of broad spectrum or persistent chemicals may result in soil contamination, fungicidal resistance, or other harmful effects. However, despite much early optimism, as yet there are few established practical applications of biological control. Of the 15 biological controls profiled by Cook and Baker (1983) only four involve direct application of a biological control agent. The other 11 involve indirect efforts, and though they involve some form of biological mechanism, it is generous to interpret them solely as "applications" of biological control. Table 16.1 lists biological control agents that have reached commercial development and utilization. Some of these may no longer be used but there may be other materials in process by industry.

Parris (see Chapter 1 for reference) cites Aristotle as recommending "biological methods" of plant disease control as early as 384–322 B.C. but gives no specific details. These methods could be crop rotation, soil tillage, and roguing diseased plants, since these cultural practices are construed by some plant pathologists to be biological control measures themselves. Cook and Baker devote extensive space to historical accounts of what they consider to be biological control of plant diseases.

Biological control has been defined and conceptualized in various ways by different workers. Most general, as well as scientific and agricultural, dictionaries define it as "the reduction or elimination of plant or insect pests by means of parasitic organisms or of animals that prey on them." This definition follows closely the "classic" definition used by entomologists, such as DeBach (1964), who consider biological control a natural phenomenon consisting of "the action of parasites, predators, and pathogens in maintaining another organisms density at a lower average than would occur in their absence." Some other entomologists take a much broader view of biological control as the use of "any living organism that can be manipulated by man for pest control" (Beirne, 1967).

TABLE 16.1. Biological Control Agents That Are or Have Been Commercially Available

Control Organism	Disease Controlled
Viruses	
Mild tomato mosaic virus	Tomato mosaic
Mild citrus tristeza virus	Citrus tristeza
Mild papaya ringspot virus	Papaya ringspot
Bacteria	
Agrobacterium radiobacter K84	Crown gall
Pseudomonas fluorescens	Bacterial and fungal diseases of cereals, cotton and other crops
Streptomyces sp.	Fungus diseases of wheat, cucumber, carnation and crucifers
Fungi	
Trichoderma viride	Damping-off, root diseases, wood decays
T. harzianum	
T. hamatum	
Fusarium oxysporum	Fusarium wilt
Peniophora (Phlebia) gigantea	Annosus root rot
Pythium oligandrum	Damping-off
Paecilomyces lilacinus	Nematodes

Although the term biological control had been used by plant pathologists in the early 1900s there was no attempt to formally define the term until 1963 when Garrett (1965) opened a symposium by describing a biological control agent as "a living microorganism or macroorganism other than the diseased or damaged plant acting as host and the pathogen or pest causing the disease or damage." This narrow definition of a control agent would exclude many present examples of biological control in the broad context of Cook and Baker, such as cross-protection, hypovirulence,and host resistance. However, Garrett went on to define biological control as "any condition under which, or practice whereby, survival or activity of a pathogen is reduced through the agency of any other living organism (except man himself), with the result that there is a reduction in incidence of the disease brought about by the pathogen. Biological control can be brought about either by introduction or augmentation in numbers of one or more species of controlling organisms, or by a change in environmental conditions designed to favour the multiplication and activity of such organisms, or by a combination of both procedures." Despite the seeming contradiction of his opening statement with his proposed definition, it is clear from discussion that followed that Garrett did not consider "rotation, fertilizer practices, and adjustments in crop husbandry" to be biological control methods. Different participants at the symposium offered definitions or interpretations of biological control that ran the gamut from the narrow classical use of parasites and predators to everything else, including chemicals.

Baker and Cook produced the first comprehensive book dealing exclusively with biological control in 1974. This, and a revision (with authorship reversed) in 1983, gave broadest possible interpretation to biological control. To them "biological control is the reduction of inoculum density or disease-producing activities of a pathogen or parasite in its active or dormant state, by one or more organisms, accomplished naturally or through manipulation of the environment, host, or antagonist, or by mass introduction of one or more antagonists." This definition does not leave much out. An abbreviated version used by other plant pathologists is "the reduction of the amount of inoculum or disease producing activity of a pathogen accomplished by or through one or more organisms other than man" (Baker, 1985). If the phrase "other than man" excludes cultural practices and use of resistant cultivars included by Cook and Baker, many examples of successful biological control are eliminated. A recent view by Deacon (1983) equates biological control to microbial control and considers that the Cook and Baker definition "does little more than exclude some cases of chemical control."

It is not the purpose of this chapter to decide which of the various definitions or concepts is correct, but for simplicity only those controls involving direct application of an organism are included. Many of the cultural measures mentioned in Chapter 15 indirectly involve microorganisms and are considered to be biological controls by some writers. Despite the many definitions accorded biological control, it is a strategy for reducing disease incidence or severity by direct or indirect manipulation of microorganisms. The principle may be eradication or protection, depending on specific tactics involved in promoting biological control. Antagonists that produce antibiotics kill pathogens and eradicate them from a substrate. Some microorganisms occupy niches and exclude pathogens from becoming established, thereby protecting plants from infection.

As with many popular topics new terms are coined; the word "biocontrol" is one. This phonetically pleasing, and convenient, word appears to be a contraction of "biological control." "Biocontrol" appears in few dictionaries but is defined in the *World Book Dictionary* as "the control of the physical and mental processes of animals or humans by means of electrical signals transmitted into the central nervous system." By this definition the term is not applicable to plant pathogens. However, it is used more frequently in plant disease writing and like many words this new meaning may become accepted with usage.

Mechanisms by which biological control is accomplished have been presented and discussed in many ways but generally fall into two categories. One takes place within host plants and includes such specific activities as cross-protection, competition for sites, production of inhibitors such as phytoalexins or antibiotics, and hypovirulence. The second is some degree of antagonism directed at pathogens from outside the plant.

This includes antibiosis, competition for nutrients, and predaceous exploitation, including parasitism.

C. von Tubeuf in a 1914 publication titled "Biologische Bekampfung von Pilzkrankheiten der Pflanzen" apparently became the first on record to use the term "biological control" for plant disease. From 1901 until the 1930s, he used hyperparasitism in attempts to control white pine blister rust with purple mold, *Tuberculina maxima*, a parasite of *Cronartium ribicola*. Potter, in 1908, prevented bacterial soft rot of turnips and blue and green mold decays of citrus with culture extracts of the causal organisms, namely, *Erwinia carotovora*, *Penicillium italicum,* and *P. digitatum*, respectively. This was not antibiosis in the present sense but involved toxins produced by microoganisms.

TRADITIONAL FORMS OF ANTAGONISM

Direct Parasitism

Parasitism is direct utilization of one organism for food by another. The action is usually subtle and the parasite works within the host.

Hyperparasites are organisms parasitic on other parasites. Fungi parasitic on other fungi are referred to as mycoparasites. Many hyperparasites, some considered for biological control, occur on a wide range of fungi. *Darluca filum* (now *Sphaerellopsis filum*) was described by Saccardo as a parasite of some rust fungi, especially *Puccinia* and *Uromyces*, and *Cicinobolus* (now *Ampelomyces*) was described as parasitic on powdery mildews. Little interest developed in these hyperparasites except from a mycological view. *Tuberculina maxima* also was described by Saccardo as a parasite on pycnial and aecial stages of some pine stem rusts. When eastern white pine planted in Europe in the 1700s and 1800s became severely damaged by white pine blister rust, interest developed in using *T. maxima* as a control agent. By 1930 von Tubeuf reported that the aecial stage of *Cronartium ribicola* had been eliminated in some regions in Germany. However, attempts to inoculate white pine blister rust cankers in the United States were marginally successful and inadequate natural spread of the hyperparasite failed to produce any significant level of control. Attempts to use a related fungus, *Tuberculina persicina*, to control cotton rust have also failed because spread of the hyperparasite was too slow to interfere with the life cycle of the rust fungus.

Weindling in 1932 observed *Trichoderma lignorum* (now *T. viride*) parasitizing hyphae of *Rhizoctonia solani* and suggested inoculating soil with *Trichoderma* spores to control damping-off of citrus seedlings. This or other *Trichoderma* species were observed to parasitize *Rhizotonia bataticola* and *Armillaria mellea*. More recently *Trichoderma harzianum* and *T. hamatum* have been marketed as wound dressings for ornamental and forest trees and decay inhibitors for utility poles. There are many reports of *T. harzianum* controlling soilborne diseases of carnation, tomato, peanut, tobacco, and other crops, but so far commercial application has been limited.

Other fungi reported as invaders and colonizers, and presumably hyperparasites, of pathogenic fungi are *Coniothyrium minitans, Sporodesmium sclerotivorum, Gliocladium roseum*, a *Corticium* species, and an unidentified basidiomycete. Several are aggressive invaders of important sclerotial fungi, such as *Sclerotium rolfsii, S. cepivorum, Sclerotinia sclerotiorum, S. minor, Botrytis, Claviceps purpurea*, and *Rhizoctonia solani*. Yet, there has been little or no commercial exploitation of these hyperparasites.

Parasites of nematodes include viruses, bacteria, rickettsias, endoparasitic fungi, fungal parasites of eggs and cysts, and protozoans. Nematode trapping fungi parasitize nematodes after ensnaring them but because of the entrapment mechanisms, unique to these fungi, they are included under predation in the next section.

Few viruses and rickettsias have shown any promise as biological control agents. There are many reports of bacteria in nematodes but most may be symbionts and not parasites. *Pasteuria (Bacillus) penetrans* provides some control under greenhouse and field conditions. Its ability to attack important nematode pests of several crops, persist in soil for long periods, resist desiccation and temperature extremes, and to be compatible with several pesticides make it a good candidate for a biological control agent. However, its obligate nature, lack of mobility, and dependence on dispersal by water, animals, man, and cultivation are major disadvantages.

A wide range of fungi, including chytrids, oomycetes, zygomycetes, deuteromycetes, and basidiomycetes, colonize vermiform nematodes, sedentary female cyst nematodes, and eggs. Many of these appear to be merely opportunistic feeders and not good prospects as practical control agents.

Nematophthora gynophila is a recently described fungus that attacks and destroys nematodes cysts in early stages of development. It is the first well-documented example of long-term biological control of plant parasitic nematodes under natural conditions. This fungus is one of the most important parasites of the cereal cyst nematode and can kill female nematodes in less than 7 days resulting in stable decline of nematode populations. *N. gynophila* infects and colonizes female nematodes converting cysts into masses of resting spores, which are spread through soil by other organisms. The obligate nature of this fungus probably limits its potential for commercial application.

Paecilomyces lilacinus is the most promising fungus that parasitizes nematode eggs. It is most effective against egg masses such as those of root-knot and cyst nematodes because the eggs are exposed to attack and infection proceeds from egg to egg. It has given excellent control of citrus nematodes in Peru, reducing larval populations to almost zero 4 months after inoculation. It is as effective as soil nematicides, has given promising results elsewhere in the world, and is being produced commercially under the name BIOCON™ in the Philippines.

In addition to direct parasitism, *N. gynophila* produces a toxic reaction without penetrating eggs. *Penicillium analobicum* reduces golden nematode populations without parasitizing female nematodes. It produces compounds that alter permeability of eggshells and reduce nematode hatching.

Parasites of phanerogams are usually not considered disease control agents, but more likely weed control agents. *Wallrothiella* (now *Caliciopsis*) *arceuthobii*, a fungal parasite of dwarfmistletoe berries, has received some attention as a possible control agent but it probably has little practical value for dwarfmistletoe control. *Glomerella cingulata* seriously damages dwarfmistletoes in areas where there is reduced seed production and may be a better prospect than *Wallrothiella*.

Broomrape is a parasite of herbaceous plants such as sweetclover, strawberry, tomato, and tobacco. *Fusarium oxysporum* is used in Russia as a biological control agent and detroys up to 70% of broomrape seed, although its effectiveness is contingent on proper environmental conditions. *Alternaria cuscuticidae* is used also in Russia to control dodder, but its effectiveness may involve secondary invaders. The unpredictable benefits of these agents probably means that they are not very effective.

Predation

Predation also is utilization of an organisms for food by another, but the action is usually obvious and the prey attacked and consumed from without.

Nematode trapping (predaceous or nematophagous) fungi include several genera that have interesting capture mechanisms. Some have finger-like constricting loops made up of three hyphal cells that swell when a nematode passes through the loop (like inflating three long balloons tied together in a circle). The constricting loops hold nematodes while fungal hyphae penetrate the cuticle and invade the nematode body. Others have sticky knobs or form nets of hyphae that ensnare nematodes as they move through soil.

Attempts to enhance activities of nematode trapping fungi to control parasitic nematodes mostly have been disappointing. These fungi as a group are poor saprophytic competitors, intolerant of desiccation, and easily lysed. Some soils are very fungistatic to predaceous fungi.

Predaceous nematodes have either a large stoma (mouth) armed with teeth used to seize and rip open prey or swallow them whole, or a piercing tooth used to impale prey, immobilize (paralyze) it, and inject secretions to digest body contents. Several groups of nematodes are largely predaceous, but *Mononchus* and *Seinura* are genera most often mentioned. After many years of recognizing predation by these animals there still is little conclusive evidence that they are important in reducing plant nematode populations. One of their chief limitations is a general lack of specificity; also they are as likely to eat their own kind as they are plant parasitic nematodes.

Miscellaneous predaceous organisms include protozoans, springtails, and mites that feed on nematodes and other microorganisms. Most are omnivorous and have the same limitations as potential control agents as do predaceous nematodes. Furthermore, many of these organisms are present only in upper layers of soil and their impact on plant parasitic nematodes is apt to be negligible.

There has been some interest in predaceous (vampyrellid) amoeba, but they should be considered mainly a curiosity. These organisms dissolve holes in spores and hyphae of soil fungi but probably have little effect on ultimate numbers of pathogens.

Competition

Competition is an indirect effect whereby pathogens are excluded by depletion of food bases or by physical occupation of sites. Other mechanisms such as antibiosis and parasitism can also be involved since a single simple action in any biological system is unlikely. The following examples illustrate competition as biological control mechanisms.

Heterobasidion (Fomes) annosum colonizes stumps of freshly cut pine and other conifers and spreads via root grafts to adjacent trees where it causes a fatal root rot. Spraying freshly cut stumps with spore suspensions of *Peniophora* (now *Phlebia) gigantea* before infection by *H. annosum* preempts the site and the latter fungus cannot gain a foothold. Although there may be some antagonism between the two fungi the primary mechanism seems to be simply physical occupation of sites. This disease and the competing fungi are better suited to practical application of biological control than most. *Peniophora* produces masses of oidia in culture, which are packaged to prepare spore suspensions. Spraying fresh stumps with spores is standard practice in Great Britain and some other countries. Its use in the United States has been limited because borax stump treatment is as effective, cheaper, and lasts longer than *Peniophora* spores. Attempts to suspend spores in chain oil of power saws to facilitate inoculation were promising but less effective than spraying spores on stumps.

Another successful application of biological control is use of composts, a technique restricted mostly to greenhouse crops. Composts consist of a range of organic and inorganic materials that either contain antagonistic microorganisms or encourage rapid colonization by competing, and sometimes antagonistic, organisms such as *Trichoderma* species. Establishing competing organisms in composts physically and biologically precludes colonization by pathogens. Composts generally are effective against non-specific diseases such as damping-off and other seedling blights.

Antibiosis

Antibiosis is inhibition of an organism by a metabolic product (an antibiotic) from another. Many organisms, especially soil fungi and actinomycetes, produce antibiotic substances but their application to disease control has been questionable. There is no doubt that antibiotics function in localized niches in soil but so far practical application or exploitation of this phenomenon has been limited. Three examples of disease control by antibiosis include Armillaria root rot by *Trichoderma viride*, damping-off, stem and root disease by *Pseudomonas fluorescens*, and crown gall by *Agrobacterium radiobacter*.

Roots of trees infected with *Armillaria mellea* serve as food bases after trees die or are removed. Fumigation of citrus orchards affected by *Armillaria* does not kill the fungus directly. It weakens it and creates an environment that allows *Trichoderma* to colonize roots and, by direct parasitism and/or antibiosis, to destroy *Armillaria*. This is often cited as successful application of biological control but is an indirect method involving chemical control.

Several *Pseudomonas* species, especially fluorescent pseudomonads, have been associated with inhibition of several soilborne diseases. Examples include take-all of wheat and Rhizoctonia and Pythium damping-off of cotton and other plants. The active mechanism is antibiosis. Pseudomonads are one of relatively few biological (e.g., microbial) control agents that has reached commercial channels in the United States. It consists of masses of bacterial cells used to coat cotton seed and was produced under the trade name Dagger G™ but has been discontinued. A different strain of *P. fluorescens* has been tested for treatment of wheat seed to control take-all, but has not yet been applied commercially.

Crown gall, a bacterial disease of many woody plants, is a serious problem on fruit trees and grapevines. The causal agent infects through wounds, grafting and budding wounds being common points of entry. Coating new grafts with *Agrobacterium radiobacter* var. *radiobacter*, an avirulent strain, preempts infection courts and the virulent strain cannot become established. The mechanism is more than just physical occupation of sites. Some bacteria produce antibiotic substances called bacteriocins, a general term. Specific bacteriocins are produced by certain bacteria and act only against related strains. In the case of *A. radiobacter* the bacteriocin is called agrocin. There may be more than one specific bacteriocin. The one originally used for crown gall control was agrocin K84 and the bacterial strain was registered in the United States as Galltrol™. The product consists of a culture of bacteria that is suspended in a volume of water into which roots and lower stems of grafted fruit trees are dipped. Strain K84 does not work on grape crown gall because this disease is caused by a different biovar of the crown gall bacterium. In some regions the pathogen has developed resistance to this bacteriocin and a genetically modified strain, K1026, has been registered as NoGall™ in Australia. The overall importance of bacteriocin production by an avirulent strain to protect against a virulent strain is confounded by conflicting reports. Apparently *Erwinia herbicola* can protect against fire blight bacteria without producing a bacteriocin.

Hypovirulence

This term was first applied to isolates of chestnut blight fungus obtained from healing cankers in Europe. It has been defined by some workers as reduced ability of pathogens to cause disease in host plants that lack any specific resistance. Healthy chestnut trees inoculated with hypovirulent strains are protected from virulent strains. Hypovirulent strains carry some transmissible agent, possibly a mycovirus, that passes through hyphal connections (anastomoses) to virulent strains and essentially neutralizes their pathogenicity.

Hypovirulent strains have been inoculated onto trees scattered throughout forested areas in France and Italy. Natural spread accounts for inoculation and subsequent protection of other trees. Fungal spores are exuded in a gelatinous matrix and spread by insects, birds, and rain. Use of hypovirulence has not worked in the United States where chestnut blight virtually eliminated American chestnut as a commercial species within a few decades. Apparently the hypovirulent European isolates are not compatible with American isolates, which are extremely variable. Treatment of individual trees is feasible for this disease but, even with this advantage, successful application of hypovirulence as a biological control method has been limited.

Mycoviruses are found in most major fungus groups and hypovirulence has been reported in isolates of *Rhizoctonia solani, Bipolaris victoriae, Gaeumannomyces graminis*, and *Ceratocystis ulmi*, all important plant pathogens. Some studies suggest that hypovirulent strains of *R. solani* give no significant control of disease on potatoes. Dutch elm disease fungus is a likely candidate for this type of control since treatment of individual trees is feasible and acceptable. Despite the high value placed on ornamental elms and cost of other control measures such as vector control, chemotherapy, pruning, and severing root grafts, there is no indication that mycoviruses are even being considered for this use.

Cross-Protection

Cross-protection originally referred to protection of plants from viruses by prior inoculation with mild strains of the same virus. Cross-protection has since been extended to include preinfection with a mild strain of any pathogen, or even an unrelated organism or nonpathogen, to protect against infection by a more severe strain. While there are a number of potential candidates for this type of biological control only a few are now in commercial use.

The earliest and best known application of cross-protection is use of mild strains of tomato mosaic virus (sometimes considered tobacco mosaic virus (TMV) in earlier reports) to protect against infection by more severe strains. Since TMV is easily transmitted mechanically, workers in greenhouses or fields readily transmit the virus from hands and clothes to tomatoes. Inoculating tomato seedlings with a mild strain of TMV permanently protects plants against other strains. Because resistant tomato varieties are available this technique is now used primarily on tomatoes that have virtually no TMV resistance. When growers become complacent losses occasionally occur if susceptible varieties are not protected by inoculation with a mild strain.

Citrus tristeza is controlled in South America by grafting buds from trees containing mild strains of virus onto healthy citrus, especially trees on susceptible sweet orange roots. The strategy came from several observations. The disease was endemic, impossible to eradicate, and spreading rapidly. Losses were so great that some yield reduction from mild strains was acceptable. There was evidence that mild strains of the virus gave effective protection without causing undue harm. Sometimes there is a violent reaction when mild strains from one host are introduced into another citrus host. Generally, it works best when the source of the mild strain is from the same type of citrus, that is, sweet orange strain from sweet orange, grapefruit from grapefruit, lime from lime, and so on. Cross-protection did not always work when it was tried in Florida citrus groves. The reason for failure remains unknown.

Cross-protection of papaya ringspot virus has been used successfully in Hawaii and Taiwan. Papaya is propagated by seed and it is relatively simple to spray-inoculate large numbers of seedlings. Cross-protection of papaya ringspot virus in Taiwan sometimes fails when mild strains come from other regions, for instance Hawaii. Strains may be serologically indistinguishable but biologically different. Intense disease pressure from adjacent papaya plantings also contributes to incomplete protection.

Adequate numbers of plants must be inoculated with mild strains for cross-protection to be effective. Mechanically transmitted viruses such as tomato mosaic and papaya ringspot make the task simple so that an airbrush spray of a solution containing virus is all that is needed. Viruses transmitted only by insects or grafting are more difficult, time consuming, and costly. Cross- protection with these viruses is feasible only where individual plants have relatively high value, such as fruit trees.

Cross-protection using nonpathogenic strains of *Fusarium oxysporum* to control virulent wilt strains of *Fusarium* on sweetpotatoes has been reported in Japan. It has also been reported using nonpathogenic races of bean anthracnose fungus, and *Phialophora graminicola* to protect wheat against subsequent infection by the take-all fungus. However, the mechanism in these instances could also be competition for site or antibiosis, instead of cross-protection, which is generally restricted to viral interactions.

Cross-protection is not without limitations and risks. One of the dangers is that a more severe but unrecognized virus may be introduced into all individuals of a given crop. The mild strain also can spread to other hosts and may do serious damage. This is especially possible with viruses like TMV that have wide host ranges. Mild strains are mutated or attenuated forms and can revert or mutate to severe forms. This can cause considerable damage since the virus is now in all of the crop rather than just a portion of it as occurs in natural infections. An additional risk is the possibility of synergistic reactions in hosts that contain some other unrelated latent or mild virus. Synergism frequently results in greater damage than the cumulative effects of the two separate viruses.

DISEASE SUPPRESSIVE SOILS

Before biological control became a popular topic, many workers observed that some soilborne diseases, especially Fusarium wilts and a few others, were less severe or even absent from certain soils. Soon it became common to refer to soils as suppressive or conducive to disease. Suppressiveness has been recognized for Fusarium root rots of beans and cereals, take-all, Verticillium wilts, Rhizoctonia seedling blights, Pythium damping-off, black root rot of tobacco, common scab of potato, bacterial wilt of tomato, and club root of cabbage.

While there may be some differences of opinion as to what constitutes disease suppressiveness there are several general ways that suppressive soils differ from conducive soils.

1. Pathogen populations decline more rapidly under natural conditions in suppressive soils.
2. Suppressive soils require higher amounts of added inoculum to induce disease.
3. Lower germination of fungus propagules (chlamydospores, zoospores, conidia, sclerotia, etc.) and slower mycelial growth in suppressive soils.

Suppressiveness is biologic in nature and moderate heat or chemical treatments of soil eliminate the suppressive factor. Furthermore, adding small quantities of suppressive soil induces suppressiveness in conducive soils. Disease decline phenomena of several soilborne disease, best known being take-all of cereals, develop with continuous monoculture. Evidence indicates that decline results from increases of microbial populations antagonistic to pathogens. Included in antagonism are lysis and fungistasis of fungal propagules, two activities that may be interrelated. Since these activities are controlled largely by soil conditions it may be possible to enhance suppression by manipulating cultural practices. This is the basis for many of the reports of successful biological control.

FUNGISTASIS

Fungistasis is a common phenomenon characterized by reduced germination and growth of fungi in soil. Two general mechanisms, namely, deprivation of nutrients and toxic chemicals from microorganisms or plant roots, have been proposed to explain fungistasis. High levels of fungistasis have been reported in suppressive soils and suggested as a primary mechanism for suppressiveness. However, some think just the opposite, that fungistasis is a survival mechanism that permits pathogen propagules such as sclerotia, chlamydospores, and oospores to survive in absence of hosts. Whatever the relationship, fungistasis appears to be of biological origin and is enhanced by incorporating organic amendments into soil.

Lysis

Lysis is a general term for the destruction, disintegration, dissolution, or decomposition of biological materials. It has been observed in both mycelium and spores of soil fungi, including plant pathogens, but the mechanism is not well understood and its significance in disease control is uncertain. It may be one component of suppressive soils and disease decline. Lysis is often associated with depletion of soil nutrients, stimulation of germination, or colonization by saprophytic bacteria.

Autolysis also occurs in which the mechanism of lytic action comes from the fungal cell itself. The enzyme chitinase is involved in some autolysis. Several attempts have been made to exploit the action of chitinase as a lytic principle in disease control by adding chitin-containing amendments, such as crabshells, to soil. The strategy is to stimulate organisms that produce chitinase, which in turn lyse cell walls of pathogenic fungi. These attempts have been largely unsuccessful.

DISEASE DECLINE

Disease decline is characterized by diseases that gradually increase to a peak as a result of repeated cropping, a monoculture, and then decline to some stable level (Figure 16.1). Decline appears to be associated with an increase of antagonistic organisms that constitute a form of biological control. Take-all decline is the best known and understood, but similar declines have been reported for necrotic ringspot of turfgrasses, Rhizoctonia root rot of sugarbeet, and common scab of potato.

A requirement for disease decline is continuous cropping with the same host. Some nonhost crops interrupt and counteract monoculture buildup of antagonistic organisms. Rotation crops such as alfalfa in wheat rotations cause complete reversal of decline.

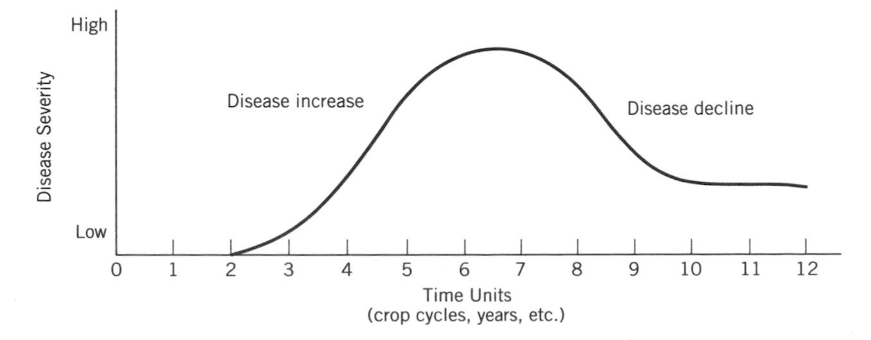

Figure 16.1. Pattern of disease development in a monoculture that results in disease decline.

Suppression is of two types, general and specific. General suppression results from activities of soil microbiota competing for nutrients and perhaps producing antibiotics. It is not transferable. Although there are varying views as to what constitutes specific suppression it is considered to be transmissable. Transmissability of suppressive or decline factors suggests that if environmental conditions controlling these phenomena are understood it is possible to maintain a crop monoculture with minimum disease loss.

MYCORRHIZAE

Mycorrhizae are symbiotic associations between fungi and plant roots. Mycorrhizae literally means fungus roots and have long been known to be important to health of forest trees. Mycorrhizae on tree roots are usually ectotrophic and consist of a fungal mat or mantle over the root surface and an intercellular network of hyphae in the root cortex called Hartig's net. Fungal components of ectotrophic mycorrhizae are usually basidiomycetes.

Endotrophic mycorrhizae are largely intracellular and are found in many herbaceous plants. These are sometimes called vescicular–arbuscular mycorrhizae (VAM) because vescicles and arbuscules develop in host cells. Vescicles are round or oval and may be propagative structures. Arbuscules are much branched structures that appear to function as nutrient-absorbing organs like haustoria. Most VA mycorrhizal fungi are phycomycetes in the genus *Glomus*.

Mycorrhizae are considered to be normal components of healthy roots and plants that lack them may be the exception. General benefits of ectomycorrhizae include increased surface area of roots with fungal hyphae extending into adjacent soil; increased uptake of nutrients such as nitrogen, phosphorus, potassium, and calcium; increased carbon dioxide production, which in turn increases solubility of elements such as calcium, magnesium, iron, and potassium; and production of growth-promoting substances such as thiamin and nicotinic acid. Disease control benefits derive from healthier, more vigorous plants.

Mycorrhizae also provide direct protection against infection by root pathogens. Three mechanisms have been proposed for this action in ectomycorrhizae: (1) mycorrhizal mantles provide mechanical barriers to invasion, (2) the fungal symbionts produces antibiotics, and (3) fungal and root excretions such as carbohydrates support protective microflora in the rhizosphere.

Non-antibiotic-producing ectotrophic mycorrhizal fungi such as *Pisolithus tinctorius*, one of the more common fungal symbionts, provide barriers against infection of pine roots by *Phytophthora cinnamomi*. The mantle is effective as a barrier only when intact over root surfaces. Hartig's net forms an internal barrier that prevents spread of pathogens beyond areas of mycorrhizal development in root tissue.

Antifungal and antibacterial substances are produced by many mycorrhizal fungi. Such fungi are included in the following genera; *Amanita, Boletinus, Boletus, Cantharellus, Cenococcum, Clitocybe, Clitopilus, Collybia, Cortinarius, Hebeloma, Inocybe, Lactarius, Lepiota, Leucopaxillus, Paxillus, Rhizopogon, Rhodophyllis, Russula, Scleroderma, Thelephora,* and *Tricholoma*. Antibiotics appear to translocate or diffuse from mycorrhizal roots to adjacent nonmycorrhizal roots, which also escape infection by *Phytophthora cinnamomi*.

Populations of rhizosphere microorganisms are 10 times or more prevalent near mycorrhizal roots than nonmycorrhizal roots, probably in response to nutrients excreted by mycorrhizae. Many of the rhizosphere microorganisms are fungi; mostly nonpathogenic fungi on mycorrhizae, while pathogens such as *Pythium, Fusarium*, and *Cylindrocarpon* predominate in nonmycorrhizal rhizospheres.

VA mycorrhizae lower the incidence or severity of Fusarium wilt of tomato, black root rot of tobacco, pink root of onion, and root knot on several crops. However, they have a higher incidence of Verticillium wilt of cotton and Phytophthora root rot of soybean. Since VA mycorrhizae do not form mantles nor do they produce antibiotics their mode of action in reducing plant diseases is unknown. They may produce sugars, amino acids, or other compounds that stimulate beneficial organisms that inhibit plant pathogens.

Mycorrhizae of both types are practical for application to plant disease control. They provide plant growth benefits and can be established by inoculation. Most ectotrophic mycorrhizal fungi can be grown in culture and inoculation is fairly simple. VA mycorrhizae cannot yet be grown in culture and require special inoculation techniques using infested soil or macerated infected roots.

COMMERCIAL USE OF BIOLOGICAL CONTROL

Despite all of the enthusiasm and optimism about biological control of plant diseases, using introduced microorganisms has resulted in relatively few commercial applications. This does not suggest that the general principles or strategies of biological control are faulty. It is likely that technologies are incomplete or imperfect, that industry is reluctant to develop this technology, or growers hesitate to apply these new methods.

Biological control agents must satisfy several requirements before they are likely to be developed and marketed. These can be considered in four general categories: economic, environmental, technical, and regulatory. First there must be a need that generates adequate markets, just as there must be a demand for any successful product. Products must be technically feasible. A major criticism of many biological control systems is that they work from a scientific standpoint but compete poorly with other control measures. Control agents must be highly consistent performers. Few growers are willing to risk a crop on an uncertain procedure if something better is available. The product must offer some advantage in performance and cost. But the greatest economic handicap of most biological control agents is their specificity. Too often they target only a single or a few related pathogens. This means that range of application is extremely narrow, without sufficient promise to cause a rush of investors.

Environmental demands are among the main reasons that biological control has gained much of its present support. However, if a potential biological agent appears to jeopardize environment or safety it will not be accepted. Agents should be indigenous microbes so that danger of introducing potentially damaging organisms is nil. Organisms should have saprophytic abilities so that they can maintain populations during periods when target pathogens are absent or they should have resting stages that ensure survival.

Saprophytes also have the advantage of being produced without hosts. They are also likely to be more stable than strictly parasitic organisms. One problem with cross-protection is that viruses are unstable and often mutate to virulent forms. Some bacteria may not survive long enough in commercial formulations to allow storage for more than short periods (short shelf life). This is a significant disadvantage compared to conventional fungicides which can be stored for years under suitable conditions.

Products should be adaptable to conventional and available technology. The need for expensive equipment to apply biological control agents is an obstacle to their use. Those requiring extra or special operations such as budding operations in cross-protection of citrus tristeza, or inoculation of individual trees with hypovirulent pathogens, will be limited to crops of high value.

Biological control agents are subject to the same registration requirements of EPA as pesticides. In the United States release of living microorganisms as "biofungicides" requires authorization from APHIS, either in the form of a permit or a statement that a permit is not required. When organisms have been modified in some way by DNA manipulation they are subject to additional restrictions. For example, introduction of a genetic marker in *Pseudomonas fluorescens* considered for take-all control required extraordinary precautions and statements that a permit was not needed for field testing. In another case, modified *P. fluorescens*, introduced experimentally into an elm tree for Dutch elm disease control caused a minor furor. Such cases generate considerable publicity, intentionally or otherwise. It is unlikely that organisms attacking or adversely affecting plant pathogens will have a bad effect on humans or animals. Exceptions might include endophytes that produce harmful reactions in animals, or fungi that produce mycotoxins. Nevertheless, the general public perceives modified organisms as posing an extraordinary threat. The Toxic Substance Control Act (TSCA) allows EPA to regulate production and application of chemical substances that present an unreasonable risk of injury to health or environment. Microorganisms deliberately altered to contain genetic material from dissimilar organisms are defined as chemicals under TSCA.

The scientific procedure involved in developing biological control agents is fairly straightforward and involves at least six steps.

1. Select potential biological agents by mass isolation of organisms from either target pathogens or their environment, or cultivating a single known organism.
2. Screen agents to further determine that they have properties requisite for acceptable biological control agents, such as production of propagules, longevity, and compatibility with pesticides and other chemicals.
3. Characterize agents to identify those features that make particular organisms candidates for development. Cook and Baker have done this in their Chapter 9 titled "Antagonistae Vitae," a valuable addition to their first book.
4. Maintain a required level of quality control. Many reports in the voluminous literature reflect great variability and inconsistency of biological systems. It would be worse if some standard is not maintained.

5. A delivery system to apply the agent must be developed. This is probably as important as finding effective control agents for unless a grower can apply agents there might as well be none.
6. Finally, products must be tested under field conditions. One of the chief criticisms of scientific reports of disease control measures, of all types and not just biological control, is that they may work in the laboratory and greenhouse but when applied in the field they fall short of hopes and expectations. Field tests are sometimes manipulated with extra or extraordinary procedures that exagerate beneficial effects.

Because biological control is relatively new and includes direct involvement of living organisms, other than host and pathogen, it is beset by more problems than are standard control measures. Growing pains perhaps. Many are difficult to formulate commercially for efficient application or consistent benefits. They may require special handling since some agents mutate and lose their efficacy. Or they may not be compatible with pesticides, fertilizers, growth regulators, or adjuvants used in agriculture.

Perhaps the greatest impediment to commercial development of a biological control agent is the limitation on patenting microorganisms. The organism itself cannot be patented, but strains having unique properties such as specific antibiotic production, the process of producing or applying, or a particular use can be patented. It is likely that most future biological control organisms will be modified by genetic engineering in some way rather than released as wild types. This enhances their chances for a patent. Without a patent and license to protect its investment commercial institutions are unlikely to be interested in undertaking development and marketing costs of biological control agents.

REFERENCES

Baker, R., Biological control of plant pathogens: Definition. In Hoy, M. A., and D. C. Herzog (eds.), *Biological Control in Agricultural IPM Systems*. Academic Press, New York, 1985. Pp. 25–39.

Beirne, B. P., Biological control and its potential. *World Rev. Pest Control* 6:7–20 (1967).

*Bruehl, G. W., *Soilborne Plant Pathogens*. Macmillan, New York, 1987.

*Burge, M. N. (ed.), *Fungi In Biological Control Systems*. Manchester University Press, Manchester, UK, 1988.

*Campbell, R., The search for biological control agents against plant pathogens: A pragmatic approach. *Biol. Agric. Hortic*. 3:317–327 (1986).

*Chet, I. (ed.), *Innovative Approaches to Plant Disease Control*. John Wiley, New York, 1987.

Cook, R. J., and K. F. Baker, *The Nature and Practice of Biological Control of Plant Pathogens*. American Phytopathological Society, St. Paul, MN, 1983.

Deacon, J. W., *Microbial Control of Plant Pests and Diseases*. American Society of Microbiology, Washington, D.C., 1983.

DeBach, P. (ed.), *Biological Control of Insect Pests and Weeds*. Reinhold, New York, 1964.

Garrett, S. D., Toward biological control of soil-borne plant pathogens. In Baker, K. F., and W. C. Snyder (eds.), *Ecology of Soil-borne Plant Pathogens.* University of Calififornia Press, Los Angeles, 1965. Pp. 4–17.

*Hornby, D. (ed.), *Biological Control of Soil-Borne Plant Pathogens.* CAB International, Wallingfors, UK, 1990.

*Hoy, M. A., and D. C. Herzog (eds.), *Biological Control in Agricultural IPM Systems.* Academic Press, New York, 1985.

*Klingmuller, W. (ed.), *Risk Assessment for Deliberate Releases.* Springer-Verlag, New York, 1988.

*Marks, G. C., and T. T. Kozlowski (eds.), *Ectomycorrhizae: Their Ecology and Physiology.* Academic Press, New York, 1973.

*Mukerji, K. G., and K. L. Garg (eds.), *Biocontrol of Plant Diseases.* 2 Vols. CRC Press, Boca Raton, FL, 1988.

*Scher, F. M., and J. R. Castagno, Biocontrol: A view from industry. *Can. J. Plant Pathol.* 8:222–224 (1986).

* Publications marked with an asterisk (*) are general references not cited in the text.

17 Pathogen-Free Seed

Planting pathogen-free seed is one of the first steps farmers can take to ensure healthy crops. Certification programs ensure that seeds are true to type (i.e., a variety is what it is claimed to be), have a high germination rate, have a good seed size to produce vigorous seedlings, and are free of pathogenic organisms. The latter is an important measure to prevent introduction of pathogens into fields and regions. Some methods of producing pathogen-free seed are discussed in this chapter.

Seeds comprise the largest group of agricultural commodities. Wheat, rice, corn (maize), and barley dominate the crops grown throughout the world. Legumes (pulse crops) are lower in total production, but are important seed crops worldwide. These include many kinds of beans, peas, lentils, soybeans, chickpeas, and lupines. Although peanuts (groundnuts) are legumes and important commodities in warm areas of the world, they are generally classed as oil crops instead of pulse crops.

Seeds vary greatly in size from almost microscopic seeds of orchids to very large seeds of coconut. It is surprising that seedborne pathogens have not been reported on orchids, the largest plant family. The many compilations do not record any seedborne pathogens on orchid seeds but this probably reflects the fact that only one orchid, *Vanilla*, is used for food, and that most commercially grown orchids are propagated vegetatively. In addition, extremely small size and low food reserves in orchid seeds make them poor substrates for fungi and bacteria.

Seeds are important for several reasons. Many can be used directly or converted to other products; they yield proportionally higher amounts of dry matter in relation to total weight than fruits and vegetables, which have higher water contents, they can be stored for long periods, and they can be used for propagation. Their use in propagation carries risk of spreading plant pathogens and an important strategy in disease control is production and use of pathogen-free seed. Seed pathology is a distinct applied branch of the seed industry involving two broad areas. Seed pathology covers the entire range of subjects including causal agents, infection processes, transmission, morphological and anatomical considerations, epidemiology, and control. Seed health testing involves primarily detection and identification of seedborne pathogens.

Plants now grown originated in isolated and specific locations and hence widespread movement of seeds has distributed pathogens to many parts of the world. Of the approximately 200,000 species of plants 300 species are widely grown for food and 16 of these comprise 90% of the world's food supply. These include wheat, rice, corn (maize), barley, potatoes, soybeans, sugarcane, sorghum, oats, millet, sweetpotatoes, cassava, sugar beet, rye, peanuts, and bananas. All but five of these are propagated from true seed. In addition to these food crops there are many forage and ornamental crops such as alfalfa, clovers, and grasses that are grown from seed. Furthermore, even though a plant is propagated primarily by vegetative means in commercial agriculture, for example strawberry, potato, pineapple, cassava, and banana, seeds are still important in breeding programs to improve plants for agricultural purposes. Figure 17.1 shows areas of the world where certain crops originated. It is noteworthy that North America was largely devoid of major crop plants. Its contribution to world agriculture has been primarily small fruits such as strawberry, raspberry, blueberry, and cranberry.

Seed pathology and seed health programs have fostered a number of institutes devoted to this branch of plant pathology. The first and best known is that in Denmark established by Paul Neergaard as the Danish Government Institute of Seed Pathology for Developing Countries. This institute has been deeply involved in training plant pathologists and seed technologists from all parts of the world. It has been particularly active in assisting developing countries to develop and implement seed health testing programs for use in seed certification and plant quarantine.

Seed production is a high value branch of agriculture. A crop grown for seed may produce twice the return or more than the same crop grown for market. Seed production is a major industry in regions where climates are warm and dry and less conducive to fungal and bacterial diseases. Statistical reporting of seed crops by the USDA was discontinued in the 1980s and it is now difficult to obtain seed production information. However, some estimates are available. Table 17.1 gives amounts of various seeds exported from the United States in 1989-1990. This does not indicate amounts of seed used for domestic plantings but the value of exported seed alone exceeds half a billion dollars annually. Table 17.2 shows the annual value of seed crops in Washington state for the period of 1983–1987. The tables illustrate the high value of seed crops. Other states and regions specialize in production of different types of crop seeds such as sunflower, soybean, corn, sorghum, and peanut from central and southern United States.

TYPES OF SEEDBORNE PATHOGENS

Pathogens are associated with seeds at one of four levels.

1. Pathogens distributed coincidentally with seeds but not organically attached to them. Such pathogens often are in plant debris or soil mixed with seeds. Examples are sclerotia of ergot with cereals, *Sclerotinia* with bean seeds, and *Verticillium* mycelium on plant residues mixed with alfalfa seeds. This type of association probably constitutes the largest group of seed transmitted pathogens.
2. Pathogens on seed surfaces. These pathogens are usually readily controlled by chemical seed treatments. Examples are smut spores on cereal seeds and bacterial spot of tomato and pepper.

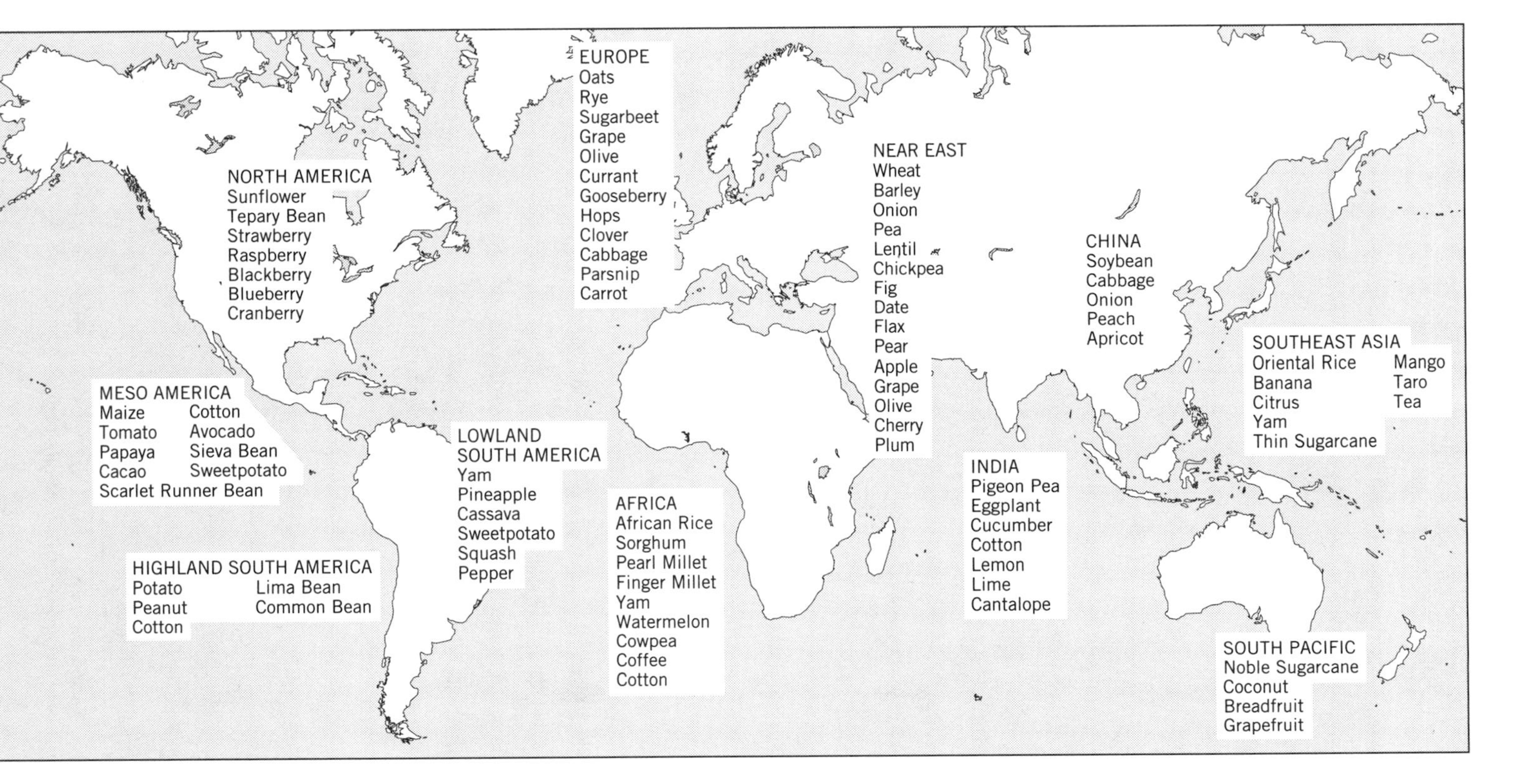

Figure 17.1. Areas of early domestication and centers of origin of crop plants.

TABLE 17.1. Seed Exports from the United States (1989–1990)

Crop	Yield (lb.)	Value ($)
Forage crops	150,355,951	104,350,368
Legume vegetables	101,449,077	36,139,509
Vegetables (except legumes)	66,775,529	134,869,137
Field crops	908,968,927	261,183,969
Flower seeds	776,855	12,399,323
Tree and shrub seeds	490,805	3,153,418
Others	9,844,204	13,112,237
	Total export seed value	565,207,561

Source: Compiled from 1991 *International Seed Directory*, Seed Trade News, Eden Prairie, MN.

TABLE 17.2. Average Annual Seed Crop Production in Washington State (1983–1987)

Crop	Acres	Farm Value ($)
Peas	11,142	3,064,000
Beans	2,403	1,090,000
Clover	3,200	1,487,000
Alfalfa	18,000	10,537,000
Turfgrass	25,000	12,500,000
Field grasses	4,500	3,150,000
Vegetables		
Beet and swiss chard	540	838,770
Cabbage, broccoli, etc.	1,778	1,855,508
Carrot, parsnip, dill, etc.	4,747	4,380,644
Onion	320	528,990
Radish	4,376	2,968,537
Spinach	2,689	1,812,512
Miscellaneous	120	59,800
Total vegetables	14,570	12,445,000
Total Washington	133,283	65,273,000

Source: From Gabrielson, R. L., and G. Q. Pelter, Vegetable seeds for the nation and the world. *Washington's Land and People* 3 (l):2-7 (Spring 1989)

3. Pathogens inside seed coats but not inside the seed itself. The pathogen may be passively present having entered through breaks or natural openings or actively present by colonizing seed coats and sometimes producing fruiting structures. Examples of this group are *Fusarium* species in asparagus seeds, *Phoma lingam* in cabbage seeds, and *Ascochyta lentis* in lentil seeds.
4. The most specialized and difficult pathogens to eliminate are those inside seeds and established within embryos or endosperm. Examples are loose smut of barley, endophytes in fescues, and some seedborne viruses.

Effects of seedborne pathogens on plant health vary widely. Some pathogens such as *Gloeotinia temulata*, the cause of blind seed of fescue, kill seeds as they develop. Others such as *Phoma medicaginis*, the cause of spring black stem of alfalfa, kill developing seedlings shortly after germination. Several internally borne pathogens such as loose smuts of cereals show no adverse effects until plants reach maturity. *Epichloe typhina* and some other endophytes have no outward effect in some grasses. In most cases, however, seedlings developing from infected seeds have lower vigor and are subject to other diseases.

Many compilations have been made of seedborne diseases. Some of these lists include any pathogen ever associated with seed. Others are more restrictive and include only pathogens of a certain type such as viruses or on specific crops such as beans, peas, or legumes as a group. Some of these lists can mislead by suggesting more than a simple coincidental occurrence of pathogens with seeds. Table 17.3 gives some examples of pathogens in different associations with seeds.

TABLE 17.3. Examples of Diseases Where Pathogens and Structures Are Present in Various Associations with Seeds

Coincidental mix
Microslerotia of *Verticillium*
Sclerotia of *Rhizoctonia, Sclerotium, Sclerotinia, etc.*
Ergot sclerotia
Nematode cysts
Smut balls
Rust pustules
Blind seed of ryegrass
Tobacco mosaic virus
Fusarium solani chlamydospores with bean debris
On the seed surface
Smut spores
Rust spores
Tomato bacterial canker
Bean halo blight
Yellow slime of wheat

(continued)

TABLE 17.3. Continued.

Fusarium spores on asparagus and china aster
Stem nematode
Phoma medicaginis var. *pinodella* on peas
Angular leafspot of cotton
Angular leafspot of cucumber
Bacterial spot of tomato
Shallow penetration of seed coat
Black leg of crucifers
Downy mildew of lima bean
Tomato bacterial canker
Rice white tip nematode
Karnal smut of wheat
Bean anthracnose
Within the seed
Under the seed coat
Fusarium spp.
Black rot of cabbage
Late blight of celery
Bacterial blight of carrot
Black chaff of wheat
Heterosporium on nasturtium
Ascochyta in lentil and chickpea
Tobaco mosaic virus
In the embryo
Loose smut of wheat and barley
Anther mold of red clover
Epichloe typhina and other endophytes
Lettuce mosaic virus
Pea seedborne mosaic virus
Barley stripe virus
Peanut stripe virus

SEEDBORNE PATHOGENS

Fungi

As described above, fungi have been associated with seeds in several ways. Many are loosely associated with seeds and present in or on contaminating plant debris or soil. Relatively few are internally transmitted, but many are transmitted on seed surfaces. Downy mildew fungi produce oospores in plant tissues that become mixed with seeds.

Some powdery mildew fungi appear to infect seeds but are not visible until seeds germinate and produces seedlings.

Bacteria

Many bacteria produce gelatinous exudates in plant tissues. This material smeared on seed surfaces dries to form a durable adhesive film in which bacteria can survive for long periods. These pathogens are commonly transmitted on the seed surface. Many are also internal. Seed-transmitted bacterial diseases include bacterial blights of bean, bacterial blight of pea, black rot of crucifers, and black chaff of wheat.

Viruses

Approximately 10% of the known viruses are seedborne to some extent. Tobacco mosaic virus can be present in plant debris mixed with seeds, on seed surfaces, or in seed coats. Many, such as bean yellow mosaic, barley stripe mosaic, and lettuce mosaic, are in embryos or endosperm and can be transmitted in a high percentage of seeds. Seedborne viruses are especially prevalent and important in legumes.

Nematodes

Most nematodes are soilborne, but a few, such as stem nematode, rice white tip nematode, wheat gall nematode, grass seed nematode, and coconut wilt (red ring) nematode, can be carried with or on seeds. None is transmitted internally in viable seeds. Nematode cysts can contaminate seed lots, but since most cysts are produced on roots this association is not common.

DETECTION OF SEEDBORNE PATHOGENS

Much of the work involved in seed pathology tests seeds for presence of certain pathogens, and is an important component of seed inspection and certification programs. These tests are many and varied, sometimes involving general searches for any obvious signs of disease and othertimes very specific searches to detect latent or invisible pathogens.

Direct examination of seeds with or without staining detects many pathogens. These examinations may simply involve placing seeds in a dish or on a white surface and observing for discoloration, surface defects, fungal structures, or other indications of pathogens. Sclerotia of *Claviceps* or *Sclerotinia* mixed with seeds, smut balls, and nematode galls require little or no magnification to be seen. Pycnidia, acervuli, sporodochia, cirrhi, and other fungal structures require low power magnification. A survey for black point in wheat seed fields involved examining seeds incubated on water agar by means of a stereomicroscope. Many fungi, including *Helminthosporium* and *Alternaria*, that developed on these seeds were readily identified using a stereomicroscope at 30× magnification.

Scanning electron microscopy can reveal fungal and bacterial cells lodged in crevices on seed surfaces (Figure 17.2). However, mere observation cannot always determine if these are pathogens or simply saprophytic transients. Transmission electron microscopy is more complicated and requires more preparation, but is a standard technique for direct observation of viruses in plant material.

Barley or wheat embryos suspected of containing loose smut are extracted, softened by boiling in KOH, sectioned or squashed, and examined for dark mycelium of the fungus. Grass seed endophytes are detected by partially digesting seeds in nitric acid and staining with anilin blue. The material is then examined microscopically to detect the small diameter, convoluted mycelia characteristic of endophytic fungi.

Standard isolation techniques using damp blotter paper, water agar, potato dextrose agar, or other nonselective medium is a common procedure for detecting fungi and bacteria in many seed testing programs. Many organisms, especially fungi and nematodes, can be identified without further treatment. Most, however, require isolation and culturing for identification. Some advanced level of knowledge and training is essential. Often these organisms are sent to taxonomists of bacteria and fungi.

Selective or semiselective media are used to detect some bacterial and fungal pathogens in or on seeds. These media usually contain specific growth factors or inhibitory agents that favor specific organisms or groups of organisms. Seeds are either plated directly, with

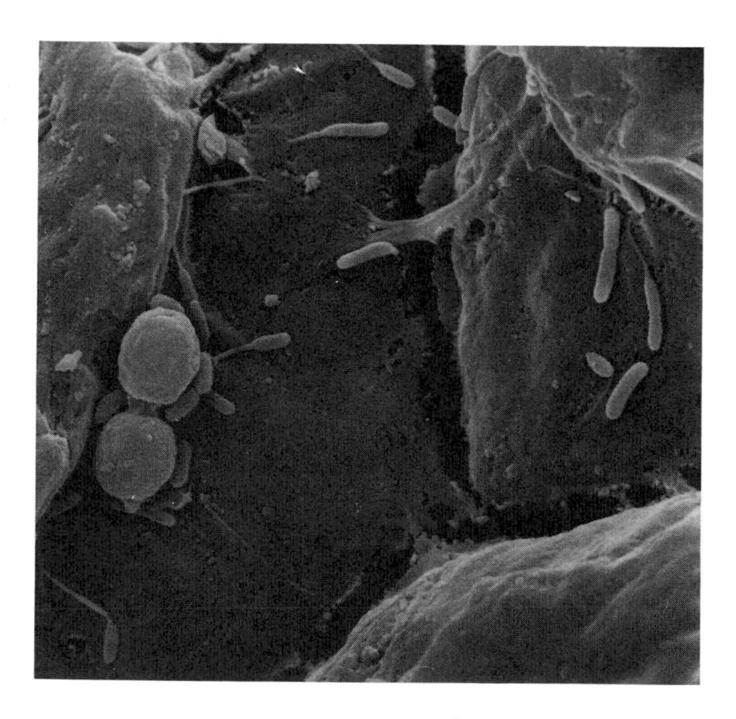

Figure 17.2. Fungal spores and bacteria in a crevice in an asparagus seed coat (7000X). (From D. A. Inglis, The Association of Two *Fusarium* Species with Asparagus Seed. M.S. Thesis, Washington State University, Pullman, 1978. With permission.)

or without surface sterilization, or soaked in sterile water, the soak or wash water clarified by centrifugation, and the pellet used to inoculate selective media.

Selective media are used to detect several species of *Fusarium* on wheat, corn, and other grain seeds, *Phoma* on sugar beet seeds, and *Pyricularia* and *Drechslera* on rice seeds. Selective media are also available for isolating plant pathogenic bacteria such as *Agrobacterium*, *Clavibacter*, *Erwinia*, *Pseudomonas*, and *Xanthomonas*.

Indexing or grow-out tests (sometimes called "growing-on" tests) require growing large numbers of plants from a seed lot to detect low levels of infected seeds. The test sometimes involves only pregermination of seeds and inoculation of suitable indicator plants. These tests aim at detecting one infected seed out of 10,000 or more and are costly because of the space required to grow and maintain large numbers of plants and time involved in conducting the tests. Several examples illustrate this method.

Grow-out test. Before serological tests were available all lettuce seed lots for use in commerial lettuce areas of California were indexed for lettuce mosaic virus by a grow-out test. This test required 30,000 seeds from each lot to establish a zero index or tolerance. The seeds were grown in insect-proof greenhouses in groups of 200 seeds in each of 150 boxes per seed lot. The plants were removed and examined for mosaic symptoms after 20 days. Cost of the test came from acreage assessments of each lettuce grower, and all growers in an area used those seedlots with a zero index.

Seed soak assay. Bacterial blights of beans can be detected in water in which seeds have been soaked. Seeds from test lots are surface disinfested in hypochlorite solution and then soaked in sterile water for 18–24 hours. The soaking period allows for multiplication of bacteria from within seeds. Soak water is decanted and vacuum infiltrated into presoaked seeds from the same seed lot. Inoculated seeds are incubated 12–13 days and seedlings examined for symptoms. Isolations are made from lesions on seedlings; once the bacteria are identified they are tested for pathogenicity on greenhouse-grown bean plants. This technique extracts and increases relatively low seed inoculum to a level that can be detected. There are several modifications of this technique involving shorter soak times and plating of leachate on selective media, and direct inoculation of bean seedlings with leachate.

Direct seed assay. A modified direct seed assay was used in the Pacific Northwest to index pea seed for pea seedborne mosaic virus before a suitable serological test (ELISA) was developed. This technique was necessitated by lack of time and space to grow pea seedlings for direct observation or index seeds directly by inoculation onto susceptible pea varieties. In this method 300 pea seeds selected from each seed lot were soaked in 10-seed units in water for 24–48 hours. Imbibed seeds were then macerated and used to inoculate leaves of *Chenopodium amaranticolor*, an indicator plant. Presence of a chlorotic spot on the inoculated leaf seven days after inoculation indicated presence of the virus. Infected seed lots were discarded.

Blotter technique. An outbreak of black leg of cabbage in central and eastern United States in 1971 was traced to infected seed from Washington state, a major crucifer seed-producing area. A single infected seed in a seed lot can result in high levels of infection on seedlings in transplant beds because of spore dispersal by water. A blotter technique was developed for detecting the pathogen in crucifer seeds. To eliminate infected lots or provide for their treatment 10,000 seeds are indexed from each lot. If the seed lot is relatively small, breeder seed for example, only 1000 seeds are indexed. Seeds are surface disinfected with hypochlorite to eliminate contaminating fungi that grow over the seeds and mask the presence of black leg infections. Surface sterilized seeds are placed on blotter paper moistened with 0.2% 2,4-dichlorophenoxyacetic acid to inhibit seed germination. The trays of seeds are incubated at 20–25°C under light to stimulate sporulation of the pathogen. Trays are examined twice, at 7 and 11 days, for a slight pink discoloration of the blotter around seeds. This color results from fungal spores exuded from pycnidia on seeds.

Ergosterol extraction has been used to detect *Drechslera graminea* in barley seeds. Ergosterol is a common sterol in many fungi. The amount of ergosterol present is a reliable indicator of infection by the fungus. Sterols are extracted from seeds by standard chemical procedures and the amount of ergosterol determined by liquid chromatography based on UV absorbance. The technique probably does not distinguish between *Drechslera* and other fungi like *Ustilago nuda* that may be present in seeds.

Serological tests

Specific antibodies produced in animals, usually rabbits, react with their specific complementary antigen. The antigen is usually a virus or in a few cases plant pathogenic bacteria. There has been little success in developing antisera to fungi because of the production of antibodies to extraneous proteins that mask specific reactions. The reaction of antigen (the material being tested) with antiserum is detected in a variety of ways and there may be modifications and combinations of different methods. All of the methods require some degree of purification of antigen so that antiserum will not react with extraneous proteins. Some of these tests can give false positive readings and require caution in interpreting results. In the ELISA test, for example, the end reaction is a yellow color which can vary in intensity.

Enzyme-linked immunosorbent assay (ELISA) has become one of the most widely used serological tests for detecting viruses, especially for many viruses undetected by other serological methods. It also is used to detect blight bacteria in bean seeds. ELISA is rapid and sensitive, but complicated, and requires considerable labor. It involves coating wells of a polystyrene plate with antiserum for a specific pathogen and rinsing off the excess. The antigen to be tested is then coated onto walls of the cells and excess rinsed off. An enzyme is then added to the wells and rinsed off. Finally this is followed by a substrate that binds to the enzyme. The enzyme and substrate used for plant pathogen ELISA usually are from an animal system while those used to detect animal viruses are from plants. This prevents antibodies from reacting with normal plant (or animal) proteins. The function of the enzyme and substrate is to produce or enhance the color reaction.

ELISA has been used to detect barley stripe mosaic, and as a result of seed testing incidence of this virus in Montana was reduced from 40% of seed lots infected in 1966 to less than 1% in 1985. ELISA is also used to detect bean common mosaic, alfalfa mosaic, soybean mosaic, tobacco ringspot virus in soybean, lettuce mosaic, and blight bacteria in bean and cassava seeds.

Radioimmunosorbent assay (RISA) is similar to ELISA but uses radioactive gamma globulin as the substrate and measures the amount of radioactivity adsorbed to the antiserum–antigen sandwich. This method has been used to detect lettuce mosaic and soybean mosaic viruses. It is a quantitative assay, the amount of radioactivity being proportional to virus concentration.

Serologically specific electron microscopy (SSEM) is used to detect a number of seedborne viruses. Among these are bean common mosaic, lettuce mosaic, tobacco ringspot, soybean mosaic, and pea seedborne mosaic. A copper electron microscope grid is coated with dilute virus antiserum and then floated on an extract of the seed to be tested. Virus particles react with the antibodies and adhere to the grid. The grid is then treated with a shadowing reagent (uranyl acetate), washed, dried, and examined by transmission electron microscopy for virus particles.

Double diffusion is a technique where a circle of several wells around a center well is cut into an agar surface. Antiserum is placed in the center well and antigens (seed extracts) to be tested in encircling wells. Antigen and antiserum diffuse toward each other. Where there is a positive reaction a white precipitation line forms in the agar between antiserum (center) and antigen (outer) wells indicating a relationship between the two components. This method has been used for several viruses in seeds, including barley stripe mosaic, soybean mosaic, and cowpea mosaic. A modification using blotter paper disks instead of wells is used as a routine test for barley stripe mosaic in some areas.

Indirect immunofluorescence (IF or IIF) staining is used to detect common blight bacteria in bean seeds. Surface sterilized seeds are soaked for 24 hours and the leachate containing bacterial cells is concentrated by centrifugation. Cells in the concentrate are smeared on a microscope slide, treated for 30 minutes with antiserum, and finally with a second antiserum conjugated with a fluorescein stain. This precedure (abbreviated here) binds the fluorescing material to suspect bacterial cells. When examined microscopically using a mercury vapor lamp to excite the stain, cells that fluoresce brightly are considered bean blight bacteria. Smears can be counterstained with rhodamine to avoid misleading readings in plant fragments, but this is not always necessary. The main advantage of this method is that it allows distinction of bacterial cells from other fluorescing objects on the slide.

CONTROL OF SEEDBORNE PATHOGENS

Physical and chemical seed treatments were described in earlier chapters but these are directed at seeds already carrying pathogens. This chapter is concerned primarily with measures that preclude presence of pathogens on or in seed which eliminates the need for seed treatment. Production of pathogen-free seed is usually accomplished by combinations of several general procedures including locating seed production areas in climates unfavorable to disease development, management practices that reduce disease incidence, seed certification, and quarantines. Quarantines were discussed in Chapter 7.

Location of Seed Production

Many seed companies in the United States locate production fields in the west where growing seasons are relatively dry. Moisture is the major factor contributing to infection by foliar fungi and bacteria. Locating seed production fields where irrigation is necessary gives growers control of water management. There is a general recommendation in the United States to use bean and pea seed produced west of the Continental Divide where hot and dry summers inhibit bacterial diseases and others. Even the Puget Sound area of Washington state with its maritime climate has relatively dry summers. This is a major crucifer seed area producing about 80% of the crucifer seed and 30% of the world's supply. Southern Idaho produces a significant amount of bean, pea, and sweet corn seed. Oregon produces a large amount of grass seed in the Willamete Valley. California produces a variety of seeds in inland regions where climates are dry. The Piedmont region of southeastern United States produces cotton seed relatively free of fungal contaminants because of low rainfall during critical growth stages. Similar situations prevail in Europe where the cleanest seed comes from dry areas.

Even in relatively dry regions some local sites should be avoided where microclimates might favor disease development. Low lying or shaded areas, soils that remain exceptionally wet, and areas protected from winds where relative humidities remain high all favor bacterial and fungal pathogens.

Management Practices

Many cultural practices influence plant disease, some of which have been discussed in Chapter 15. A few of these are very important in seed production.

Crop rotation is necessary to reduce survival of pathogens in or on crop debris from one growing cycle to the next. Some pathogens survive for very long periods and crop rotation may not be a realistic control measure for such pathogens. Three to four years between susceptible crops is usually adequate to reduce populations of pathogens such as *Phoma lingam*, *Xanthomonas* species, *Pseudomonas* on peas, *Colletotrichum* on bean, and *Septoria* on lettuce, peas, and tomato. However, some pathogens such as stem nematode, *Fusarium oxysporum* in flax and peas, and wheat dwarf smut fungus may require 5 or more years without hosts for populations to drop to insignificant levels.

Distance between fields is important in many certification schemes, as is spacing between plants within a field. Crowded plantings promote disease in two ways. One is by increased contact between plants, which enables pathogens to spread from plant to plant. The other is creation of humid microclimates favorable to pathogens and disease development. Wider plant spacing allows better air movement through plant canopies and more light to reach soil surfaces, both of which create drier, warmer environments.

Weed control is important for several reasons. Eliminating them as alternative hosts for pathogens and vectors is an important factor in producing pathogen-free seed.

Seed should be harvested soon after it matures when weather is dry. In area where disease is present in only some fields these should be harvested last to avoid spreading pathogens to clean fields on harvesting equipment.

Seed storage can also influence survival of pathogens in or on seed. Both length of storage (aging of seed) and conditions of storage are involved. Some bacterial pathogens like those that cause angular leaf spot of cucumber, leaf and fruit spot of tomato and pepper, leaf spot of sesame, leaf spot of celery, and wildfire of tobacco and soybean survive for less than 2 years. Storing seed for this time is a feasible means of eliminating these pathogens from seed. Most seedborne viruses, fungi, and nematodes survive in or on seed too long to consider storing seed as a means of control.

Pathogens that require longer storage periods also are not likely to be controllable by this method because of reduced germination of seed. There are many reports that aged seed is detrimental to germination and seedling survival. Effect of aged seed on seedling vigor and relative susceptibility to facultative (weak) parasites is also well established.

Conditions under which seed is stored also affect pathogen survival. Cucumber angular leaf spot bacteria may survive for more than a year (but less than two) but if seed is stored for several days at temperatures as high as 50°C with high relative humidity the bacteria are killed or their numbers greatly reduced. Similarly, several fungi including *Drechslera* and *Fusarium,* are killed in barley seeds which have 14% moisture content when stored at 20°C for several months but survive in seeds with 10-12% moisture. Seeds with moisture content higher than 14% rapidly lose germinability.

Certification

Seed certification is a procedure that regulates quality of seed placed on the market. More than 30 kinds of seeds are certified in the United States. There are four categories of seed in certification programs. *Breeder* seed is the initial source or nucleus of a genotype and is controlled by the originating or sponsoring plant breeder or institution. It produces *Foundation* seed grown under supervision of the certifying agency and used to produce *Registered* seed, which also must be approved by the certifying agency. *Certified* seed can be produced from breeder, foundation, or registered seed and is approved and certified by the appropriate authority.

A number of qualitative parameters are involved in seed certification, but this book is primarily concerned with relative freedom from plant pathogens. Use of certified seed is an important method for reducing plant disease losses. The main consideration for control of seedborne diseases through certification is the degree that disease levels are allowed, or tolerated, in a seed lot. Disease tolerances are most stringent (lowest) in foundation and registered seed and somewhat more relaxed (higher tolerances) in certified seed. The level of tolerance is usually based on whether pathogens are exclusively seedborne or coincidentally carried with seeds. Diseases that are exclusively or mostly seedborne can often be eliminated by using pathogen-free seed.

Tolerance is also determined by the correlation between level of seed infection and amount of disease that develops in the field. Diseases such as halo blight of bean, black leg and black rot of crucifers, or lettuce mosaic can be extremely serious, even when levels are as low as one seed per 10,000. These diseases require zero tolerance to be controlled effectively by seed certification. Other diseases such as smuts of wheat, white mold of sunflower and other crops, and Ascochyta blight of peas require fairly large amounts of inoculum on or with seed to result in appreciable field disease. These have tolerances of 5% or more.

Some other requirements for seed certification were discussed in Chapter 7 and are only briefly reviewed here. At least two inspections are required but others may be required under some circumstances. Amount of disease allowed declines with subsequent inspections, so that second inspections permit less disease than the first. There may also be stipulations as to cropping sequences and time intervals between successive crops. Type of culture might by regulated, such as disallowing sprinkler irrigation for seed crops. Buffer zones between seed crops and other seed or commercial fields of the same crop may also be required.

REFERENCES

*Agarwal, V. K., and J. B. Sinclair, *Principles of Seed Pathology*. 2 Vols. CRC Press, Boca Raton, FL, 1987.

*Agarwal, V. K., C. N. Mortensen, and S. B. Mathur, *Seed-Borne Diseases and Seed Health Testing of Rice*. CAB International Mycological Institute, Kew, Surrey, UK, 1989.

*Anon., International symposium on seed pathology. *Seed Sci. Technol.* 11:455–1364 (1984).

*Baker, K. F., Seed pathology. In Kozlowski, T. T. (ed.), *Seed Biology*. Academic Press, New York, 1972. Pp. 317–416.

*Kaiser, W. J., Testing and production of healthy plant germplasm. Tech. Bull. No. 2. *The Danish Institute of Seed Pathology for Developing Countries*, Copenhagen, Denmark, 1987.

*Kulik, M. M., New techniques for the detection of seedborne pathogenic viruses, viroids, bacteria and fungi. *Seed Sci. Technol.* 12:831–840 (1984).

*McGee, D. C., Seed pathology: Its place in modern seed production. *Plant Dis.* 65:638–642 (1981).

*Nasser, L. C., M. M. Wetzel, and J. M. Fernandes (eds.), *Advanced International Course on Seed Pathology*. Brasilia, D.F. Abrates, 1987.

*Naumova, N. A., *Testing of Seeds for Fungous and Bacterial Infections*. Translated from Russian by Israel Program for Scientific Publications, Jerusalem, 1972.

*Neergaard, P., *Seed Pathology*. 2 Vols. Macmillan Press, London, 1979.

*Richardson, M. J., *An Annotated List of Seed-Borne Diseases*. Commonwealth Mycological Institute, Kew, Surrey, UK, 1979.

*Saettler, A. W., N. W. Schaad, and D. A. Roth (eds.), *Detection of Bacteria in Seed and Other Planting Material*. APS Press, St. Paul, MN, 1989.

* Publications marked with an asterisk (*) are general references not cited in the text.

18 Pathogen-Free Vegetative Propagative Material

Many plants, both dicots and monocots, are propagated vegetatively even though direct seeding is less costly than transplanting. Vegetative propagation requires more skill than seeding but is used for several reasons, chiefly to preserve highly desirable genotypes and produce uniform plants. This method avoids phenotypic heterogeneity that results from true seed. Furthermore, some plants produce little or no viable seed and must be propagated by other means. In many woody plants, such as ornamentals and fruit trees, vegetative propagation permits combinations of rootstocks and scions that result in greater disease resistance, increased vigor, size control (for example, dwarfing rootstocks), or earlier fruiting.

Vegetative propagation is any means of increasing planting stock other than by true seed and uses budding, grafting, stem, root, and leaf cuttings, tissue and meristem culture, suckers, rhizomes, tubers, bulbs, and corms. However, danger of transmitting plant pathogens is much greater than with true seed since any systemic or even local infection in mother plants can be transmitted to all daughter plants. Viral and mycoplasmal diseases constitute the greatest danger because they are often systemic in plants and sometimes produce no symptoms, hence are difficult to detect.

Some plants that can be seeded directly in fields are instead grown in nurseries and then transplanted. Plants such as asparagus and nutmeg are started from seed and then selected for sex. Female plants are essential for nutmeg production and male plants are preferred for asparagus. They are then increased by cuttings and crown divisions for commercial plantings. Other plants such as tobacco, tomato, cabbage, and onion are started from seed in nurseries or greenhouses and transplanted to fields. This maximizes use of costly seed, enhances seedling survival, improves plant stands, and brings plants into production earlier than when grown directly from seed. However, this method is usually not considered vegetative propagation since there is no increase in plant numbers. Table 18.1 lists some plants propagated vegetatively.

Several crops are subject to important diseases transmitted via vegetative propagation but not by true seed. Table 18.2 lists some of the crops grown in Washington state and pathogens transmitted through vegetative parts.

TABLE 18.1. Structures Used in the Vegetative Propagation of Crop Plants

Propagative Structure	Plants Propagated
Bulbs and bulbils	Onions and garlic, daffodil, tulip, lily
Corms and cormels	Gladiolus, banana, taro
Crown division	Asparagus, hops, caneberries, iris, strawberry, bamboo, chives
Tubers	Potato, yams, Jerusalem artichoke
Rhizomes	Bamboo, ginger, mint
Cuttings	Orchids, many woody and herbaceous ornamentals, pomegranate, breadfruit, spice pepper, fig, sugarcane, grape, sweetpotato, blueberry, cranberry, cassava
Bud and scion grafts	Deciduous fruit trees, citrus, roses, avocado, rubber, nuts
Suckers	Pineapple, banana, agave, strawberry, caneberries

TABLE 18.2. Pathogens Transmitted on Vegetatively Propagated Crops of Washington

Crop	Pathogens transmitted
Potato	Viruses, *Verticillium*, *Rhizoctonia*, ring rot and blackleg bacteria, root-knot nematodes
Fruit trees	Viruses
Grapes	Viruses
Strawberries	Viruses, nematodes, red stele fungus
Caneberries	Viruses
Blueberries	Viruses
Mint	*Verticillium*
Hops	Viruses, downy mildew fungus
Asparagus	Viruses, *Fusarium*
Ornamental bulbs	Viruses, nematodes, fungi
Ornamental trees and shrubs	*Phytophthora*

PRODUCING PATHOGEN-FREE PROPAGATIVE MATERIAL

Pathogen-free is a sweeping term loosely used in agriculture. It implies absence of any pathogen in propagative material, but this generally is not true, particularly for latent viruses. Most programs involve relative freedom from one or more specific pathogens. The basic strategy in all programs designed to produce pathogen-free stock is to locate or create horticulturally desirable individuals free of specific pathogens, maintain pathogen-free nuclear stock, and make this material available for propagation. Most diseases of concern are caused by viruses, but a few bacterial, fungal, and nematode pathogens transmitted in vegetative material and difficult to control with fungicides, resistance, or other means are also targeted.

The main methods used to produce pathogen-free material are certification, chemotherapy, thermotherapy, and micropropagation, all of which involve some type of pathogen or disease detection. The first step after locating promising candidates is to determine that they are disease-free. In some cases this is by observation in the field, but disease is not always apparent and elaborate tests are necessary. These candidates must be indexed by some method.

Indexing

Disease indexing is any means that reveals the presence of otherwise invisible or inconspicuous pathogens. Indexing can involve chemical and colorimetric tests, indicator plants, serological tests, or standard culturing techniques. Indexing is done in laboratories, greenhouses, growth chambers, or fields, depending on the procedures. Most indexing aims at detecting latent viruses in plant material and is used to locate virus-free plants that provide nuclear stock for propagation.

Indicator Plants (Bioassay) Many viruses produce no obvious symptoms on economically important host plants such as fruit trees, grapevines, and potatoes. To detect many of these viruses they must be transmitted by some means to indicator plants that exhibit visible symptoms. Transmission can be by rubbing plant sap on leaves of indicators, use of aphids or other insect vectors of viruses, or budding and grafting suspect shoots onto indicator plants.

Some viruses of herbaceous plants produce local lesions or systemic symptoms on indicator plants, including *Chenopodium amaranticolor*, *C. quinoa*, *Gomphrena globosa*, and *Nicotiana glutinosa*. For example, carnation is indexed for mottle, vein mottle, and ringspot on *C. amaranticolor*, for ringspot and carnation latent virus on *C. quinoa*, and mottle and mosaic on *G. globosa*. Potato virus X is also indexed on *G. globosa*. Viruses in strawberries are indexed by grafting onto *Fragaria virginiana* or *F. vesca*. Most fruit tree viruses can be transmitted only by grafting. Indicator hosts include Elberta peach for peach yellows, little peach, red suture, peach rosette, and peach mosaic. Other peach varieties are used for other viruses. Bing sweet cherry is an indicator for mottle leaf, rusty mottle, rasp leaf, twisted leaf, and several more. Shiro plum is an indicator for line pattern and Italian prune for prune dwarf. Indicators for pome fruit viruses can also be quite specific. Radiant crabapple is used to detect apple stem pitting virus and Lord Lambourne apple is a good indicator for rubbery wood virus. Sweet orange is an indicator for several citrus virus or virus-like diseases including blind pocket, concave gum psorosis, crinkly leaf, cristacortis, greening, leaf curl, ring spot, and stubborn. The more viruses a given indicator plant detects and differentiates the more valuable it is because it reduces the need for many different indicators and the number of budding or grafting manipulations.

A double inoculation technique has been used to detect the potato spindle tuber viroid using tomato as the indicator plant. There are mild and severe strains of the spindle tuber viroid. The severe strain produces extreme symptoms in tomato in less than 20 days. The mild strain takes longer to produce symptoms. Since there can be partial or temporary cross protection, the presence of the mild strain is detected by first inoculating tomato plants with juice from test tubers and 10–20 days later inoculating plants not showing symptoms with the severe strain. If the mild strain was present in test tubers no additional effect will be seen, because the plants have been cross-protected. However, if tubers are viroid-free test plants will exhibit severe symptoms. Because indexing for pathogens requires much time and labor, faster and easier methods are used when available.

Chemical and Colorimetric Tests Various chemical or staining tests such as phloroglucinol-HCl and methyl red-HCl have been used to indicate the presence of viruses in woody tissues, but in general these have not been very successful. Gram stain is used to detect potato ring rot bacteria in potato tissue and fluorescent staining of leaf petioles has been used to detect little cherry virus. Results of these tests are often faulty and require skilled and careful interpretation.

Certain citrus pathogens can be recognized by chromatographic techniques used to separate compounds associated with infection. Gentisoyl glucoside is present in tissues infected with greening or stubborn MLOs, but it is more abundant in greening tissue. Scopoletin and umbelliferone are present in exocortis-infected tissues. These tests are generally less reliable than indexing on indicator plants.

Serological Tests Serological methods of detecting pathogens, especially viruses, were briefly described in Chapter 17. These same methods, and particularly ELISA, are very useful with vegetative propagative material.

Culture Indexing Propagative material can be tested for fungi and bacteria by transferring portions of candidate plants to selective or nonselective culture media. Carnation cuttings selected for propagation are cut one internode longer than needed for propagation. The lowest internode is surface disinfested, cut into several thin sections, and transferred to nutrient broth or agar medium. The cultures are incubated for 2 weeks. Any evidence of growth during that time is cause for rejecting the cutting from the program. A similar technique is used to detect fungi and bacteria in chrysanthemum cuttings. Sections of rose rootstocks cultured on selective media that enhance microsclerotial production by *Verticillium* aids in identifying the fungus. Potato tuber pieces placed on a selective medium reveal the presence of blackleg bacteria by formation of pectolytic zones around tissue pieces. Culturing also detects citrus stubborn spiroplasma.

METHODS OF ELIMINATING PATHOGENS FROM VEGETATIVE MATERIAL

Chemotherapy

Treatment of diseased plants with chemicals to eliminate pathogens has been attempted many times but with few exceptions chemotherapy has not been successful. The chemicals are either ineffective, too phytotoxic, or too expensive for practical use. Chemotherapy of some fungal and MLO pathogens was discussed in Chapter 9. In addition, 2-thiouracil, an inhibitor of nucleic acid metabolism, and more recently ribavirin, a broad-spectrum antiviral agent, have been tried for controlling plant viruses. So far, however, chemotherapy of viruses has been mostly experimental.

Thermotherapy

The use of heat to eliminate pathogens is an old technique and was discussed in detail in Chapter 11. Hot water therapy usually kills pathogens in plant tissue and is generally used for pathogens other than viruses such as mycoplasmas in fruit tree scions and sugarcane cuttings. Hot air treatment inhibits replication of viruses while allowing some growth of pathogen-free tissues. This method combined with shoot tip or meristem culture has been used to eliminate viruses from many types of plants. Two-thirds or more of the European and United States programs producing pathogen-free propagative materials use heat treatments.

Hot air is the usual method of treatment and causes less injury than hot water. Temperatures for treating growing plants range to about 40°C, and treatment lasts 2–8 weeks. Dormant plants tolerate slightly higher temperatures, but probably no higher than 50°C, and then only for short periods. Success has been very good with some viruses and poor with others but even when only one or two virus-free plants can be obtained they can form the foundation for virus-free planting stock.

Micropropagation

Micropropagation includes a number of techniques used to regenerate plants from small portions of tissue. These range from tissue culture techniques involving only a few undifferentiated cells to tip culture methods in which differentiated growing tips are rooted directly or grafted onto rootstocks. These techniques are means to rapidly increase pathogen-free foundation plants in various certification schemes. There are differences in use of terms by various workers. Tissue, meristem, and tip culture are frequently used interchangeably and there is some overlapping of plant tissue size categories.

Micropropagation has several advantages including indefinite preservation of nuclear stock by *in vitro* cold storage. There also can be rapid proliferation of nuclear stocks, earlier production of daughter plants, earlier introduction of disease-free varieties, and rejuvenation of valuable clones.

Growing points of many infected plants are often free of viruses or other pathogens. One effect of heat treatment allows plant tissue to develop without concomitant development of pathogens. Pathogen-free apical meristems can be excised and subsequently propagated. As a rule, the smaller the tissue taken the better the chances of excluding pathogens.

Many countries use indexing, meristem culture, and rapid multiplication methods to produce propagative materials. Pathogen tests are performed before rapid multiplication and include testing for viruses, viroids, bacteria, and fungi. After multiplication most testing is to detect viruses.

Tissue cultures are started by growing a few undifferentiated cells to form a callus. The callus must be treated with hormones to initiate cell differentiation and root and stem primordia. After several stages of treatment and development, small plantlets are produced that are propagated like any other plant. Chrysanthemums are propagated by tissue culture to obtain plants free from tomato aspermy, mosaic complexes, stunt, chlorotic mottle, and a latent virus. Small shoot tips less than 0.1 mm are aseptically excised from non-heat-treated plants and tested for bacteria, fungi, and viruses. Strawberry plants free of anthracnose crown rot are reportedly produced by tissue culture and made available to industry as small rooted plants. The plants are grown in summer nurseries to produce plants for field planting.

Other than these few examples tissue culture has not been used to any extent for producing pathogen-free stock because of time and cost. Unfortunately, there is a high incidence of mutation in tissue cultures, which detracts from horticultural properties of the progeny.

Meristem culture has been applied indiscriminately to pieces of plant tissue ranging from 0.1 mm to more than 1 cm. This is called meristem culture by some and tip culture by others. Meristem culture should be restricted to the apical dome, along with up to four leaf primordia, and tissue pieces 0.1–0.5 mm long. This tissue is often grown under special lights on tissue culture media sometimes containing napthaleneacetic acid (NAA) or other growth hormone to stimulate tissue differentiation, rooting, and shoot development. Plantlets are transferred to some growth medium such as sphagnum, sand, or vermiculite and maintained for propagation.

Meristem culture has been used to free strawberries from plant viruses by aseptic excision of apical shoot meristems 0.3 mm or less, from runners previously subjected to 40°C for 4–6 weeks. Potatoes were freed of PVX and PVS by treating plants at 36°C (day) and 33°C (night) for 2 weeks. After 6–8 weeks shoot tips 15 cm long were cut from each plant and meristems 0.3–0.6 mm long excised and placed in sterile culture media. Roots and shoots generally developed within 2 months and rooted plantlets transplanted to soil when at least 3 cm or longer. The plantlets were tested for virus and a clean one selected for propagation by stem cuttings.

Tip culture or tip cuttings involves removing growing tips, usually 0.5–2 cm long, and rooting them directly in sand, vermiculite, peat moss, or other rooting medium, or grafting onto suitable rootstocks.

Meristem or tip cultures have been used to generate pathogen-free propagation material for many crops including fruit trees, asparagus, carnation, dahlia, orchids, sweetpotato, ornamental bulbs, and sugarcane.

Stem cuttings can be used to propagate material indexed directly, or obtained by meristem or tip culture. Stem cuttings of carnation, chrysanthemum, and geranium are rooted directly in potting media to produce mother plants for further propagation. Figure 18.1 is a schematic diagram of a process for selecting, indexing, and propagating plants of this type.

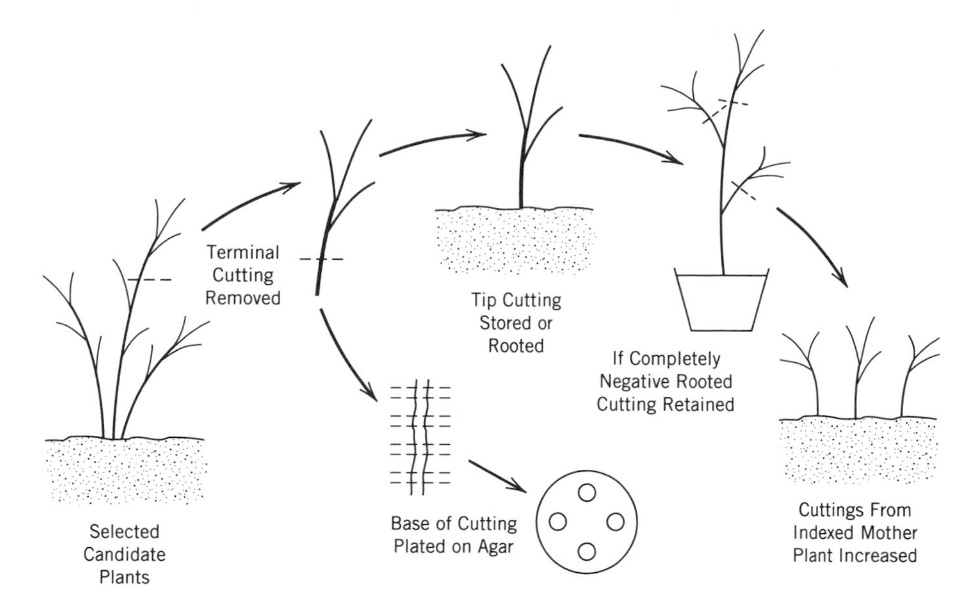

Figure 18.1. Schematic diagram of stem cutting procedures for vegetative propagation of herbaceous plants.

Potatoes are also started from indexed stem cuttings. These are grown in culture media to produce pea-sized microtubers in leaf axils of plantlets, somewhat like aerial tubers. Stem cuttings can also be rooted in sand or similar substrate (other than soil) to produce small minitubers in about 7 weeks. Whole microtubers are used to produce new plants, and minitubers are divided to yield several plants. Scotland was the first country to require all seed lots entered for certification to be derived from nuclear stocks developed from virus-tested stem cuttings.

Nucellar embryony is unique to *Citrus* and a few related genera. It is characterized by production of extra embryos from cells of the nucellus, which surrounds the embryonic sac. It is common, sometimes 100%, in some citrus including grapefruit, sweet orange, lemon, lime, and mandarins. The extra embryos develop from mitotic division of somatic cells. Since there is no involvement of pollen the progeny are essentially clones of mother trees even though propagules develop from ovarian tissue. This means that reproduction is without the genetic variation of seedlings. Since many important citrus viruses such as psorosis, tristeza, exocortis, cachexia, and vein enation are not seed transmitted, this method produces nucellar seedlings that are genetically identical to parent trees but are free of viruses carried through budwood.

A major problem is distinguishing nucellar seedlings from zygotic seedlings. Several methods have been used. These include morphological differences, such as irregular growth or juvenile growth, or use of radioactive pollen so that fertilized (zygotic) seedlings can be detected. The most satisfactory method is isozyme analysis, a kind of genetic fingerprinting. Nucellar seedlings have characteristics only of the female parent.

Certification Programs

Certification provides official assurance in a formal statement that certain conditions and standards apply to specific materials including horticultural propagants. The procedures for certification programs for vegetative materials are similar to those for seed certification. Most certification programs are aimed at viruses, although a few bacteria, fungi, and nematodes are also included. The list of disease-free certification programs is long but some of the important ones throughout the world are those concerned with viral diseases of apple, cacao, citrus, peach, potato, and sugarcane.

The United States has no federal certification programs as do other countries. Canada has a federally administered potato certification program and the UK has a strawberry program. In the United States the majority of certification programs are administered by state departments of agriculture or crop improvement associations, and a few by state agricultural experiment stations. Some highly specialized crops such as ornamentals that are produced by a relatively few large operations are not certified as such but quality, including freedom from pathogens, is maintained by private companies or industry organizations.

Certification programs are formally established, usually by legislative action, and define objectives and requirements of programs. Regulations are formulated designating certifying authorities (usually state departments of agriculture) and specifying tolerances. Certifying agencies then establish procedures for programs. These include number and types of inspections, such as field, harvest, storage, or random samples. There are provisions for roguing diseased plants between inspections, and separation of propagation–production areas from commercial plantings. They may also require pathogen testing or auxillary grow-out tests such as greenhouse or winter testing.

Certification programs are established in the United States for potatoes, fruit trees (pome fruits, stone fruits, avocado, and citrus), grapevines, strawberry, raspberry, sweetpotato, and some specialty crops such as asparagus, garlic, hops, and mint. Many of these programs contain two distinct elements. First is an improvement program or process to obtain pathogen-free nuclear stock. This material is maintained at some protected location and subjected to frequent retesting to guarantee continued freedom from pathogens. It is the source of foundation stock grown to produce registered and certified planting material. These procedures require a great deal of time, facilities, and money and are usually carried out by state agricultural experiment stations, often in cooperation with federal and private agencies. The second step is actual production of certified planting stock in regulated commercial nurseries or farms supervised by certifying agencies.

Two large certification programs in the United States and possibly worldwide are for potatoes and fruit trees. These are discussed in detail to illustrate application of strategies and technologies described above.

Potato Seed Improvement and Certification

In 1904 W. A. Orton of the USDA observed potato seed improvement programs in Europe and recommended that a system of official inspection and certification be established in each seed-growing state with emphasis on freedom from disease. The first official potato seed certification conference was held in Philadelphia, Pennsylvania in 1914 and included representatives from Canada, Germany, Ireland, the USDA, and 12 states. By the 1920s most U.S. and Canadian potato-growing areas had seed certification programs.

Seed potato certification represents voluntary agreements between seed growers and certifying agencies. Each agency develops certification standards outlining eligibility requirements for inspection, fees charged growers for registration, inspection, and certification services, grade requirements, types of tests, and rules for selling certified seed. Twenty states have seed potato programs and procedures vary widely from state to state. All have visual field inspections of some type; 9 require testing for specific pathogens such as ringrot bacteria and 12 include greenhouse or other winter testing.

Seed Potato Improvement Canada has one of the more intensive seed potato improvement programs and has developed nuclear stock (pre-Elite) free of several viruses, such as potato viruses X and S (PVX, PVS). The program includes over 200 accessions including the more important commercial varieties such as Russet Burbank, White Rose, Kennebec, and Norgold, which account for more than half of Washington state's seed potato acreage.

There is some debate about the value of completely virus-free seed stocks. Some potato varieties such as Russet Burbank are believed to be universally infected by PVX. Comparisons of PVX-free potatoes with yields from infected seed tubers have been inconclusive and has not led to a rush to plant PVX-free seed in the United States.

The Canadian procedure serologically tests mother plant (pre-Elite I or Nuclear I are other terms used) to ensure freedom from viruses. All plants from first-year tubers (Elite I or Nuclear II) are visually inspected and serologically tested. Second-year tubers (Elite II or Foundation I) produce plants that are field inspected and 10% tested serologically. Tubers from these plants comprise the third generation (Elite III or Foundation II), which is also field inspected but only 2% tested serologically. The fourth tuber generation, foundation seed, is field inspected and 30-50 plants per acre tested serologically. These tubers produce certified virus-free seed potatoes.

Once disease-free nuclear seed is produced it must be increased in some way. This is done by micropropagation, usually stem cuttings. Material from the Canadian program is sent to countries around the world, often for research purposes. About 90% goes to Canadian Provinces and the United States. The rest goes to Australia, Central and South America, Europe and North Africa, New Zealand, Pacific Rim Countries, and the Russian Republics (formerly USSR).

Certification Material to be used by foundation seed growers is increased in several ways, tuber indexing, tuber unit planting, hill selection, or mass selection. In *tuber indexing* a single eye is removed from each mother tuber and planted in greenhouses. Diseased plants are destroyed along with the mother tuber. Those that pass this first test are increased by *tuber unit* planting where each tuber is cut into four pieces and planted about 12 inches apart in rows. Wider spaces separate each tuber unit. If any of the plants in a unit are diseased the entire unit is destroyed. Tubers from units that are apparently free of disease are harvested and indexed during winter months. Those that pass winter indexing are increased to provide seed for foundation plantings. During increase periods plantings are rogued intensively and screened for the presence of viruses. Fields are generally rogued three times during growing seasons.

Hill selection consists of selecting and labeling promising plants during the growing season. Tubers from these plants are winter indexed and those that pass are increased through tuber unit planting. *Mass selection* is similar to hill selection except that instead of tubers from desirable plants being held separately they are stored in bulk.

Tolerances permitted for various diseases vary somewhat between states. Table 18.3 shows the range of tolerances allowed in seed in western states certification programs and those allowed in Washington state. There are zero tolerances for bacterial ring rot and northern root-knot nematode in all 20 state potato seed certification programs.

Requirements for certified seed are somewhat less rigid than for foundation seed. Certified seed can be produced from any higher category of seed, including breeder, registered, or foundation. Certified seed cannot be grown for recertification except under certain circumstances such as absence of registered or foundation seed and then only with approval of the owner or controlling authorities. Justification for disallowance is readily seen in seasonal increase of potato leafroll virus. A study of leafroll in 70 potato fields in the Columbia Basin region of Washington state revealed an average of 2.2% leafroll infected seed tubers at planting but 37% of harvested tubers were infected, a 17-fold increase in just one growing cycle. Some other viruses such as PVS and PVY showed similar increases from planting to harvest.

Seed potato certification has been successful. Percentage of seed potato acreage passing certification in the United States increased from about 40% in the early 1920s to 90% in the 1970s. The Washington State Potato Seed Lot Trials have shown a dramatic effect on leafroll reduction. This program was started as an extension demonstration project in the 1950s to show farmers what quality of seed they were buying. Seed producers are asked to submit 300-tuber samples and these are planted without cutting so that each seed lot constitutes one row. In late June or early July the plants in each row are examined by experts from Washington State University and the

TABLE 18.3. Disease Tolerances for Foundation and Certified Seed Potatoes

	Range in Western States	Washington State			
		Foundation Inspection		Certified Inspection	
Disease	(%)	1	2–3	1	2–3
Leafroll	0.05–0.2%	0.2	0.1	0.4	0.2
PVX	4.0–2.0				
Mosaic	0.2–2.0	1.0	0.5	2.0	1.0
Blackleg	0.25–5.0	2.0	1.0	4.0	2.0

Washington State Department of Agriculture. Each diseased plant is marked with a colored flag. A few days after plot readings a field day is held where farmers view the results. Individual rows are identified by name and address of the seed producer. This allows potato farmers to identify sources of their seed and be more selective, essentially forcing seed producers to clean up their stock. Table 18.4 compares level of leafroll from different sources soon after the program started with recent results. In the 25 elapsed years average occurrence of leafroll-infected seed lots dropped from 32 to 2%.

Fruit Tree Certification

Citrus There are citrus registration and certification programs in California, Florida, Texas, and Arizona in the United States and Brazil, Colombia, Sicily, France, Morocco, Israel, Spain, South Africa, and Australia.

The California program, initiated by the University of California Agricultural Experiment Station, selects budwood from candidate trees and indexes for 10 or more virus or virus-like diseases. Budwood that passes initial screening goes into a foundation block tree that is indexed and reindexed for tristeza and several other viruses. Budwood from the foundation block tree goes to three maintenance locations. One is the U.C. protected block of prime bud lines where one tree is held in a screenhouse and two are planted outdoors to check on performance. A second becomes the growers registered tree, which is also indexed. The third is held in a screenhouse in a growers protected block where it is indexed. Budwood from all of the above sources also goes to nursery increase blocks where trees are visually inspected and to nursery certified blocks where trees are produced for sale to citrus growers.

TABLE 18.4. Leafroll Incidence in the Washington Potato Seed Lot Trials from Various Seed Sources (1961–1986)

	1961		1986	
Origin	Total No. Samples	No. with Leafroll	Total No. Samples	No. with Leafroll
Montana	52	3	68	0
Idaho	48	7	30	0
Alberta	5	2	6	0
Nebraska	2	1	1	0
Oregon	10	6	12	0
North Dakota	7	4	5	0
British Columbia	37	19	7	2
Washington	23	18	11	1
Total	184	60 (32%)	140	3 (2%)

Deciduous Fruit Trees California, Oregon, Washington, Maryland, Michigan, Minnesota, Missouri, New York, Pennsylvania, and Tennessee have certification programs for pome (apples, pears, quinces) and stone (cherries, prunes, plums, apricots, peaches, nectarines) fruits. The main source of foundation stock for most of the states indicated is the Interregional Research Project (IR-2) financed from regional research funds appropriated through USDA. Facilities and personnel are located at Washington State University Irrigated Agriculture Research and Extension Center, Prosser, Washington.

The primary objective of the IR-2 project is to obtain apparently virus-free valuable stock of stone and pome fruits. Where none can be found virus-free individuals are developed by thermotherapy and meristem culture. Virus-free foundation material is maintained in semiisolated repositories or screenhouses and material distributed to research workers and nursery industry. Pome fruit trees maintained for foundation stock generally require isolation zones of only a few hundred feet from noncertified plants of the same kind. Stone fruits require greater distances (i.e., greater isolation) because some viruses are transmitted on pollen grains.

Most indexing is of a double-budding technique, which cuts a year from standard budding procedures. In double-budding an indicator bud is placed above two test buds on a seedling rootstock. Seedlings are topped just above indicator buds to force growth. Test buds (inoculum) are not intended to grow, only knit, which takes about 7 days, so that any viruses present can move into indicator buds. By this technique readings that take 1 or 2 years in the field can be made in 4–10 weeks in greenhouses. Because of the possibility of reinfection of virus-free budwood from infected roots, rootstock material also goes through the program.

The Interregional Virus-Free Deciduous Fruit Tree Repository at Prosser maintains about 40 lines of apricot, 212 peach, 37 nectarine, 135 plum and prune, 25 sour cherry, and 72 sweet cherry. There are also 180 fruiting apple lines plus 150 other apple lines used for rootstock, ornamentals, cider, virus indicators, or interstock, 170 fruiting pears plus 32 other pears, and 8 quince. Two additional National Germ Plasm Repositories are at Corvallis, Oregon, and Davis, California. Oregon holds pears, hops, mint, strawberries, caneberries, and filberts, and California holds prunes, pistachios, olives, and grapes.

Production and maintenance of virus-free stock is the main responsibility of the IR-2 program. Figure 18.2 shows the process of maintaining virus-free nucleus trees. When budwood is distributed for propagation the Washington State Department of Agriculture (WSDA) becomes the certifying authority. The WSDA Nursery Inspection Program sets trees out in nurseries. Nurseries propagate the trees and WSDA indexes them. If virus-free, trees are certified for sale by nurseries.

Other Certification Programs Several other vegetatively propagated crops have some type of certification program in the United States. These are sweetpotato (9 states), strawberries (12), raspberries (9), and grapes (7). Except for sweetpotato most of these programs require virus testing of the nuclear stock and a few require some indexing of foundation plants. Certification of sweetpotato roots is based entirely on inspection of plants in fields for Fusarium wilt, and roots in storage for black rot and internal cork virus. Roguing of Fusarium wilt plants must be supervised by an inspector.

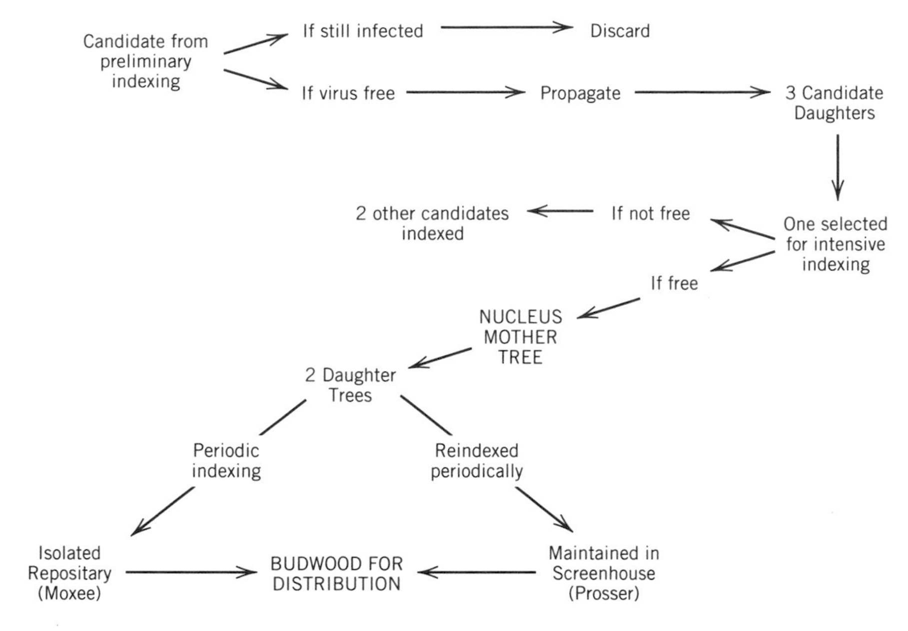

Figure 18.2. Flow chart for propagating and maintaining virus-free budwood in IR-2 program.

REFERENCES

*Anon., Micropropagation of selected rootcrops, palms, citrus, and ornamental species. *FAO Plant Production and Protection Paper* 59, 1984.

*Anon., Symposium: Applications of tissue culture and micropropagation techniques to potato production. *Amer. Potato J.* 65:161–235 (1988).

*Baker, K. F. (ed.), Development and production of pathogen-free propagative material of ornamental plants. *Plant Dis. Report. Suppl.* 238:57–95 (1956).

*Fridlund, P. R., The IR-2 program for obtaining virus-free fruit trees. *Plant Dis.* 64:826–830 (1980).

*Fridlund, P. R. (ed.), *Virus and Viruslike Diseases of Pome Fruits and Simulating Noninfectious Disorders.* Washington State University Cooperative Extension SP0003, 1989.

*Hansen, A. J., An end to the dilemma—virus free all the way. *Hortscience* 20:852–859 (1985).

*Hollings, M., Disease control through virus-free stock. *Annu. Rev. Phytopathol.* 3:367–396 (1965).

*Lawson, R. H., Controlling virus diseases in major international flower and bulb crops. *Plant Dis.* 65:780–786 (1981).

*Mink, G. I., Control of plant diseases using disease-free stocks. In Pimental, D. (ed.), *Handbook of Pest Management in Agriculture*, Vol. I, 2nd ed. CRC Press, Boca Raton, FL, 1991. Pp. 363–391.

*Nyland, G., and A. C. Goheen, Heat therapy of virus diseases of perennial plants. *Annu. Rev. Phytopathol.* 7:331–354 (1969).

*Reuther, W., E. C. Calavan, and G. E. Carman (eds.), *The Citrus Industry*, Vol. IV. *Crop Protection*. University of California, Riverside, 1978.

*Shepard, J. E., and L. E. Claflin, Critical analysis of the principles of seed potato certification. *Annu. Rev. Phytopathol.* 13:271–293 (1975).

* Publications marked with an asterisk (*) are general references not cited in the text.

19 Resistance

Resistance is the ideal way to control plant diseases, if satisfactory levels of durable resistance can be incorporated into culturally desirable crop plants. Resistant varieties save time, effort, and money otherwise spent fighting plant diseases. Resistance reduces or eliminates two economic losses, direct reduction of yields and additional cost of controls. The environment also benefits because there is no need for pesticide applications.

Resistance to plant diseases embraces a wide range of biological phenomena including true genetic resistance, physiological, anatomical, or mechanical resistance, tolerance, and escape. Resistance also conveys a range of concepts as to degree of freedom from disease. Immunity and resistance are often used interchangeably, even though immunity is absolute and resistance relative.

Differences in susceptibility of plants to diseases were recorded by Theophrastus as early as 300 B.C. It is probable that even then, farmers saved seed from the best plants for replanting and by this selection process obtained some disease resistance. Darwin (see reference, Chapter 1) compiled many reports of disease variation in several important cultivated plants although specific diseases were not identified. This was soon after potato late blight epidemics ravaged Europe and efforts focused on control. Darwin hoped to gain government support for work being conducted by James Torbitt of Belfast to overcome the "potato disease" (late blight) by "raising fungus-proof varieties of the potato." Darwin said the plan consisted of "rearing vast numbers of seedlings from cross-fertilized parents, exposing them to infection, ruthlessly destroying all that suffer, saving those which resist best, and repeating the process in successive seminal generations." This procedure is exactly the same as that used today to develop resistant varieties by selection.

Among the earliest directed efforts at producing disease-resistant plants was hybridization of the common potato with other tuber-producing *Solanum* species from Central and South America. These early attempts at interspecific hybridization were not very successful. Crossing common potato (*S. tuberosum*) with Darwin potato (*S. maglia*) or Uruguay potato (*S. commersonii*) resulted in few fertile seeds and hybridization was thought to have little practical use. Most benefits came from crosses between varieties of cultivated potatoes. One of the earliest late blight resistant potatoes, Magnum Bonum, was introduced in 1876, from a cross of Early Rose with Victoria. However, modern techniques and different sources of germ plasm make hybridization, followed by selection, a more successful method of producing disease-resistant varieties. By 1937 blight-resistant varieties had been produced by crossing *S. tuberosum* with *S. demissum* and other *Solanum* species.

Rediscovery of Mendel's laws in 1900 started an avalanche of genetic improvement of crop plants including incorporation of disease resistance. Biffen, whose pioneering work in breeding for disease resistance is acknowledged in Chapter 1, mentioned that as early as 1815, use of disease-resistant varieties had been proposed. A few years before Biffen's work, workers in Australia in 1889 declared that susceptibility to rust is hereditary in wheat.

Biffen wanted to demonstrate heritability of certain characteristics in wheat and included resistance to yellow (stripe) rust as one component. He used Michigan Bronze, a variety of hexaploid common bread wheat (*T. aestivum*), which he considered to be the most susceptible wheat in existence to yellow rust, and Rivet, a subspecies of tetraploid poulard wheat (*T. turgidum*). Diploid and tetraploid wheats are generally more resistant to rusts than are hexaploids. Ratios of susceptible to resistant progenies of this cross were approximately 3:1 suggesting to Biffen that "liability" (i.e., susceptibility) is dominant over "immunity" (i.e., resistance).

Early breeding for stem rust resistance started in the United States after the destructive epidemic of 1904. Kanred, a selection of a wheat brought from Russia in 1906, was grown in Kansas in 1916 and discovered to be stem rust resistant. By 1924 it was grown on more than four million acres, but it began to lose its resistance within a few years. Ceres was developed in 1926 and by 1933 was grown on five million acres in the United States and Canada, but stem rust race 56 ruined it in 1935. The successor to Ceres was Thatcher, offspring of a cross between a bread wheat, Marquis, and a durum variety. Thatcher was resistant to stem rust races of that time but susceptible to leaf rust and scab. This example and others in cotton, flax, and banana illustrate that development of a variety resistant to a disease may simply exchange one problem for another because of new pathogenic races or different pathogens. Living with the lesser of evils may be a consequence.

About 1900 W. A. Orton of the USDA observed that some asparagus varieties were partially or highly resistant to rust. By selective breeding he obtained varieties with near immunity to rust, improved size, quality, and productivity. Orton used the selection method to obtain resistance to Fusarium wilts in cotton and cowpeas but used hybridization to incorporate wilt resistance in watermelon.

Rivers, Centerville, and Sensation were early wilt-resistant varieties of Sea Island cotton that originated from what Orton called "mutations" and improved by selection. This resistance did not always extend to other diseases. The variety Rivers was very susceptible to bacterial blight, Centerville nearly immune.

Orton said the wilt and root-knot resistant cowpea variety Iron was of "chance origin" and crosses between this and other varieties produced additional resistant varieties. No source of Fusarium wilt resistance could be found in watermelons, but Orton obtained resistance by crossing watermelon with the highly resistant citron-melon. This resulted in a true disease-resistant hybrid, the Eden watermelon. From this hybrid, Orton obtained Conqueror, a watermelon of good quality, but its round shape was disdained in favor of oblong melons, and, as Orton noted, "styles in watermelon changed." However, the genetic base for wilt resistance was created.

H. L. Bolley selected a few flax plants surviving in severe Fusarium wilt-infested soil in North Dakota and by individual and mass selection produced a series of wilt resistant lines. He observed that resistance is relative and even the most resistant selections would wilt severely at high temperatures. An early selection, North Dakota Resistant (NDR) 22, was developed from a single plant and was very resistant to wilt but susceptible to rust. This was followed by NDR 52, which was slightly resistant to rust and later by NDR 73, which was less resistant to wilt than either "22" or "52" but very resistant to rust. By 1908 NDR 114 was derived from a previous selection to obtain both wilt and rust resistance.

Development of Fusarium wilt (yellows)-resistant cabbage began in Wisconsin in 1909. Surviving cabbage plants were selected from severely diseased fields and planted to produce seed. Heads were self- or cross-pollinated and seed from these selfings and crossings were planted in infested and noninfested fields to compare yields. Results were dramatic. Average yields of susceptible (commercial) varieties were about 1 or 2 tons per acre while yields of selected strains were as high as 19 tons and averaged about 12 tons.

The first resistant cabbage varieties were Wisconsin Hollander in 1916 and Wisconsin All Seasons 4 years later. From their first introduction it was noticed that Wisconsin Hollander became diseased at temperatures above 24°C while Wisconsin All Seasons remained healthy. J. C. Walker discovered that two types of resistance were involved. Hollander resistance was quantitative and polygenic (type B), that is, it had varying levels of resistance controlled by more than one gene. All Seasons had both type B and type A resistance, which is qualitative, monogenic, and does not vary with temperature. Type A resistance is either expressed maximally or is absent, no varying levels in between, and is controlled by one dominant gene. Considering the frequent occurrence of pathogenic (i.e., physiologic) races of Fusarium wilt fungi it is remarkable that type A resistance has remained stable and not compromised by pathogenic races.

Production of new resistant varieties and circumventing pathogenic races that develop in response to resistance genes in host plants are principle strategies in developing and deploying disease-resistant varieties. This chapter outlines some methods for obtaining disease resistance. Genetics and plant breeding are highly developed disciplines and involve more than incorporating resistance into plants. It is not the purpose of this chapter to attempt a thorough treatise on plant breeding but to show various ways that this technology is applied to controlling plant diseases.

LEVELS OF DISEASE RESISTANCE

Resistance and susceptibility are polar phenonena; a continuum on a single scale, and as one increases the other decreases. Figure 19.1 illustrates this inverse relationship between susceptibility and resistance. Sometimes what is considered resistant on a population basis is relatively high in the amount of disease that is allowed. For example, the National Alfalfa Variety Review Board evaluates alfalfa varieties for agronomic characteristics including resistance to diseases. Varieties are considered resistant (R)

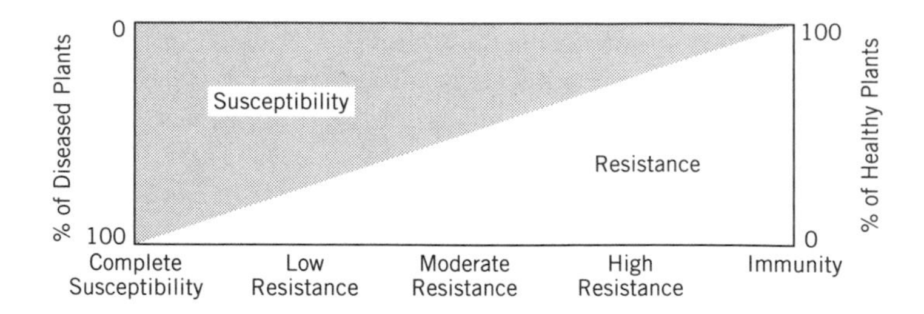

Figure 19.1. The polar relationship beween susceptibility and resistance.

if more than 30% of seedlings in a population are resistant to a specific disease. In other words, two-thirds of individual plants can succumb to a disease yet the variety is considered resistant. Proportionally, genotypes are susceptible, but in practice phenotypes are resistant because the surviving 30% can make a crop. Correlation between varying degrees of resistance (amount of disease) and productivity (yield) is illustrated graphically in Figure 19.2.

Resistance is inherent ability of plants to remain relatively unaffected by disease. There is a direct correlation between disease and yield reduction in susceptible plants (Figure 19.2A). Moderately resistant (i.e., slightly or moderately susceptible) plants

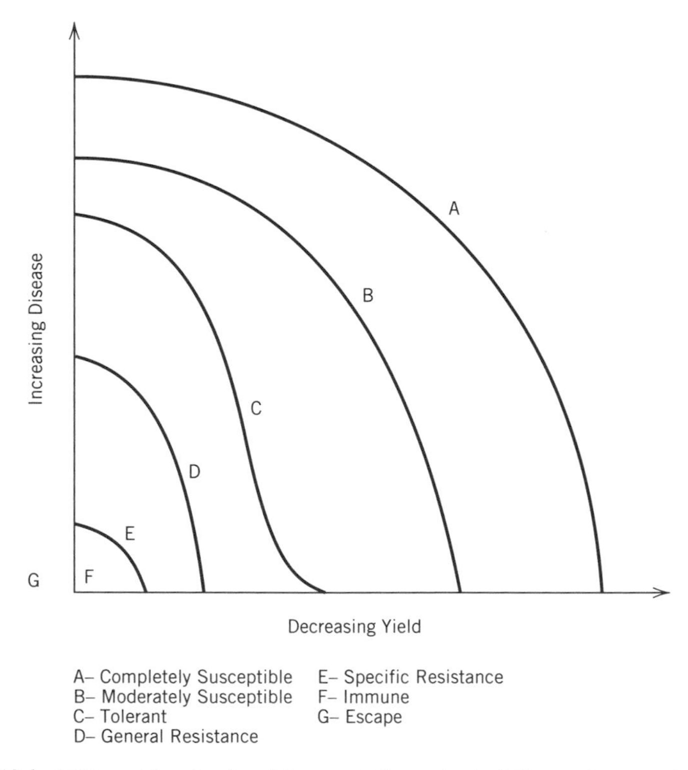

Figure 19.2. Different levels of resistance relating extent of disease to crop yield.

have recognizable amounts of disease but proportionally better yields than susceptible plants (Figure 19.2B). Two kinds of resistance are recognized and have been characterized in a number of ways. General (horizontal) and specific (vertical) resistance are terms commonly used to reflect reactions of plants to pathogenic races. General resistance imparts moderate levels of resistance to all known races of a pathogen with only slight reductions in yield (Figure 19.2D). Specific resistance results in minimal disease and decrease in yield (Figure 19.2E), *as long as no new overriding (compromising) pathogenic races occur.*

Tolerance is ability of plants to become diseased and possibly show it, but not have yields appreciably or proportionately affected (Figure 19.2C). Some workers maintain that tolerance is to pathogens; others argue that tolerance is to diseases. Some believe that tolerance can be to both. From a practical standpoint it does not matter which is tolerated as long as it is a predictable and useful characteristic.

Immunity is absolute freedom from infection or disease. Plants cannot be infected under any circumstances (Figure 19.2F). This is the ideal and ultimate control, but unfortunately it is extremely rare within a given plant species. It is common, however, between plant species and essentially reflects the fact that an immune plant is a nonhost. For example, *Pinus ponderosa* and other two and three-needle pines are immune to white pine blister rust, which affects only five-needle pines. For the most part immune species are not useful sources of germplasm for plant improvement, too often because of genetic incompatibility or undesirable qualities.

Escape occurs when inherently susceptible plants do not become infected because of phenology, absence of inoculum, or conditions unfavorable for infection. The escape phenomenon illustrated by point G in Figure 19.2 is not comparable with true resistance because escaped plants could just as easily be at point A.

Klendusity is a special type of disease escape owing to morphologic features of plants. This term has been misused to include characteristics not originally intended by the author of the term, or included in its derivation or definition. Klendusity was introduced to replace "disease-escaping," which was considered to be an awkward expression. The term originally was applied to raspberry varieties that escaped mosaic virus infection because aphid vectors could not reach the epidermis through layers of hairs on leaves and canes. The definition generally includes escape from infection by an otherwise susceptible variety as a result of any type of mechanical hindrance to infection such as thickened cuticle, closed flowering, and smaller leaf size. One worker appropriately recognized reduced target area of foliage on certain western conifers as a klendusic factor in escaping dwarfmistletoe infection. However, the same worker also believed removal of dwarfmistletoe seeds by snow, wind, rain, or some other external force constituted klendusity. These actions do not qualify as klendusity though, since they are not the result of host plant morphology.

CHARACTERISTICS OF RESISTANCE

Several terms describe and compare characteristics of disease resistance in broad and narrow senses. Commonest of these are *general* (or nonspecific) and *specific* resistance. Van der Plank (1968) introduced the terms *horizontal* and *vertical* to describe resistance in relation to pathogenic races of disease organisms. Horizontal resistance is variable, that is, quantitative due to several genes, but that operates against all races of a pathogen (Figure 19.3A). Vertical resistance is usually complete, that is, qualitative due to a single gene that acts against individual races (Figure 19.3B). The underlying mechanism that determines whether resistance is general or specific usually is the number of genes contributing to resistance and its physiological action in hosts. Some exceptions exist such as the Type A resistance to Fusarium wilt in cabbage and stripe rust in Nugaines wheat where general resistance is controlled by one or two genes.

General resistance is also called nonspecific, multigene, polygenic, minor gene, adult or mature plant, field, uniform, durable, incomplete, or partial resistance. Certain mechanisms in plants help to resist colonization by pathogens, which results in reduced infection, pathogen development, and/or reproduction. Resistance in plants is variable, often not expressed in seedlings, but increases as plants mature; moreover, it may be

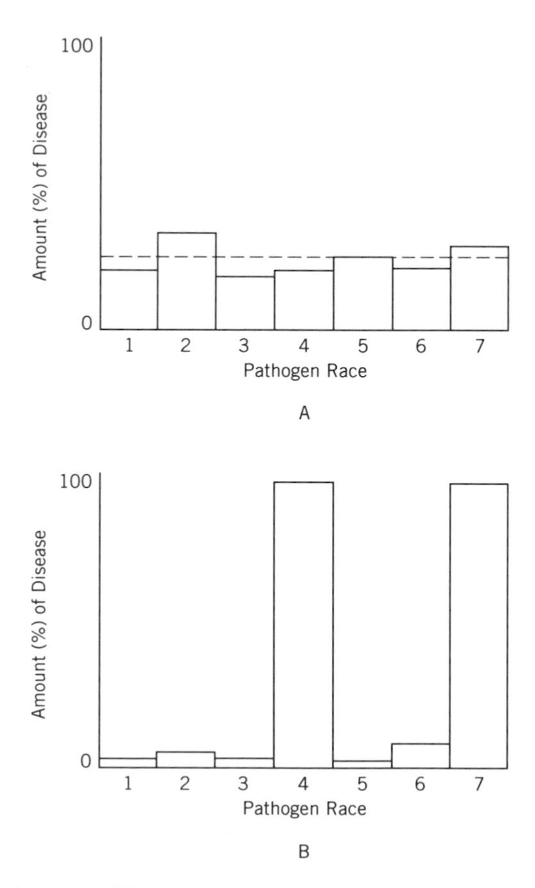

Figure 19.3. Development of disease in crop varieties with (A) general resistance or (B) specific resistance in response to seven different races of a pathogen.

influenced by temperature. Polygenic resistance is controlled by many genes, often with small but additive effects. This resistance is fairly stable and unaffected by changes in virulence genes of pathogens (i.e., new races).

Specific resistance alludes to race-specific, single gene, monogenic, major gene, seedling, and differential resistance. It produces immune or hypersensitive host reactions that resist invasion by pathogens. Vertical resistance is highly efficient and specific against some races (qualitative) but may mask extreme susceptibility to races not present at a particular time or place. Control by one or a few genes with major effects makes specific resistance vulnerable to sudden failure due to new pathogenic races.

THE MEANING OF RESISTANCE

What does resistance mean? As Bolley and many others have wisely observed, resistance is relative and hence a subjective value. In some cases environment is a contributing factor: temperature as already mentioned is probably the most important. Resistance evaluations can be influenced by disease pressure applied to plants. Some varieties hold up well at low or moderate disease levels but resistance fails when disease pressure is high. Several examples illustrate different levels of disease resistance and how these levels are defined.

A resistance continuum frequently consists of several categories, including very susceptible, susceptible, moderately susceptible, intermediate, moderately resistant, resistant, and very resistant. The Washington State Crop Improvement Association annually publishes a listing of leading wheat varieties grown in the state. The resistance categories are based on comparisons of varieties and breeding lines in research plots and empirical field evaluations. Table 19.1 is a condensed version for a recent year showing levels of resistance assigned to different varieties. This table also demonstrates that assessing resistance is variable and depends on which diseases are of concern in a particular region. Ten white winter wheat varieties were selected to compare disease resistance in similar types.

As it applies to common smut (Table 19.1) the real meaning of some ratings might be questioned, since common smut tests involve a limited range of pathogenic races and the real values for common smut are probably more like those of dwarf smut. Dwarf smut is less prevalent and the need for resistance is not as great in all varieties, depending on where a variety is grown. The great success of seed treatment in controlling common smut has reduced the need for resistance to this disease. Stem rust is not a major problem in the Pacific Northwest, except in late maturing varieties, so there is no great need for stem rust resistance in this region. Flag smut has yet to become an economic problem in the Pacific Northwest except when highly susceptible varieties such as Paha are grown.

Cephalosporium stripe reaction is probably true tolerance since plants often appear to be equally infected but yields of some varieties are notably higher than those of susceptible varieties. Varieties from certain breeding programs, such as Stephens, Malcolm, Hyslop, and McDermid, are highly susceptible to this disease and apparently have a common genetic base.

TABLE 19.1. Disease Resistance Ratings of 10 Soft White Winter Wheat Varieties Grown in the Pacific Northwest

	Resistant Rating for							
Variety	Common Smut	Dwarf Smut	Leaf Rust	Stripe Rust	Stem Rust	Flag Smut	Cephalosporium Stripe	Strawbreaker Foot rot
Daws	R[a]	S	MS	MR	S	MS	3[b]	S
Dusty	R	S	MS	R	S	—	5	VS
Eltan	MR	S	MS	MR	S	—	6	MS
Hill 81	R	S	MR	MR	S	—	4	S
John	MR	S	S	S	S	—	4	S
Kmor	R	R	MS	R	S	—	5	S
Lewjain	R	R	MS	MR	S	S	6	MS
Madsen	R	S	R	R	R	—	4	R
Malcolm	R	S	S	R	S	MS	1	MS
Stephens	R	S	MR	R	S	MS	1	MS

[a] Disease resistance categories: R, resistant; MR, moderately resistant; MS, moderately susceptible; S = susceptible; VS = very susceptible.
[b] *Cephalosporium* stripe tolerance index: 1= very susceptible; 10 = very tolerant.

Resistance to strawbreaker foot rot is a recent advance, derived from *Aegilops ventricosa*, which also is a source of resistance to other cereal diseases, including rusts. Some varieties are developed to fill certain agricultural niches. The variety John was selected for snow mold resistance. It is grown in restricted areas where diseases like rust and common smut are not major problems or smut is controlled by seed treatment.

Roses are popular ornamental plants and subject to several serious diseases like rust, blackspot, and powdery mildew. Varieties resistant to these diseases have been developed and are widely grown. Some varieties may react differently to specific diseases in different areas, but even comparisons made at the same location may give different impressions, depending on how ratings are expressed.

Roses are evaluated in several test gardens and ratings included in the Pacific Northwest Plant Disease Control Handbook. Before 1991 disease reactions were given as numerical ratings with 0–2 being "highly tolerant," 3–5 "moderately tolerant," and 6–10 "seriously affected." Since 1991 disease ratings are given simply as resistant, moderately resistant, or susceptible. Resistance or tolerance have been used interchangeably to express these ratings. Numerical scales allow for finer comparisons between varieties, even though these ratings are based on empirical evaluations. From a practical standpoint, however, simple categories of resistant, moderately resistant, and susceptible are more useful. A grower can easily determine which of the three categories will best suit disease situations in an area.

Table 19.2 compares disease ratings for eight hybrid tea rose varieties included in pre-1991 evaluations and in post-1991 tests. The numerical ratings do not always reflect placement of varieties in the three resistance categories. For example, only two varieties, Pascali and Red Devil, are rated as "resistant" to any disease, namely rust; all other ratings are "moderately resistant." Using numerical ratings Pascali would have only a "moderately tolerant" rating to rust while its black spot rating is "highly tolerant." If numerical ratings of 0–3 are equivalent to "resistant" then all varieties should be classed as resistant to black spot, yet none is. Disease ratings cannot be considered absolute or inviolable because of these inconsistencies. They are only comparative guides to performance of varieties in an area.

OBTAINING RESISTANCE

Resistance to certain diseases is obtained in several ways with selection or hybridization; a combination of these methods is most commonly used.

Selection

Selection or selective breeding involves exposing plant populations to high levels of disease and selecting individual plants that survive. These plants are further tested, often by artificial inoculation, to ensure they have not simply escaped infection. Those

TABLE 19.2. Comparison of Disease Ratings of Hybrid Tea Rose Varieties Using a Numerical Scale (0–10) or Disease Categories (R, M, S) from the 1990 and 1991 *Pacific Northwest Plant Disease Control Handbooks*

	Pre-1991[a]			1991[b]		
Variety	Rust	Powdery Mildew	Black Spot	Rust	Powdery Mildew	Black Spot
Chicago Peace	5	4	2	M	M	M
Dainty Bess	3	4	0	M	M	M
Irish Gold	1	2	1	M	M	M
Mister Lincoln	7	6	0	M	M	M
Pascali	3	3	0	R	M	M
Peace	6	3	2	M	M	M
Red Devil	0	1.5	3	R	M	M
Tiffany	4	4	0	M	M	M

[a] Pre-1991 Numerical disease ratings: 0–2, highly tolerant; 3–5, moderately tolerant; 6–10, seriously affected.

[b] Post-1991 disease ratings: R, resistant; M, moderately resistant; S, susceptible.

plants that are resistant are used as parents to improve levels of disease resistance. Since the parents involved are of the same genetic lineage and not separate species, varieties, or races, this process should not be considered hybridization, although hybridization has different meanings and interpretations. Successful selection requires genetically variable populations that contain both resistance and susceptibility. Selection might be considered a "Darwinian" process in that selection promotes survival (or expansion) of fittest individuals. Genetic lines used in selective breeding are often maintained, or owned, by private companies.

Selection of plants for desirable traits has probably been practiced since the beginning of agriculture. One of the earliest records of selection for disease resistance (or tolerance) occurred in Iowa in 1898. Wheat varieties from many sources were being compared for winter hardiness, yield, and other characteristics. From a planting of 40 varieties of spring wheat individual plants were selected for short growing season. The purpose was to obtain wheat that would mature in short time to minimize losses from stem rust.

Hybridization

Hybridization is genetic manipulation of plants to combine disease resistance with desirable agronomic or horticultural features. Hybridization can be considered a "Mendelian" process since it involves identification and incorporation of specific genetically controlled characteristics. Perhaps the simplest explanation of hybridization is illustrated by crossing a plant with good horticultural properties (H) but poor disease resistance (r) with a plant that is less desirable horticulturally (h) but has good disease resistance (R). In the process of combining genes from these two parents it is hoped that some of the progeny will have both desirable features (HR). In other words, Hr × hR = HR, the only one of four possible combinations to have both good disease resistance and desirable horticultural characteristics.

Several things make hybridization more difficult than selection. First, sources of resistance must be found. Genetic resistance is frequently found in regions where plants, and often pathogens, originated. Plant collectors from many countries have gathered germplasm in the form of wild plants throughout the world. These plants are maintained in repositories, both as seed and growing plants. In the United States these collections are controlled by the Plant Introduction Division of the USDA. There are more than 25 Plant Introduction Stations or Clonal Germplasm Repositories in the United States. These contain more than 1500 plant genera and provide sources of disease resistance and agronomic characteristics for plant breeders.

Of major consideration in hybridization is the possibility that undesirable characteristics may be linked with resistance factors. Sometimes edible qualities of hybrids are unsatisfactory, such as off-color flour or poor milling properties of wheat with leaf rust resistance, or poor flavor in tomatoes bred for resistance to curly top. Growth habit may be unacceptable. Breeding for resistance to Dutch elm disease usually involves crossing susceptible American elms that have a desirable upright (i.e., vase-shaped) growth habit with resistant oriental elms that have a short, round shape that is aesthetically less pleasing. Most resistant offspring of crosses have crowns shaped like the resistant parent. Despite more than 50 years of breeding for Dutch elm resistance very few acceptable varieties possessing both disease resistance and desirable form have been produced.

An additional complication that discourages hybridization is the disease spectrum that may be present. Hybridization usually involves a gene or genes against a single pathogen, but if other diseases are present and pose equal danger then it does little good to release a variety resistant to only one disease. Most alfalfa varieties are resistant to bacterial wilt but in areas where stem nematode and Verticillium wilt are present, varieties must have resistance to all three diseases to be acceptable.

Delay in adopting disease-resistant varieties for the above reasons is exemplified by the reluctance of the banana industry in Central America to grow the more Fusarium wilt-resistant Cavendish varieties. Gros Michel was considered the best banana in the world, and until the 1940s it was the only banana variety exported from Central America, and to a lesser extent, Malaysia and West Africa. The problem with Gros Michel is high susceptibility to Fusarium wilt (Panama disease).

Dessert and cooking bananas are triploids derived from *Musa acuminata* (AA) and *M. balbisiana* (BB). Dessert bananas have the AAA genome and cooking bananas generally are AAB or ABB. In general *M. balbisiana* is less susceptible to various plant diseases than *M. acuminata*. In areas where bananas are used for local consumption, varieties are selected primarily for adaptation and these often include bananas containing a B genome. However, since bananas grown in Central and South America are primarily for export, the best banana available, namely Gros Michel, was the only one accepted and it lacks disease resistance of hybrid triploids.

There is some variation in resistance to Fusarium wilt within the AAA genome, which includes two major groups, Gros Michel and Cavendish varieties. Certain of the Cavendish group such as Lacatan and Dwarf Cavendish are immune or highly resistant to Fusarium wilt, but are very susceptible to other diseases such as leaf spot (Sigatoka). By 1947 Lacatan had replaced Gros Michel as the main export banana in Jamaica. Lacatan and Dwarf Cavendish have now largely replaced Gros Michel in Central America. Susceptibility to leaf spot is unimportant because effective fungicides are available to control it; also, higher yields and resistance to wind damage offset any disadvantages. However, it is difficult to say how long it might have taken to accept these varieties if banana producers had not run out of land on which to establish new plantations to avoid Fusarium wilt.

Sometimes varieties are inadvertently released that are highly susceptible to diseases previously of little importance. The wheat variety Paha was developed for release in the Pacific Northwest and resistance against only rust and common smut were considered. Since flag smut had never been a general problem it was not considered in breeding programs. However, when planted in field plots in some parts of the test region, Paha became severely infected with flag smut. Popular varieties such as Gaines might have 5% or less flag smut infection while Paha was 95% infected.

Cougar bluegrass was developed in Washington state because of its deep green color and short growth habit, two desirable characteristics in lawn grasses. Helminthosporium blight (melting out) had not been a serious problem in the Northwest. However, Cougar was supersusceptible and some newly seeded lawns were killed in their first year. As a result, popularity of Cougar bluegrass was short-lived.

Grafting

Many woody perennials are grafted or budded for propagation. This permits insertion of disease-resistant or -tolerant roots, interstems, or tops. Grafting leaf blight-resistant tops onto desirable latex-producing trunks to produce disease-resistant rubber trees was described earlier. Collar rot is a disease on apples in several regions of North America and Europe. Some clonal rootstocks selected for size control at the East Malling research facility in the United Kingdom vary in resistance to collar rot. Those designated as MM104 and 106 are very susceptible and should be avoided where the disease is likely to occur. Others including MM111 and EM2, 7, and 26 are moderately resistant and EM9 is highly resistant.

Cork oak is very susceptible to Phytophthora ink disease and 50 to 60% of cork oaks in the Black Sea coastal area are affected. The disease can be controlled by grafting cork oak onto roots of Iberian oak, which is not affected.

Verticillium wilt is a serious disease in young sweet cherry trees in the Pacific Northwest. If trees survive until they are 8–10, chances of severe Verticillium wilt is greatly reduced. Choosing the right rootstock can be an important factor in survival. Mazzard roots are more severely affected and killed more rapidly than Mahaleb. The latter should therefore be used where Verticillium wilt is a threat.

Induced Mutations

Many induced mutants have been used commercially to obtain growth characteristics such as reduced plant size, altered or improved flower and fruit color, larger and more uniform fruit size, and earlier fruiting and ripening. Mutations have been induced primarily by chemicals such as colchicine, methyl or ethyl-methanesulfonate, and diethyl sulfate, or by ionizing radiations such as X- or gamma rays.

There are many reports of disease resistant mutants but few of these have developed into commercial varieties. Most resistant mutations have been induced by gamma radiation. Selection methods are then used to further screen these mutants for field application. Gamma radiation-induced resistance has been reported for several diseases including stem and leaf rust of wheat; Ascochyta blight, wilt, stunt, and root rot of chickpea; potato late blight, leaf roll, and wart; sugarcane red rot and smut; grape downy mildew; citrus canker; apple powdery mildew and scab. None of these has yet reached commercial application.

Resistance to Verticillium wilt of peppermint and Scotch spearmint has also been obtained by radiation of planting material. The application of radiation-induced mutations is best documented for peppermint and was carried out by M. J. Murray of the A. M. Todd Company. Rhizomes of the primary commercial variety, Black Mitcham, were irradiated with 5–6 thousand roentgens of gamma rays. From this irradiated material 138,575 plants were grown in Verticillium infested soil and subjected to severe disease pressure. These were increased to more than 6 million plants from which 58,724 wilt-free selections were field tested in replicated plots. Twelve highly or moderately wilt-resistant strains were identified and six of these were tested regionally in Indiana, Oregon, and Washington. Eventually two selections were made that combined high disease resistance with desirable growth habit, yield, and oil quality. These became available as Todd's Mitcham and Murray Mitcham. These selections are not highly resistant and become diseased in heavily infested soil. However, they extend the life of mint plantings in lightly infested fields whereas standard varieties do not survive.

It is believed that high radiation dose inactivate preformed meristems and force development of adventitious shoots derived from single cells. This process ostensibly increases chances that cells carrying disease-resistant traits can be identified without being obscurred or masked by nonresistant tissue that is better fit to develop and grow.

This procedure has also been used to develop *Verticillium* resistant strains of Scotch spearmint. Two lines, Skot-227 and Skot-770 (named for their developer C. B. Skotland), have been released to Oregon and Washington mint growers. These varieties yield 25–50% higher than the original varieties. However, if levels of resistance are no better than they are in peppermint or the proportion of resistant selections is no higher than two out of over six million or more, this technique will likely have limited use.

Induced Resistance

Induced resistance is nonheritable resistance governed by external factors and is initiated by balanced or enhanced nutrition, stimulation of phytoalexins, chemotherapy, and cross- protection, which was discussed under biological control and more accurately called protective action. Some methods, such as improved nutrition, have practical application and are used to control a range of diseases in many crops. Utilization of phytoalexins, and to a large extent chemotherapy, is yet to be realized on a practical scale.

Nutrition

Three elements, namely, nitrogen, phosphorus, and potassium, affect relative resistance of plants to diseases. Nitrogen, for example, has two opposing effects on susceptibility. Adequate and balanced nitrogen nutrition improves plant vigor and reduces damage from facultative parasites that attack unthrifty plants. Early blight of potato and Armillaria root rot are diseases seen typically on tissue that is senescent or lacks vigor. Nitrogen also stimulates production of succulent juvenile tissues more vulnerable to attack by rusts, powdery mildews, and bacterial diseases such as fire blight of pears.

Phosphorus, together with potassium, hastens cell maturity, thickens cell walls, and increases tissue hardiness. These elements are reported to increase resistance to many diseases, especially those caused by facultative parasites. There are some contradictory reports of influence of particular nutrients on disease resistance, but these often result from nutrient imbalances rather than direct effects of the specific element. It is difficult to separate direct nutritional effects of nitrogen and phosphorus from indirect effects on soil pH and resultant influence on microbial activity or availability of other nutrients.

Phytoalexins

Phytoalexins are protective chemicals produced in certain plant tissue in response to invasion by some agents other than pathogens. The idea that host plants can be stimulated to produce compounds that repel invasion is not new but no mechanism was suggested to explain this until 1940 when it was demonstrated that potato tubers infected by the late blight fungus were resistant to decay by a different pathogen. The mechanism was believed to be localized production of inhibitory compounds subsequently named phytoalexins, which means something produced in plants that ward off attack. Subsequent research showed that inhibitory compounds accumulate in fluid around spores of fungi that are usually nonpathogenic to a particular host, for example the stone-fruit brown-rot fungus placed on bean tissue.

Phytoalexins are induced by organisms and chemical and physical agents. However, some enthusiasts define phytoalexin in a narrower sense as low-molecular-weight, antimicrobial compounds synthesized and accumulated in plant tissues after exposure to microorganisms. This has been equated to a type of cross-protection.

Phytoalexins are often given names derived from the plant in which they are produced, for example, pisatin in peas, phaseolin in beans, orchinol in orchids, and betavulgarin in beets. Phytoalexins are distinct from compounds normally present in resistant plants in that they are induced by external agents. They are also localized in plant tissues and hence do not involve the whole plant, suggesting that application to disease control will be limited. Furthermore, some phytoalexins are very phytotoxic and application to plants would be self-defeating.

Chemotherapy

Application of chemicals to impart disease resistance has been tried many times but with little success. Treatment of mycoplasma diseases with tetracycline antibiotics or vascular diseases with systemic fungicides as described in Chapter 9 does not produce resistance but only contains or eliminates already established pathogens.

Fosetyl-Al, a registered fungicide, has little or no direct effect on target fungi nor is it converted to toxic metabolites in plants. In other words, it is neither fungicidal nor fungistatic. However, it appears to enhance production of capsidol, a phytoalexin, in tobacco plants inducing hypersensitive-like reactions to *Phytopthora nicotianae* similar to those exhibited by genetically resistant seedlings. A similar action is induced by dichlorocyclopropane against rice blast. This chemical has no antifungal activity but stimulates phytoalexin production in rice plants and renders them resistant to blast.

Genetic Engineering

Genetic engineering is a recent catch-phrase in the scientific arena and is generating considerable interest in commercial areas. Genetic engineering is aimed at inserting foreign genes into organisms by some means; so far it has been applied mostly to microorganisms. Much has been written about genetic engineering, but major applications of transgenic plants have yet to be made.

The strategy behind genetically engineered plants for disease resistance is sound; simply insert a gene for resistance from some foreign source into desired plants. The technology is available. An important method used to insert genetic material into plants employs tumor-inducing plasmids of crown gall bacteria as vectors. One nonvector method uses a "gene gun" to fire DNA-coated metallic particles into plant protoplasts. Presumably, some DNA becomes part of nuclei. In both methods treated cells are propagated by tissue culture techniques to produce plants.

Forecasts in scientific and commercial literature indicate that alfalfa, rice, potato, tomato, tobacco, cantaloupe, and other plants with engineered resistance to viral diseases are in the making. However, these are not yet available for commercial agriculture. This does not mean that genetic engineering is without merit, but that too much may have been expected too soon. There appears to be the same governmental obstacles to field testing transgenic plants that there is to biological control agents that delays development and use.

MECHANISMS OF RESISTANCE

The mechanics of resistance at cellular or tissue levels are not understood in many cases, but some mechanisms have been postulated or demonstrated. Resistance mechanisms are described in various ways and probably do not function singly but more likely in combinations.

The major mechanisms that have been proposed are either passive or active. Passive mechanisms involve preexisting resistance that is normally present in healthy plants. Included are such characteristics as cuticle thickness, presence of phenolic or other fungitoxic (or fungistatic) compounds, and cork or lignin barriers. Active resistance (i.e., induced or provoked resistance) develops in response to parasitism. This includes production of fungitoxic substances such as phytoalexins, resins, and gums, wound meristems that produce cork or suberin barriers, and thickened or altered cell walls including formation of tyloses, lignitubers, and papillae, which prevent spread of pathogens. Other active mechanisms of resistance include physiological or functional reactions that impede entry or reduce development of pathogens, such as stomatal structure or function, closed flowering habit, absence of nectaries, hypersensitivity, and slow rusting phenomenon. These proposed mechanisms can be combined into three simple categories, mechanical, chemical, and functional.

Mechanical Resistance

Cuticle is a plant's outermost barrier to invasion. Thickened cuticle resists penetration and renders older leaves of citrus more resistant to anthracnose than young leaves. Thick cuticle is also associated with resistance to direct penetration by powdery mildews, some rusts, gray mold, and coffee berry pathogen. Thicker epidermal cells with higher silica content are believed to account for brown spot resistance in rice.

The role of gums and resins in disease resistance is problematic; in fact, these materials may result from disease and not be resistance mechanisms. Wound gums fill intercellular spaces and perhaps create barriers to further penetration of tissues in diseases such as Cytospora canker of stone fruits and citrus melanose. Resin formation in conifers is associated with resistance to annosus root rot. Conifers with high resin production such as spruce and pine are reported to be more resistant than Douglas fir or larch, which produce less resin. This correlation is substantiated in part by reports in the western United States that nonresinous conifers such as hemlock and true firs are damaged more by annosus root rot than are resinous conifers.

Modified cell walls are considered to be active disease defense mechanisms. Tyloses, for example, are balloon-like formations in xylem cells that develop from xylem parenchyma. They are common in vascular wilts and are thought to contribute to resistance. Tyloses develop sooner and more extensively in Lacatan, a Fusarium wilt-resistant banana variety, than in the very susceptible variety Gros Michel.

Lignin and suberin deposits reduce cell to cell spread of late blight fungus and soft rot bacteria in potato tubers. They may also cause sufficient cell wall thickening to block plasmodesmata and thus prevent virus movement from cell to cell. Rapid lignification in leaf tissue restricts lesions of cucumber scab or gummy stem blight and results in fewer, smaller lesions. Lignification is slower in susceptible varieties.

Barrier zone formations occur as nonspecific resistance mechanisms in vascular wilts of trees. These zones promote recovery and survival of infected trees by walling off invaded tissues with atypical xylem rings. This mechanism is reported in Dutch elm disease, Verticillium wilts, and Fusarium wilt of mimosa. Similar containment is reflected by the "compartmentalization of decay in trees" (CODIT) concept. Lignin and other polyphenols deposited in and adjacent to wounded or diseased tissue prevent spread of fungi beyond the original invaded cylinder.

Cork barriers consist of new tissue that develops in response to infection and that restricts pathogens. This occurs in citrus scab, fire blight of rosaceous plants, white pine blister rust, common scab of potato, and many leaf spot diseases. An abscission layer and periderm produce corky layers that fall away, or remain attached as scabs.

Lignitubers and papillae are deposits of dense material in and around cell walls, formed in response to infective structures. They are implicated as nonspecific resistance mechanisms in take-all and strawbreaker foot rot of wheat, powdery mildew of cereals, pink root of onion, and Verticillium wilts.

Chemical Resistance

Certain compounds have been implicated in disease resistance but the role of most has not been clearly defined. Their involvment is usually based on the presence in plants and *in vitro* demonstration of some fungitoxic action. These chemicals include those that are present before invasion and those produced in response to invasion, such as phytoalexins.

Catechol and protocatechuic acid are classical examples of chemicals associated with resistance to disease. These compounds are present in scales of smudge-resistant red and yellow onions but are absent in susceptible onions with colorless scales. Chlorogenic acid in plant tissues is related to resistance in Verticillium wilt of potato, pear scab, and bull's eye rot of apples. Caffeic acid is associated with sweetpotato black rot, and juglone with resistance to branch wilt of walnut. These chemicals are all phenolic compounds. Borbonol is a preformed, fungitoxic, nonphenolic ring compound in avocado species that are resistant to Phytophthora root rot. It is present in resistant *Persea borbonia* and *P. caerula* but not in susceptible *P. indica* or only in small amounts in *P. americana*.

Functional Resistance

Small or closed stomata are less subject to penetration by fungi causing cereal rusts, brown rot of stone fruits, sugar beet leaf spot, and other pathogens that enter through stomata. Stomatal morphology can also contribute to resistance. Mandarin types of citrus, for example, are more resistant to citrus canker because overarching ridges project from stomatal guard cells and prevent entry of canker bacteria.

Closed (cleistogamous) flowers of cereals are less subject to infection by loose smut or ergot. In contrast, rye, triticale, and some barley varieties that are cross-pollinated have open flowers and hence are more susceptible to loose smut and ergot than cereals like wheat that are self-pollinated and bear closed flowers. Because there are no extrafloral nectaries on bracts and leaves of certain lines of cotton there is a klendusic type of resistance to fungal boll rots.

Hypersensitivity and slow rusting are difficult to categorize, but probably qualify as functional types of resistance. Hypersensitivity is a pronounced local reaction to invasion by pathogens resulting in small areas of killed cells. This reaction is most frequently associated with rust diseases of cereals and grasses and is characterized by a range of pathogen responses and symptom expression. Infection-type classes for rusts reflect hypersensitivity and range from 0 to 5, in which 0 is an immune reaction where no rust pustules develop but at times accompanied by small flecks of dead tissue. Classes 1 and 2 are very resistant and moderately resistant, respectively, and bear small to medium rust pustules surrounded by either dead tissue (class 1), or bands of chlorotic and dead tissue (class 2). Classes 3, 4, and 5 are degrees of susceptibility differing only in pustule size and having chlorotic zones or none around pustules (no hypersensitivity).

Hypersensitive reactions also occur in a number of other diseases caused by obligate parasites such as downy and powdery mildews. Reactions similar to hypersensitivity can occur in diseases caused by facultative saprophytes, such as potato late blight, bacterial blight of cotton, and some fungal leaf spots.

Slow rusting is a phenomenon reported in wheat, barley, and corn, in which development of rust sori in fields and on individual plants is much slower than on other varieties. In some instances this type of resistance is considered partial resistance and in others adult plant resistance. Sometimes it is said to be specific and at other times nonspecific or general. Whatever the mechanism it is better described than understood.

TOLERANCE

Tolerance to disease is a pragmatic and useful form of disease control. The concept, however, is a subject often debated. Some consider tolerance and resistance as synonymous. One worker selecting for snow mold resistance in wheat admits that the mechanism is probably tolerance but prefers to use resistance. Some say tolerance is a vague, ill-defined concept. Others consider it a valid term when applied to infection by latent viruses, but not for other pathogens. Then there is a middle ground, wherein tolerant plants become diseased but produce nearly normal crops. The concept often centers around foliar diseases in which symptoms and signs appear to the same degree on two groups of plants, but one group yields better than the other and is thus said to be tolerant. Table 19.3 compares relative pathogen and plant growth in tolerant hosts with those that are susceptible, intolerant, or resistant.

TABLE 19.3. Pathogen and Plant Growth Characteristics in Plants with Different Levels of Resistance

Degree of Resistance	Pathogen Growth	Plant Growth
Susceptible	Good	Poor
Intolerant	Poor	Poor
Resistant	Poor	Good
Tolerant	Good	Good

One of the earliest examples of disease tolerance was seen in selecting sugar beets for "resistance" to curly top. Beets were planted next to dessert breeding grounds of leafhopper vectors and surviving plants selected for breeding and propagation. These "resistant" plants sometimes showed symptoms of disease and often contained virus at levels similar to those of severely affected plants. In other words, plants tolerated both disease and pathogen. Repeated selections resulted in a many-fold increase in sugar beet yields. Table 19.4 shows the progress made through early selections for curly top "resistance." These selections have been greatly improved, but even today the sugar beet industry in curly top areas is dependent on disease-tolerant varieties.

A major deficiency of tolerance in comparison with true resistance is that tolerance allows continued production of pathogens. Tolerant varieties have little impact in slowing disease epidemics since they serve as reservoirs of pathogens and vectors. It was believed that curly top might be less serious on other crops such as tomato, melons, and beans in Washington state when sugar beet production was terminated. However, there are so many wild hosts that serve as reservoirs for both virus and vector that no major change was seen although the disease may be more sporadic than before. Another limitation in breeding for tolerance is difficulty in demonstrating its presence. To demonstrate tolerance in plants there must be some means of controlling diseases, such as fungicides for foliar pathogens, soil fumigants for root diseases, or placing plants in screened cages to protect them from vectors.

TABLE 19.4. Yields from Early Sugarbeet Selections under Severe Curly Top Pressure

Variety	Description	Root yields (T/acre)
Old type	Susceptible	0.7
US 1	First "resistant" variety	6.3
US 33	Moderate but uniform "resistance"	8.4
US 12	Selection from US 1	11.3
US 22	Selection from US 1	14.3
Improved US 22	Selection from US 22	16.6

BREAKDOWN OF RESISTANCE

Some types of resistance often fail suddenly. Failure is attributed to development of pathogenic races (i.e.,physiologic races) that render varieties obsolete. Because of extensive breeding of wheat and other cereals for disease resistance most information about pathogenic races involves pathogens of cereals, especially rusts. However, many races also exist especially for pathogens of smuts, late blight of potato, Fusarium wilts of peas, bananas, and other crops, southern leaf blight of corn, bean anthracnose, and many more. Most plant pathogens probably have the potential to produce pathogenic races but host varieties resistant to all extant races must be available to identify new races. A range of host plants having different reactions to different races and used to identify races are called *differential hosts*.

Pathogenic races were first demonstrated in bean anthracnose. Farmers in New York state complained that resistant bean varieties became susceptible to anthracnose. M. F. Barrus of Cornell University obtained isolates of the pathogen from diseased plants that previously were considered resistant and inoculated a range of resistant and susceptible bean varieties. Reactions were always definitive, either disease or no disease, indicating that differences were not in degree of virulence but inherent and absolute. Table 19.5 shows how differential bean varieties are used to distinguish four races, alpha, beta, gamma, and delta, of bean anthracnose. This table also illustrates race determination for leaf rust of wheat and Fusarium wilt of peas.

The genetic explanation of formation of pathogenic races evolved from the gene-for-gene theory developed by H. H. Flor. He studied inheritance of pathogenicity in the flax rust fungus and concluded that for each resistance gene in the host there is a matching or complementary virulence gene in the pathogen. This concept applies best in systems where host resistance is conditioned by major genes (i.e., specific or vertical resistance). In addition to flax rust other host–parasite systems where a gene-for-gene relationship has been suggested include leaf, stem, and stripe rusts, loose smut, common and dwarf smuts, and powdery mildew of wheat; stem rust, loose smut and Victoria blight of oats; coffee rust; apple scab; late blight and wart of potato; and tomato leaf mold.

A widely planted variety that holds up year after year is considered to have stable or durable resistance. Stability (durability) is a desirable feature of resistance and is usually obtained with general or nonspecific resistance. However, the well-known example of stability from type A resistance to cabbage yellows is an exception. Perhaps there are others.

Failure of vertical or race-specific resistance can be sudden and absolute. This requires continuous effort to develop new varieties to replace those lost as resistance fails. Wheat varieties grown in the Pacific Northwest provide a good illustration of this. It is estimated that varieties with race-specific resistance to smut or rust last only 3 to 8 years before new pathogenic races render them obsolete. Figure 19.4 shows the rise and fall of four predominant wheat varieties. Elgin and Elmar failed because of races of common smut. An effective seed treatment for smut (HCB) came into use in the mid-1950s; failure of a variety after that was due to stripe rust. Gaines, Nugaines, and subsequent varieties with a nonspecific (i.e., horizontal) type of resistance to stripe rust

TABLE 19.5. Differential Hosts Used to Distinguish Pathogenic Races of Bean Anthracnose, Leaf Rust of Wheat, and Fusarium Wilt of Peas

	Reaction to Indicated Races Bean Anthracnose Race			
Differential Variety	Alpha	Beta	Gamma	Delta
Michelite	S	R	R	S
Dark Red Kidney	R	S	S	S
Perry Marrow	R	R	S	S
Sanilac	R	R	R	S

				Leaf Rust Race Group					
Differential Variety	1	2	3	4	5	6	7	8	9
Malakoff	R	R	R	R	S	S	S	S	S
Webster	R	R	R	R	R	R	S	S	S
Loros	R	R	V	S	R	S	R	S	S
Mediterranean	R	S	S	S	S	S	S	S	R
Democrat	R	S	S	R	S	S	S	R	R

	Pea Wilt Race			
Differential Variety	1	2	5	6
Little Marvel	S	S	S	S
Darkskin Perfection	R	S	S	S
New Era	R	R	S	S
WSU 23	R	R	R	S

have been reasonably durable. However, varieties such as Moro and Tyee have single gene resistance and pathogenic races soon appeared. When these races appear they also attack other varieties with the same source of resistance. For example, Faro and Weston have the same gene for rust resistance as Moro, and their future was also jeopardized by appearance of the Moro race.

STRATEGIES FOR MANAGING RESISTANCE FAILURE

When a single genotype is grown over a wide area, loss of resistance can be dramatic and devastating. A recent example is the southern corn leaf blight epidemic in the United States in 1970. It was discovered in the 1930s that hybrid corn could be produced without laborious and costly detasseling by incorporating male sterility in female parents. A useful cytoplasmic gene for male sterility was discovered in Texas and became identified as *Tcms*. By 1970 most hybrid corn produced in the United States contained the *Tcms* gene. About 1961 plant breeders in the Phillipines observed that lines with this cytoplasmic gene were very susceptible to southern corn leaf blight.

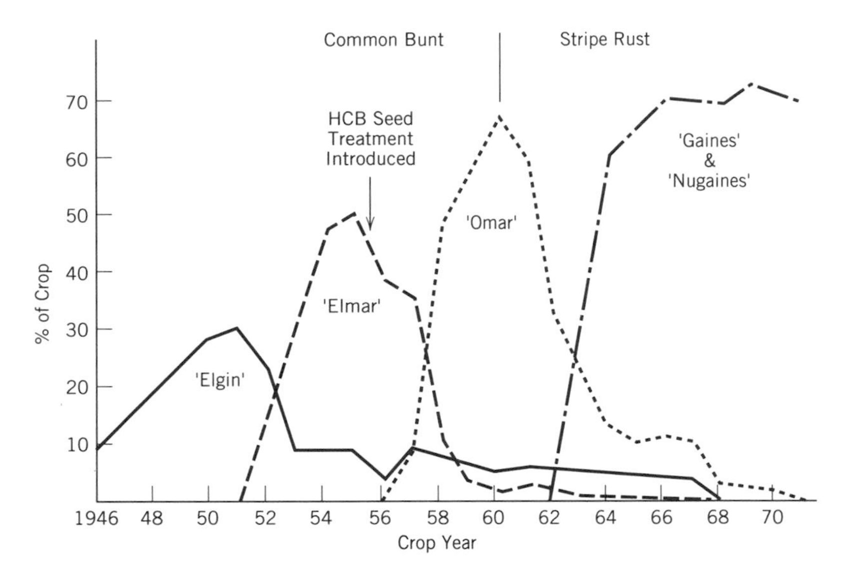

Figure 19.4. Major wheat varieties in the Pacific Northwest that failed because of pathogenic races of common smut and stripe rust.

Until 1969 southern corn leaf blight was considered an unimportant disease, but that year the disease was severe in Iowa, Illinois, Indiana, and Minnesota, the heart of the corn belt. By then two races of leaf blight fungus were recognized. Race "O" was much less virulent than race "T," which attacked *T* cytoplasm. The "T" race could develop at lower temperatures than the "O" race and reproduce more rapidly. In 1970, blight started in Florida in February and by September had spread throughout the eastern half of the country. Losses were estimated at one billion dollars and there was much anxiety about the coming year. However, no epidemic developed the following year. Corn seed producers in Brazil increased seed stocks that carried other male sterility cytoplasm (*Ncms*), which is not affected by the "T" race. In 1971 corn yields were back to normal.

A similar dramatic breakdown involving widespred use of a single gene occurred earlier with Victoria oats and a different Helminthosporium blight. Victoria had been developed for resistance to crown rust. By the mid-1940s virtually all oats grown in the midwest carried the Victoria gene for rust resistance and were severely damaged by blight. Fortunately, a resistant variety, Bond, was being used for breeding and some of its progeny was increased to end the epidemic.

These examples illustrate the importance of having genetic diversity to manage and contain plant diseases. Such diversity is realized in several ways, the most common being nonspecific major gene resistance, gene stacking or pyramiding, multilines, and gene deployment.

Nonspecific major gene resistance is based on a few closely related major genes (oligogenic) rather than a single major gene (monogenic) characterized as vertical resistance, or polygenic horizontal resistance. This results in genotypes that are more durable than vertical resistance and have higher resistance levels than horizontal resistance.

Gene pyramiding or gene stacking is accumulation of several genes into single lines or varieties to provide long-lasting disease resistance. Genes introduced to form pyramids or stacks can be either major or minor genes, or both, or any other type of gene that imparts disease resistance. This strategy has been proposed mainly for cereal rusts, in which the largest inventory of resistance genes is available. It has also been suggested for powdery mildew of wheat and northern leaf blight of corn.

Multilines and varietal mixtures avoid genetic monocultures. They involve planting mixtures of isogenic lines or varieties that have similar agronomic or utilitarian properties but different sources of disease resistance. When Stephens wheat became popular in the Pacific Northwest because of its high yields, growers often hedged against its lack of winter hardiness by planting mixtures with more cold hardy varieties such as Daws. In the same way the white winter wheat varieties Crew and Rely are physical mixtures of 10 similar wheat lines having different types of resistance to stripe rust. If a rust race arises that attacks one of the 10 components that line is replaced with another. This virtually precludes catastrophic crop failures from stripe rust.

Miramar 63 was a multiline mixture developed by CIMMYT, the international wheat development program in Mexico, by crossing a Brazilian wheat with about 600 varieties and lines. From these crosses selections were made to incorporate 10 of the best lines giving resistance to both stripe and stem rust into a physical mixture. Within 2 years stem rust was attacking two components of the mixture but total losses were less than the 20% expected, probably because of compensating effects by the other eight lines. The two susceptible lines were removed and replaced by four new lines to produce a new multiline, Miramar 65.

Use of multilines has been most effective against cereal rusts where development of races is rapid and spread extensive. A large diversity of multiline components held in reserve provides for almost instant synthesis of cultivars with general resistance.

Multilines have been suggested as a means of combatting rice blast. Some workers believe this is not feasible since in the tropics where two rice crops are grown in a year the fungus can generate races faster than plant breeders can develop isogenic replacments for susceptible lines. Furthermore, need for repeated backcrossing limits yield potential in new lines to that of the recurrent parent and improved yield is usually a major objective of plant breeding.

Gene deployment restricts spread of pathogenic races by utilizing different genes for resistance in a patchwork fashion over large areas. Advantage of deployment of different genes is apparent in the central wheat-growing region of the United States where wheat is grown in a continuous band from northern Mexico into southern Canada. Stem rust follows as wheat crops develop from south to north. If a single wheat genotype for rust resistance is grown over much of this region, a race of rust that attacks that gene can spread without interference. Planting varieties with different genes for resistance in different zones within the region prevents a race from moving into an adjacent zone. For example, variety A is planted in Texas and Oklahoma, variety B with a different gene for resistance is planted in Kansas, a third gene for resistance in variety C is planted in Nebraska, and a fourth variety D in the Dakotas. If a race that attacks variety B appears, it will be confined to the Kansas region. This pattern has been used in the central United States to deploy different genes for resistance to crown rust of oats.

An important requirement for effective gene deployment over political and economic boundaries is the need for cooperation among plant breeders to orchestrate gene selection and deployment.

This concept also works on a smaller scale. Figure 19.4 shows that a single variety, Omar, comprised two-thirds of wheat grown in the Pacific Northwest in 1960. The severe stripe rust epidemic that followed resulted largely from extensive planting of this single variety and emergence of a pathogenic race of the fungus. Introduction of the variety Gaines, which has nonspecific resistance, ended that epidemic. Several years later when a pathogenic race was discovered that could attack the specific single gene resistance of Moro, that variety was not grown in sections where the new race had been found. This restrictive deployment of a genetic line allowed Moro and similar varieties to be grown for an additional period.

DISEASE ESCAPE

Disease escape is the bane of plant breeders. Plants can escape disease either because pathogens are absent or environmental conditions are unfavorable for infection. The latter is sometimes called disease avoidance and in this book has been treated as a strategy under the general principle of protection. Survivors in a plant population subjected to severe disease pressure simply may fail to become infected rather than be inherently resistant. Modified behavior of insect vectors such as avoiding shaded areas can result in plants escaping viral infections. In populations of diseased plants these individuals appear to be resistant.

Selective screening for pink and speckled snow mold resistance in wheat must involve date of planting as a relevant factor in disease development. Early seeding favors survival because plants develop large crowns with abundant food reserves. Very late seeding produces small plants that escape severe impact by the pathogens. Plant breeders can regulate degree of selection pressure in this instance by altering seeding dates.

Commercial nurseries can also be deceived by disease escape. Stewart's Bartlett pear is an example of an apparently disease-resistant clone being marketed and then found to have no resistance. Fire blight is a severe disease and Bartlett pear is both a favorite market variety and very susceptible to fire blight. In 1959 a pear grower in central Washington found a disease-free branch on an otherwise severely blighted Bartlett tree. Inoculations made on the branch in the field indicated that the branch was blight tolerant and it was suggested as a replacement for the completely susceptible Bartlett. By 1964 Stewart's Bartlett was widely advertised and being sold for this purpose. Ten trees purchased from a commercial nursery were planted in the eastern United States as part of a USDA pear breeding program at a location where trees are spontaneously infected by blight. At such a location the Stewart strain was as susceptible as standard Bartlett, or more so. Subsequent tests in Washington state showed that trees propagated with buds from the Stewart tree were as susceptible as any other pear. However, inoculations on the branch still on the original tree demonstrated that tolerance was still there. No explanation was ever given for this phenomenon, but it certainly appears to be some form of disease escape.

REFERENCES

*Anon., *Plant Mutation Breeding for Crop Improvement.* 2 Vols. International Atomic Energy Agency, Vienna, 1991.

*Bailey, J. A., and J. W. Mansfield (eds.), *Phytoalexins.* John Wiley, New York, 1982.

*Borojevic, S., *Principles and Methods of Plant Breeding.* Elsevier, New York, 1990.

*Broetjes, C., and A. M. van Harten, *Applied Mutation Breeding for Vegetatively Propagated Crops.* Elsevier, New York, 1988.

*Clarke, D. D., Tolerance of parasite infection in plants. In Wood, R. K. S., and G. J. Jellis (eds.) *Plant Diseases: Infection, Damage and Loss.* Blackwell, London, 1984. Pp. 119–127.

*Grierson, D., (ed.), *Plant Genetic Engineering.* Blackie and Son, London, 1991.

*Huber, D., The use of fertilizers and organic amendments in the control of plant disease. In Pimentel, D. (ed.) *CRC Handbood of Pest Management in Agriculture.* CRC Press, Boca Raton, FL, 1991. Pp. 405–494.

*Jones, D. G. and B. C. Clifford, *Cereal Diseases: Their Pathology and Control.* 2nd ed. John Wiley, New York, 1983.

*Lenne', J. M., and D. Wood, Plant diseases and the use of wild germplasm. *Annu. Rev. Phytopathol.* 29:35–63 (1992).

*Nelson, R. R., (ed.), *Breeding Plants for Disease Resistance.* The Pennsylvania State University Press, University Park, 1973.

*Robinson, R. S., Disease resistance terminology. *Rev. Appl. Mycol.* 48:593–606 (1969).

Van der Plank, J. E., *Disease Resistance in Plants.* Academic Press, New York, 1968.

*Wood, R. K. S., *Physiological Plant Pathology.* Blackwell, Oxford, UK, 1967.

* Publications marked with an asterisk (*) are general references not cited in the text.

20 Integrated Disease Management

Diseases can be controlled or they can be managed. Control has a more absolute connotation; it implies corrective measures after disease has occurred, and is thought by some to involve radical measures such as application of pesticides (i.e., the desperate remedies of Hippocrates). Management carries a softer intent, more like preventive medicine than radical surgery. It is based on containing damage or loss below certain economic levels. Management conveys the idea of a continuous process in which disease is one component of an agroecosystem. Management implies less dependence on chemical controls and reduction of possible detrimental effects of chemical and biological interactions on nontarget (i.e., beneficial) organisms.

Concerns about pesticides on the environment and human health brought a new era of concepts and attendant terms such as integrated pest management (IPM), integrated plant protection (IPP), integrated crop management (ICM), integrated disease management (IDM), holism, plant health, expert systems, biocontrol, biotechnology, genetic engineering, molecular biology, Master Gardener, and Phytonarian. Of all these terms, integrated disease management is probably most accurate and descriptive.

Probably no single event gave greater momentum to this new era than Rachel Carson's *Silent Spring*. This book was an indictment of the pesticide industry in general and led to drastic changes in the philosophy and practice of pest control. Even though plant pathologists had considered their use of chemicals more benign than the stomach poisons and nerve toxins used by the "bug people," all branches of pest and disease control came to be painted with the same brush. Reluctance to use the term "control" resulted from increased complaints and litigation over use of pesticides. These actions result from real or imagined damage to property or from failure to achieve some expected level of control. In either case there has been a move to avoid strong and promising words such as "control" and "recommendation."

The present era reflects a pronounced and deliberate aversion to use of chemicals as principal means of pest control. It is noteworthy, however, that of the three major groups of plant pests, namely, insects, weeds, and pathogens, control of the latter, historically, has been least dependent on chemicals. This was largely because effective materials were not available and, out of necessity, many diseases were controlled by nonchemical methods. Many integrated management concepts were crystallized by entomologists even though basic components had previously been utilized by plant pathologists.

IPM

The first highly publicized move away from the term "control" was creation of the integrated pest management concept (IPM) in entomology. This concept began, or was formalized, in the 1950s to describe methods of selecting pesticides, adjusting dosages, and timing applications to control harmful insects and mites while preserving beneficial arthropods. It was an integration of chemical and biological control methods. This somewhat restricted concept has been expanded and qualified (see pp. 41-51 in the Hoy and Herzog reference, Chapter 16) to become "a pest population management system that utilizes all suitable techniques in a compatible manner to reduce pest populations and maintain them at levels below those causing economic injury. Integrated control achieves this ideal by harmonizing techniques in an organized way, by making control practices compatible, and by blending them into a multi-faceted, flexible, evolving system." This definition has been further modified and simplified by others, and it can be argued that these simplifications do little more than compound disease control measures rather than integrate them.

Germany officially recognizes integrated plant protection (IPP) as a combination of procedures, including biological, biotechnical, plant breeding, and cultural methods that limit application of chemical plant protection agents to only those needed. An even simpler definition is that integrated control of plant disease includes all tools used to reduce dependence on a single control practice. These disregard the blending and harmonizing requisites of earlier definitions.

Since IPM is generally thought of as pertaining to insects and mites, derivatives such as integrated disease management (IDM) and integrated plant protection (IPP) attempt to identify the concept with disease management. Various presentations help make concepts of disease management, and nature of plant diseases in general, easier to comprehend by the public. Weeds and insects are tangible, usually visible, entities generally recognized as such by most people. Diseases present a problem because causal agents (i.e., the "pest") are usually "invisible." This means that when disease control is described it really means detection and management of a somewhat abstract agent (i.e., the pathogen).

Application of IPM principles is described in publications from several states and regions. *Integrated Pest Management of Tomatoes* and *Integrated Pest Management of Cole Crops* are publications of the California Integrated Pest Management Project. *Integrated Crop and Pest Management for Small Grains* came from Montana State University Extension Service. *Integrated Pest Management of Potatoes* is a product of the Western Regional Potato Project. These publications include a range of information concerning production of these crops, most of which directly relates to control, such as growth and development of plants, cultural practices, identification of pests, diagnosis of diseases, and pest and disease monitoring, all of which are essential for implementing effective disease control programs, integrated or otherwise. In most cases the integration process is not clearly defined and implementation is left to growers. This is necessary because conditions vary widely in agriculture; it also demonstrates the value of public and private advisory specialists to counsel growers and homeowners as to the best combinations of measures for particular situations.

PLANT HEALTH MANAGEMENT

Healthy plants function within a range above and below an optimum level. This range or zone of normal plant development is sometimes called the latitude of health. When plants cease to function within this latitude they are considered to be unhealthy, abnormal, or diseased. Thrifty, vigorous plants are less subject to attack by many pathogens and for this reason maintaining healthy plants is the front line of defense against disease. Plant health was a primary concern of Emil Rostrup and his student, J. L. Jensen (of hot water treatment fame), in the 1800s. This was a uniquely Danish concept according to several historians of plant pathology.

Despite this early recognition of the concept, attention paid to plant health by plant pathologists has been spotty. In many instances no mention is made of plant health although it must be assumed to be the very heart of disease control. On its 75th anniversary in 1983 The American Phytopathological Society (APS) published a volume of contributed papers collectively titled *Challenging Problems in Plant Health*. Some papers were developed around a central theme of plant health, others did not mention plant health, and a few used the phrase only in the title. Almost 10 years later the APS began a series of Plant Health Management publications. The first of these, *Wheat Health Management,* is an in-depth treatise of various practices involved in producing crops of healthy wheat. Presumably, other major crops will receive similar treatment.

The United Kingdom has a government agency called the Plant Health Service that sends reports on disease interception and diagnoses by electronic mail. This speeds up reporting procedures and allows rapid summation of pertinent information. These reports appear to be similar to Plant Disease Surveys of the USDA. Most developed countries provide similar services by agencies with various names, some suggesting plant health and others reflecting plant protection. The APS publication *Plant Disease* started a new section called "Plant Health Strategies" in September 1983 and initially drew a number of appropriate papers. However, contributions have been sporadic and mixed; some on diagnosing and some on monitoring and management.

HOLISM

The term holistic health found its way into plant pathology writings without much consideration of the origin of this term. One use defines it as the "whole health of wheat managed through a combination of biological, physical, and chemical approaches integrated for maximum overall effects" (Cook and Veseth, 1991), This virtually equates holistic health with integrated disease management.

The term "holism" was coined in 1926 by Jan Christian Smuts, famous South African military and political leader and philosopher, in his book *Holism and Evolution*. However, the concept probably preceded Smuts by many centuries or even millenia. Smuts developed his thesis that an organism is a living world in itself and is more than the sum of its parts. While the concept drew heavily on Darwin's ideas on evolution of organisms, Smuts did not restrict holism to biological systems, but included personality and society as holistic entities. A basic tenet of holism is that small units must develop or combine into larger wholes for survival. Four political entities directly attributed to Smuts, The Union of South Africa, The British Commonwealth of Nations, The League of Nations, and The United Nations, are tangible manifestations of holism as originally conceived.

Holism appears to have been first connected with health in its adoption by some areas of human medicine, particularly those associated with nontraditional areas such as naturopathy, herbalism, acupuncture, and psychic healing. The basic principle is one of treating whole patients rather than just afflicted organs. So the objective of holism is health of patients, wheat plants, or crops as whole entities. Like other terms that start with a specific meaning, holism and holistic may eventually come to mean whatever writers want them to mean, but for now it appears to be mainly a trendy catch-phrase. One reviewer of the holistic health movement commented that "the words *holism* and *holistic* are being tossed about as enthusiastically as a frisbee in springtime" (Alster, 1989).

THE AGROECOSYSTEM

An agroecosystem is a real and functioning system that includes all factors that influence development of crops. Probably no word more accurately reflects the influence of many external factors on disease management than the agroecosystem concept. It regulates need, application, and effectiveness of disease control programs. Influencing factors function at different levels within the system. They have been expressed and illustrated in a variety of ways, but basically crops form the base of an inverted pyramid made up of these influencing factors. All of these factors play some role in implementing and integrating disease control or management. Figure 20.1 illustrates the hierarchy of these influencing components. Each level influences components of levels below it. Most of these components have been discussed to some degree in earlier chapters and will be repeated briefly.

The schematic flow chart of an agroecosystem hierarchy also emphasizes the involvement and integration of various components in disease management programs. Producers (farmers) have little direct control over universal influences, but must adjust to them. *Climate* largely dictates what crops can be grown and the manner in which to grow them. Extremes of temperature, moisture, and daylength are critical components; they not only determine what crops can be grown but also what diseases become severe. *Economic* constraints determine to what extent growers can utilize various control measures. *Society* establishes its preferences for certain commodities and the quality standards for them. *Politics* enters into decision-making processes by regulating movement of plant products through quarantines, embargoes, price supports, and other devices. Politics may also determine whether certain plants can be grown in a particular area. The one variety cotton law of California is an example in which use of resistant varieties is not allowed unless they are of certain fiber types. Similarly, rapeseed for industrial oil cannot be grown within a certain distance of edible oil rapeseed areas because cross-pollination of the two types produces oil worthless for either use. *Ecological* considerations such as air polluting smoke from grass seed-field burning

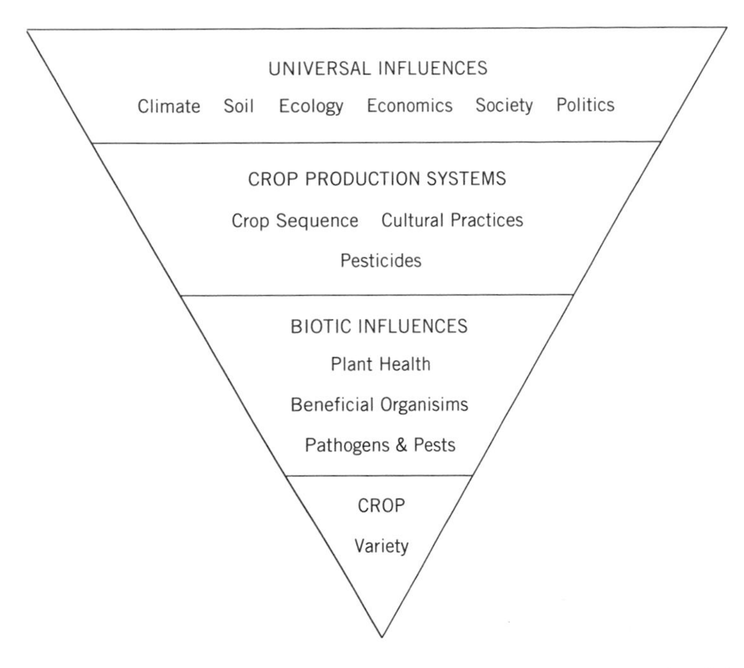

Figure 20.1. An agroecosystem hierarchy and the major components of each level.

arouse society, which in turn applies pressure to political forces. *Edaphic* (soil) factors also determine to some extent what crops can be grown and the cultural practices required to produce them. Sandy soils require frequent irrigation, soils with high water tables require drainage, and so on.

Growers have great choices in selecting crop production systems. General cropping systems differ depending on whether a crop is perennial, such as fruit trees, or annual, such as wheat. Economics is a major consideration within these systems. For example, producers of long-term crops of relatively low value, such as forest trees, will use less intensive disease control practices than citrus growers. Even within a single crop, its potential return is the determinant. A dryland wheat farmer with a potential yield of 35 bushels per acre is unable to afford as intensive a disease control program as an irrigated wheat farmer a few miles away who gets 150 bushels per acre.

Growers have many options, some of which are directly related to disease control. Cropping sequence determines whether wheat growers should resort to monocultures to induce disease decline and provide long-term control of take-all, or plan short rotations to noncereal crops to give fast but temporary control. Farmers influence plant health and microbial activity by fertilization with inorganic chemicals or animal and plant manures. Various effects on pathogens and disease were discussed in earlier chapters. Producers have many choices of cultural practices to influence biotic factors in agroecosystems.

There are two kinds of biotic influences. One is the plant and its relative health. Healthy plants are less subject to many diseases than plants on the outer fringes of good health. The other major biotic influence is the pathogen, and any attendant factors such as vectors of viruses. Biotic influences indirectly involved include weeds, insects, birds, and rodents, all of which can affect plant health. There are also many beneficial organisms such as nitrifying and nitrogen-fixing bacteria, mycorrhizae, and a variety of microbial antagonists that function in biological control.

The primary determinant in disease management is the crop. A good resistant variety eliminates the need for decisions associated with diseases. In most cases, however, varieties have little or no inherent resistance. Apart from resistance, plant vigor and density are components that can mitigate disease damage.

The foregoing factors are carefully considered in adopting and implementing disease control measures. Involvement of these components is especially critical to efficient integration of disease management with other crop management practices.

OTHER DISEASE MANAGEMENT AIDS

In addition to IPM, holism, and plant health, several other concepts are being introduced into plant pathology and are in part related to disease control. These serve important functions but are not solutions in themselves. Like those activities described earlier in this chapter these are sometimes given more credit and generate higher expectations than is realistic.

Expert systems are computerized decision-making aids similar to those used in disease predictions. The rationale for expert systems is to simulate reasoning processes of human experts. They are based on accumulation and evaluation of basic data and involve an IF–THEN reaction. *IF* certain conditions prevail *THEN* certain actions are warranted. So far expert systems have been utilized primarily as diagnostic aids. The first plant disease expert system was used in Illinois to diagnose 17 soybean diseases. It and two other systems dealing with diseases of apples and muskmelons are now being used commercially in the United States. Other systems proposed in this country are grape disease management, turf disease diagnosis, wheat disease management, and white pine blister rust hazard site prediction. EPIPRE, a prediction system described in Chapter 5, is a type of expert system for wheat disease management now being used in some areas of Europe. STAR and TUBERS are expert systems for disease diagnosis in France. Expert systems are best suited to high value crops where pesticide applications are options in disease management schemes. In this regard expert systems are extensions of disease prediction programs.

Biotechnology appears to mean different things, depending on its usage. It generally involves molecular biology and cell, tissue, and fermentation technologies applied to solution of practical problems. In relation to disease management biotechnology includes protoplast, cell, callus, and tissue culture to produce disease-resistant or pathogen-free plants. Molecular biology is used in pathogen identification, study of host–parasite relations, and induction of disease resistance or cross-protection. Genetic engineering holds promise in recombinant DNA procedures to develop disease resistant plants.

APPLYING DISEASE MANAGEMENT TECHNOLOGIES

Transfer of information from points of origin, usually research programs, to users is an obstacle in any applied discipline such as medicine, engineering, and particularly so in agriculture. The Extension Service was created in the United States in 1914 specifically to facilitate transfer of information from federal, state, and private research sources to farmers. The process by which this is accomplished has gone through many evolutionary changes in the 80 years since its inception. Extension specialists, plant pathologists among them, have used a number of innovative ways to convey useful information to farmers.

Plant pathologists have always had an identification problem, not only with respect to the general public but also with administrators and specialists in other disciplines as well. The tendency to treat diseases as one of three peas in a pod along with insects and weeds has contributed to this misunderstanding. Furthermore, the fact that plant diseases are somewhat abstract and ephemeral results of actions by usually invisible (microscopic) entities is a major reason for lack of recognition.

Two ways by which plant disease information is being conveyed to the public is through innovative Master Gardener programs and creating a new name to describe plant health practitioners. This new name, "phytonarian," was intended to capitalize on the more familiar animal health provider, namely, the veterinarian. The attempt was logical since veterinarian is derived from the Latin *veterinus*, which means cattle, and phytonarian comes from the Greek term *phyton* for plant. However, humans tend to empathize more with animals than plants and the long established usage of "veterinarian" makes this term readily identifiable. Phytonarian is as mysterious as is phytopathologist to most people and does little to improve the recognition problem.

Master Gardener programs have had considerable success in educating the general public, but except for use of innovative producers this concept has not been used to any extent in commercial agriculture. Master gardeners have practical experience in growing plants and not only present technical information in a simple straightforward manner but often are more believable and better accepted than "experts from the college."

REQUIREMENTS FOR DISEASE MANAGEMENT PROGRAMS

Planning, developing, and applying disease management programs require that several parameters be established. Most of them have been discussed in detail in earlier chapters. A brief review follows.

Identify Diseases to Be Managed

Not only must we accurately identify diseases, or more specifically causal agents, we also must understand the biology of pathogens and epidemiology. Relation of causal organisms to their environments (i.e., pathogen ecology) plays a large role in determining which specific control strategies and tactics will be effective. For example, pathogens with broad host ranges are not likely to respond to crop rotations, and facultative parasites are unlikely prospects for genetic resistance. Epidemiology, development of diseases over time and distance, is important in timing specific control measures. Effective control of monocyclic diseases such as root rots are best controlled by preplanting tactics such as crop rotation, soil fumigation, and certain cultural practices. Polycyclic diseases such as cereal rusts and apple scab can be effectively controlled during the growing season after diseases begin to develop.

It is also important to recognize presence and potential threat of other crop pests. Little benefit comes from controlling one disease only to have another unexpected problem devastate a crop. Fortunately, many cultural and chemical measures are sufficiently broad to control a range of pathogens.

Define the Management Unit

Once a disease is identified and the biologic nature of the problem determined, the next move is to assess the area affected and decide how to treat it. Levels in a plant hierarchy range from individual plants to populations (fields, orchards, rows, etc.) to plant communities (gardens, farms, forests) to biomes (regions). Management can be applied at any level within this hierarchy. Most diseases of ornamental and high value crops, such as orchard trees, are treated as individual plants. However, the greatest application of disease management is at population levels. Wheat or pea fields, greenhouse benches, nursery beds, orchards, or vineyards are common units of disease management. Sometimes units are too large to be managed efficiently or effectively. Counties, states, countries, or even continents might be subjected to quarantines or eradication programs because of diseases. Sometimes disease management areas must be reevaluated and modified to better meet realistic or reasonable goals. The Unshu orange provision of citrus canker quarantines was a reappraisal that allowed entry of some types of citrus into limited areas where they posed no real danger. The white pine blister rust *Ribes* eradication program became more selective and much smaller after economic, climatic, and ecological aspects were reviewed.

Develop Management Strategies

Strategies and tactics used to manage diseases depend on whether there are suitable resistant varieties available and if pathogenic races exist, or are likely to appear. If an agronomically acceptable disease-resistant variety is available it reduces the need for other disease management strategies. If no suitable resistant varieties are available then the nature of pathogens and their biology become important. Pathogens such as *Phymatotrichum omnivorum, Sclerotinia sclerotiorum, Rhizoctonia solani*, and *Botrytis cinerea* that attack wide ranges of hosts are protected in soil, produce sclerotia or other survival structures, can survive for long periods without hosts; some may be good saprophytes and require different strategies for control than obligate parasites that require live hosts to survive and perpetuate.

Epidemiological factors also must be considered in developing disease control strategies. If pathogens are established in a region, exclusion and eradication are precluded except on a local (i.e., field) basis. Available strategies include some forms of eradication through crop rotation, biological control, soil fumigation, protection by chemical or environmental methods, and resistance.

Monocyclic diseases such as root rots, some canker diseases, smuts, and some viral diseases are caused by pathogens from local sources. In contrast, polycyclic diseases such as rusts, powdery and downy mildews, many leaf disease fungi and bacteria, and insect-transmitted viruses originate from both local and distant sources, but as soon as infection takes place secondary cycles result in rapid increase in proportion of diseased plants. Monocyclic diseases are frequently managed by reducing local sources of inoculum with chemical seed treatments, soil fumigation, or cultural practices such as sanitation and crop rotation. Polycyclic diseases are usually managed by preventing primary cycles through eradication of alternate or alternative hosts, chemical protection, or vector control.

Genetic diversity is important to diversity of an agroecosystem. Monocultures usually reduce genetic diversity over distance (space) but multilines or gene deployment can provide genetic diversity while growing the same type of crop such as wheat or corn over large areas. Growing the same crop repeatedly reduces genetic diversity over time. Crop rotations create temporal genetic diversity by separating culture of the same crops with breaks in time (i.e., growing periods or seasons).

History of occurrence of a disease is important in developing management strategies, more so for soilborne diseases because once pathogens are established in fields they will likely remain there indefinitely. Development of disease, then, depends on environmental and host factors. These provide a basis for predicting occurrence of diseases and probable need for control measures. Strawbreaker foot rot, take-all, and Cephalosporium stripe of wheat are examples of this type of disease. Other diseases are less predictable and more sporadic since the inoculum might originate from distant places. Diseases of small grains such as leaf spots, rusts, powdery mildew, and barley yellow dwarf are mostly unpredictable.

Establish Economic Thresholds

Disease management programs must be based on sound economic principles and gains from these programs must exceed their costs. Economic thresholds, sometimes called disease loss projections, are disease levels that produce incremental reductions in crop values greater than costs of implementing disease management strategies. When fungicides became available to control strawbreaker foot rot of wheat in the Pacific Northwest cost of treatment was about 20 dollars per acre. If wheat sold for four dollars a bushel a net yield increase of five bushels was needed to break even. In this example the economic threshold is a disease loss of five or more bushels. In other words, the amount lost to disease exceeds the cost of control. This concept is illustrated in Figure 20.2. This graph has been used by several authors and it is difficult to determine who developed it originally. Regardless of origin, the graph clearly demonstrates the relationship between progress of disease, crop value, and cost of control.

One curve in Figure 20.2 represents potential value of a crop (yield revenue) at various stages of crop and disease development. An opposing curve represents total (accumulated) costs of control from a specific starting point. If there is no disease,

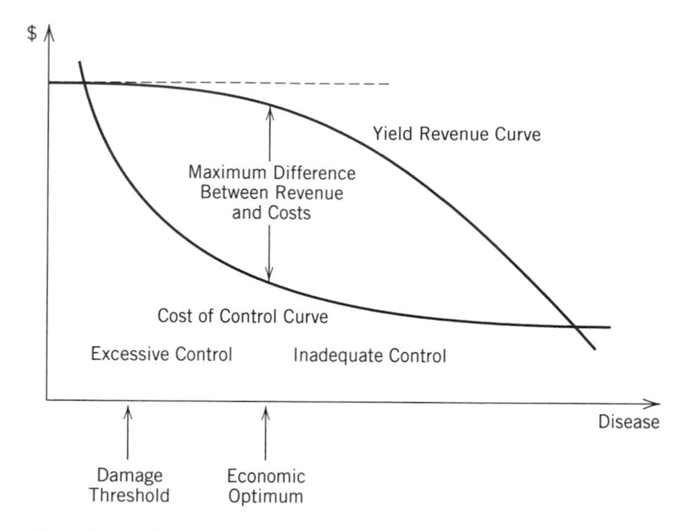

Figure 20.2. The relationship between crop revenue losses and the cost of disease control. [From Jeger, M. J., In Wood, R. K. S., and G. J. Jellis (eds.), *Plant Diseases: Infection, Damage, and Loss*. Blackwell, Oxford, 1984.]

projected yield revenues continue as a straight line. Even in early stages of disease there is no appreciable drop in the yield revenue curve. At some point potential value of a crop is lessened measurably by disease. This is the damage threshold. Such thresholds are complex and difficult to establish, a fact that has discouraged their use.

Yield revenue continues downward as disease progresses, but even in worst cases rarely reaches zero, since some harvestable product, however small, usually remains. Crops are usually abandoned before this point, however, since cost of harvest exceeds value of salvaged products.

Cost of control may exceed potential value of the crop at the beginning of disease management programs (i.e., the disease control curve). This assumes that there are startup costs such as buying tractors, spray equipment, and chemicals, which must be amortized over several crops. Cost of control declines as control action is deferred and materials, equipment, and labor are conserved. Control costs are disproportionately high before damage thresholds are reached. After this point control costs level off and eventually exceed the value of reclaimable crops.

An ideal strategy applies control measures at stages of crop and disease development where the curves are farthest apart. Cost of control can exceed maximum expected returns if excessive control measures are applied. This usually results from starting controls too soon and repeating more often than needed. As fewer control measures are applied control cost curves also drop but never reach zero even if nothing has been done since it is assumed that there are startup costs. Even minimum control costs may exceed value of crops in some cases. The ideal balance is to apply the least control to get maximum yields. This is the economic optimum of control. Too early and too many control applications are a waste, as are too little or too late.

Develop Monitoring Techniques

Monitoring techniques include disease forecasting systems and predictive models. These were described in Chapter 5. It is noteworthy and worth repeating, however, that if forecasting systems are reasonably accurate, timely and effective control measures can be expected to maximize economic benefits.

EXAMPLES OF INTEGRATED DISEASE MANAGEMENT

With few exceptions plant diseases are controlled (managed) using a variety of methods. Peach leaf curl is one exception. It is controlled almost exclusively by applying fungicides while trees are dormant. A few peach selections appear to have some resistance, but so far have had limited use. Physical protection by mechanical barriers was described in Chapter 14, but this is feasible only in special situations. Even where there is good resistance to diseases such as for common smut of wheat, there is need for additional supplemental measures to prolong resistance. Seed treatment is still recommended for smut control in the Pacific Northwest to prevent rapid buildup of pathogenic races.

Long lists could be compiled of diseases controlled by combinations of measures. Whether all of these constitute integrated disease management is a matter of interpretation. The following were selected as examples of diseases in which several control measures contribute added increments of control. One control measure is not merely a replacement or backup for another measure.

Strawbreaker Foot Rot of Wheat

Monitoring Strawbreaker is forecast on the basis of previous history of disease in fields, seeding dates, abundance of fall rain, and large plants going into winter, all of which contribute to early and extensive infection. Disease progress through winter is measured by disease thresholds in spring as determined by examining tillers for stem lesions before jointing.

Management Crop rotations to barley, legumes, or spring cereals for 3 years reduces soil inoculum levels. Late (delayed) fall seeding reduces incidence of infection by providing smaller plant canopies to intercept water-splashed spores.

Avoiding spring tillage reduces movement of soil around stems of plants and eliminates environmental niches favorable for fungal development.

Fungicide applications are economical when 10% or more tillers are visibly diseased. Several benzimidazole fungicides are registered and usually applied at rates of 0.25–0.50 lb of active material per acre.

Madsen and Hyak are relatively new resistant varieties but may not be adapted to every farmer's program. However, they provide additional weapons in the disease control arsenal.

Cephalosporium Stripe of Wheat

Monitoring The causal fungus must be assumed to be present if the disease has occurred in fields in the past. Soil type, particularly heavy clay soils, and open, cold winters where frost heaving occurs in saturated soils foretell problems with Cephalosporium stripe.

Management Crop rotations of at least 2 years between winter cereals or grasses is essential for effective inoculum reduction. Winter wheat can be rotated with legumes and spring cereals. However, some spring cereals such as barley produce volunteers that carry the fungus through winter.

Burning diseased crop residues also reduces inoculum levels but must be weighed against environmental effects such as air pollution and soil erosion. Delayed planting can reduce disease severity by producing plants with small root systems going into winter. Such plants are less subject to damage by frost heaving.

Nitrate forms of fertilizers avoid lowering soil pH, which increases disease severity.

Tolerant varieties should be grown where possible and growers should avoid highly susceptible varieties such as Stephens in soils where the disease is certain to occur.

Stripe Rust of Wheat

Monitoring Stripe rust forecasts are based primarily on two conditions. Disease development in fall provides early buildup of inoculum. Infections continue to occur during mild winters and springs.

Development of rust at flag leaf stage is a major consideration for applying fungicides. Equally important is degree of resistance in the host.

Management The basic management strategy relies on resistant varieties, even though the most durable resistance occurs during adult plant stages. Such resistance is usually not active until onset of warm weather, being typically delayed during prolonged cool springs and summers. Resistant varieties with race-specific genes are vulnerable to development of new pathogenic races. Use of varieties with nonspecific resistance or multilines like Crew and Rely is another strategy. Fungicides are used to protect varieties that lack resistance or varieties in which adult-plant resistance is not active.

Verticillium Wilt of Potatoes

Monitoring A field with a history of Verticillium wilt is a good indication that the disease will reoccur. Inoculum levels can be assayed by plating soil on selective media.

Management Field flaming before harvest reduces numbers of microsclerotia returned to soil. Soil fumigation before planting lowers inoculum levels, and controls root-knot and root-lesion nematodes, and other microorganisms that may be involved in disease complexes. These treatments combined are considerably more effective than either one alone. Crop rotation to grasses, cereals, or legumes also results in decline of soil inoculum. Sudangrass in the rotation as a green manure has added benefits of reducing *Verticillium* populations. Balanced fertilizer and irrigation programs enhance plant vigor and minimize symptom expression.

Fire Blight of Pears

Monitoring Fire blight in orchards is easily detected during dormant seasons given the retention of leaves on blighted branches. Surveys and assays detect streptomycin-resistant strains of the causal bacterium.

Disease forecasting systems are based primarily on average daily temperatures. Bacteria are active when temperatures reach 60°F and protective sprays recommended when favorable mean daily temperatures coincide with bloom periods.

Management Fire blight management calls for elimination of overwintering inoculum by sanitation pruning during dormant periods. Diseased branches must be removed well below visible margins of infection. Excised material must be taken from orchards and buried or burned.

Blossoms are the main infection courts and hence must be protected with suitable bactericides such as fixed coppers or antibiotics. If copper causes serious russetting of fruit, antibiotics such as streptomycin or tetracyclines can be used. Choice of antibiotics depends on whether streptomycin-resistant strains are present. Pears are subject to infection over a prolonged period because they produce second, or late, blossoms. These blossoms contribute little to fruit production and are removed to eliminate infection courts. Overtree sprinkling should be avoided because it intensifies blight severity. Although there are some differences in susceptibility of pear varieties to fire blight they are insufficient to be used as a management tactic.

Phytophthora Root Rots of Apples, Avocado, and Other Crops

Monitoring Phytophthora root rots affect many crop plants. Various species of this fungus are widely distributed throughout temperate and tropical regions. If a pathogen is assumed to be present, disease development is determined by environmental factors that favor the pathogen. These consist primarily of heavy, poorly drained soils.

Management Phytophthora root rots are managed by using resistant or tolerant varieties, modifying enviroment by cultural practices, and applying chemical treatments.

Soybeans and alfalfa are herbaceous plants severely affected by Phytophthora root rots. Many resistant varieties have been developed for both crops. Collar rot-resistant or -tolerant rootstocks for apples such as EM9 were discussed previously. Duke 7 and G755c are *Phytophthora*-resistant avocado rootstocks.

Cultural measures involve improving soil drainage by tiling, ditching, or growing plants on raised beds or mounds.

Fungicides, usually applied as drenches around bases of trees, are effective, but certain chemicals, such as metalaxyl, may become ineffective after several years use because of enhanced biodegradation. Fosetyl-Al is effective when applied as drenches or injected into trees.

Diseases of Greenhouse Crops

Monitoring Greenhouse environments are favorable for development of foliar diseases, for example, cucumber scab, gray mold of various plants, tomato leaf mold, and lettuce downy mildew. Monitoring for these diseases usually involves early detection.

Management Resistance is a primary control tactic for many of these diseases, but fungicides may be required to avoid or minimize appearance of pathogenic races.

Many of these pathogens are efficient saprophytes and live or survive on plant debris. Greenhouse sanitation is essential to reduce the abundant inoculum that develops on this debris. Modifying greenhouse environments by controlling heating and ventilation reduces high humidity and temperatures that favor diseases. Irrigation can be modified to use drip or subsurface irrigation instead of sprinklers, which favor foliar diseases.

Fungicides continue to be primary management tools in greenhouses even when resistance is available. The high value of these crops and sudden and extensive development of diseases in greenhouses do not permit risking severe losses that can occur with less adequate control methods.

DISEASE CONTROL IN THE FUTURE

Throughout this book I have tried to show that new strategies and technologies are constantly evolving to improve disease control. However, I have also tried to show that these innovations usually are not the panaceas their promoters often proclaim. They are simply new building blocks to improve existing structures of disease management or control. The three basic control procedures, resistance, cultural, and chemical, have a long history of successful use. Cultural practices to reduce disease losses predate history. Disease-resistant selections have been recorded for almost two centuries and chemicals even longer. It is very likely that 100 years from now we will still be managing plant diseases as we are today, only with newer tools and better mousetraps.

REFERENCES

Alster, K. B., *The Holistic Health Movement*. The University of Alabama Press, Tuscaloosa, 1989.

*Anon., Potato crop management: The need for an integrated approach. *Am. Potato J.* 67:1–68 (1990).

*Bruehl, G. W., Integrated control of soil-borne plant pathogens: An overview. *Can. J. Plant Pathol.* 11:153–157 (1989).

*Coffey, M. C., Phytophthora root rot of avocado: An integrated approach to control in California. *Plant Dis.* 71:1046–1052 (1987).

Cook, R. J., and R. J. Veseth, *Wheat Health Management.* APS Press, St. Paul, MN, 1991.

*Heitefuss, R., *Crop and Plant Protection: The Practical Foundations.* John Wiley, New York, 1989.

*Koeppsel, P. A., and J. W. Pscheidt (eds.), *Pacific Northwest Plant Disease Control Handbook.* Oregón State University, Corvallis, 1992.

*Morris, P. C., Biotechnology and plant protection. *FAO Plant Protection Bull.* 38:25–37 (1990).

*Sill, W. H., Jr., *Plant Protection: An Integrated Interdisciplinary Approach.* The Iowa State University Press, Ames, 1982.

*Travis, J. W., and R. X. Latin, Development, implementation and adoption of expert systems in plant pathology. *Annu. Rev. Phytopathol.* 29:343–360 (1991).

*Zadoks, J. C., Does IPM give adequate attention to disease control in particular at the farmer level? *FAO Plant Protection Bull.* 37:144–150 (1989).

* Publications marked with an asterisk (*) are general references not cited in the text.

APPENDIX 1 Scientific Names of Host Plants Mentioned in Text

Alfalfa	*Medicago sativa* L.
American elm	*Ulmus americana* L.
American chestnut	*Castanea dentata* (Marsh.) Borkh.
American sycamore	*Platanus occidentalis* L.
Apple	*Malus sylvestris* Mill.
Apricot	*Prunus armeniaca* L.
Asparagus	*Asparagus officinalis* L.
Aspen	*Populus* L.
Avocado	*Persea americana* Miller
Azalea	*Rhododendron* L.
Banana	*Musa acuminata* Colla
Barberry	*Berberis vulgaris* L.
Barley	*Hordeum vulgare* L.
Beet	*Beta vulgaris* L.
Black currant	*Ribes nigrum* L.
Blueberry	*Vaccinium corymbosum* L.
Cabbage	*Brassica oleracea* L.
Cacao	*Theobroma cacao* L.
Carrot	*Daucus carota* L.
Cassava	*Manihot esculenta* Crantz
Castor bean	*Ricinus communis* L.
Celery	*Apium graveolens* L.
Chickpea	*Cicer arietinum* L.
Chokecherry	*Prunus virginiana* L.
Chrysanthemum	*Chrysanthemum* L.

Citron melon	*Citrullus vulgaris* Schrad. var. *citroides*
Citrus	*Citrus* L.
Coconut palm	*Cocos nucifera* L.
Coffee	*Coffea arabica* L.
Common wheat	*Triticum aestivum* L.
Common bean	*Phaseolus vulgaris* L.
Cork oak	*Quercus suber* L.
Corn poppy	*Papaver rhoeas* L.
Corn (maize)	*Zea mays* L.
Cowpea	*Vigna unguiculata* (L.) Walp.
Cranberry	*Vaccinium macrocarpon* Aiton
Crotalaria	*Crotalaria* L.
Cucumber	*Cucumis sativus* L.
Daffodil	*Narcissus pseudonarcissus* L.
Dogwood	*Cornus* L.
Douglas fir	*Pseudotsuga menziesii* (Mirb.) Franco
Durum wheat	*Triticum durum* Desf.
Eastern red cedar	*Juniperus virginiana* L.
Eastern white pine	*Pinus strobus* L.
Eggplant	*Solanum melongena* L.
English walnut	*Juglans regia* L.
Fig	*Ficus carica* L.
Flax	*Linum usitatissimum* L.
Gladiolus	*Gladiolus hortulanus* L. H. Bailey
Gooseberry	*Ribes* L.
Grape-American	*Vitis labrusca* L.
Grape-European	*Vitis vinifera* L.
Hawthorn	*Crataegus* L.
Hollyhock	*Alcea rosea* L.
Hop	*Humulus lupulus* L.
Iberian oak	*Quercus falcata* Michx.
Jarrah	*Eucalyptus marginata* Sm.
Jimsonweed	*Datura stromonium* L.

Juniper	*Juniperus* L.
Lettuce	*Lactuca sativa* L.
Lima bean	*Phaseolus lunatus* L.
Loganberry	*Rubus loganobaccus* L. H. Bailey
Longleaf pine	*Pinus palustris* Mill.
Marigold	*Tagetes* L.
Melon	*Cucumis melo* L.
Mignonette	*Reseda odorata* L.
Mimosa	*Albizia julibrissin* Durazz.
Monterey pine	*Pinus radiata* D. Don
Oat	*Avena sativa* L.
Olive	*Olea europaea* L.
Onion	*Allium cepa* L.
Papaya	*Carica papaya* L.
Para rubber	*Hevea brasiliensis* Muell.-Arg.
Pea-Austrian winter	*Lathyrus hirsutus* L.
Pea-Garden	*Pisum sativum* L.
Peach	*Prunus persica* (L.) Botsch
Peanut	*Arachis hypogaea* L.
Pear	*Pyrus communis* L.
Pepper	*Capsicum annuum* L.
Peppermint	*Mentha piperita* L.
Ponderosa pine	*Pinus ponderosa* Laws
Poplar	*Populus* L.
Port Orford cedar	*Chamaecyparis lawsoniana* (A. Murr.) Parl.
Potato	*Solanum tuberosum* L.
Poulard wheat	*Triticum turgidum* L.
Quince	*Cydonia oblonga* Mill.
Rape	*Brassica napus* L.
Raspberry	*Rubus idaeus* L.
Red clover	*Trifolium pratense* L.
Rhododendron	*Rhododendron* L.
Rice	*Oryza sativa* L.

Rose	*Rosa* L.
Rubber	*Hevea guaianensis* Aubl.
Rye	*Secale cereale* L.
Ryegrass	*Lolium perenne* L.
Safflower	*Carthamus tinctorius* L.
Sea Island cotton	*Gossypium barbadense* L.
Sesame	*Sesamum indicum* L.
Slash pine	*Pinus elliottii* Engelm.
Sorghum	*Sorghum bicolor* (L.) Moench
Soybean	*Glycine max* (L.) Merr.
Spearmint	*Mentha spicata* L.
Spinach	*Spinacia oleracea* L.
Strawberry	*Fragaria* L.
Sugar pine	*Pinus lambertiana* Douglas
Sugarcane	*Saccharum officinarum* L.
Sunflower	*Helianthus annuus* L.
Sweet cherry	*Prunus avium* L.
Sweetpotato	*Ipomoea batatas* (L.) Lam.
Taro	*Colocasia esculenta* (L.) Schott
Tea	*Camellia sinensis* (L.) Kuntze
Tobacco	*Nicotiana tabaccum* L.
Tomato	*Lycopersicum esculentum* Miller
Upland cotton	*Gossypium hirsutum* L.
Watermelon	*Citrullus vulgaris* Schrad.
Western white pine	*Pinus monticola* Douglas ex D. Don
White clover	*Trifolium repens* L.
Whitebark pine	*Pinus albicaulis* Engelm.

APPENDIX 2
Scientific Names for Pathogens of Diseases Mentioned in Text

Alfalfa bacterial wilt	*Clavibacter michiganensis* subsp. *insidiosus* (McCulloch) Davis et al.
Alfalfa leaf spot	*Pseudopeziza medicaginis* (Lib.) Sacc.
Alfalfa root rot	*Phytophthora megasperma* Drechs.
Alfalfa rust	*Uromyces striatus* J. Schröt.
Alfalfa southern anthracnose	*Colletotrichum trifolii* Bain & Essary
Alfalfa spring black stem	*Phoma medicaginis* Malbr. & Roum. in Roum.
Alfalfa stem nematode	*Ditylenchus dipasaci* (Kühn) Filipjev
Alfalfa Verticillium wilt	*Verticillium albo-atrum* Reinke & Berthier
Alfalfa witches broom	Virus
Annosus root rot	*Heterobasidion annosum* (Fr.) Bref.
Apple European canker	*Nectria galligena* Bres. in Strass.
Apple scab	*Venturia inaequalis* (Cooke) G. Wint.
Apple powdery mildew	*Podosphaera leucotricha* (Ellis & Everh.) E. S. Salmon
Armillaria root rot	*Armillaria mellea* (Vahl.) P.Kumm.
Asparagus blue mold	*Penicillium aurantiogriseum* Dierckx
Asparagus purple spot	*Pleospora allii* (Rabenh.) Ces.& DeNot.
Asparagus rust	*Puccinia asparagi* DC. in Lam. & DC.
Aspen leaf blight	*Marssonina tremulae* (Lib.) Kleb.
Azalea flower spot	*Ovulinia azaleae* Weiss
Azalea leaf gall	*Exobasidium vaccinii* (Fuckel) Woronin
Bacterial soft rot	*Erwinia carotovora* subsp. *carotovora* (Jones) Bergey et al.
Bacterial wilt	*Pseudomonas solanacearum* (Smith) Smith

Banana bunch top	Virus
Banana cigar-end	*Verticillium theobromae* (Turc.) Mason & Hughes
Banana finger rot	*Botryodiplodia theobromae* Pat. and *Trachysphaera fructigena* Tabor & Bunting
Banana Fusarium wilt	*Fusarium oxysprorum* Schlechtend.:Fr. f.sp. *Cubense* (E.F. Sm.)
Banana leaf spot (Sigatoka)	*Mycosphaerella musicola* J. L. Mulder in J. L. Mulder & Stover
Banana Moko disease	See bacterial wilt
Banana pitting	See rice blast
Banana speckle	*Deightoniella torulosa* (Syd.) Ellis
Banana stem-end rot	*Colletotrichum musae* (Berk. & Curt.) Arx
Barley loose smut	*Ustilago tritici* (Pers.) Rostr.
Barley yellow dwarf	Virus
Bean anthracnose	*Colletotrichum lindemuthianum* (Sacc. & Magnus) Lams.-Scrib.
Bean halo blight	*Pseudomonas syringae* pv. *phaseolicola* (Burkholder) Young et al.
Bean root rot	*Fusarium solani* (Mart.) Sacc. f.sp. *phaseoli* (Burkh.) W. C. Snyder & H. N. Hans.
Bean rust	*Uromyces appendiculatus* (Pers.) Unger
Beet leaf spot	*Cercospora beticola* Sacc.
Beet cyst nematode	*Heterodera schachtii* Schmidt
Beet yellows	Virus
Blueberry mosaic	Virus
Broomrape	*Orobanche* spp. L.
Bulb nematode	See alfalfa stem nematode
Bull's eye rot	*Pezicula malicorticis* (H. Jacks.) Nannf.
Burrowing nematode	*Radopholus similis* (Cobb) Thorne
Cabbage leaf spot	*Alternaria brassicola* (Schwein.) Wiltshire
Cabbage wilt	*Fusarium oxysporum* Schlechtend. f.sp. *conglutinans* (Wollenweb.) W. C. Snyder & H.N. Hans.
Cacao swollen shoot	Virus
Cacao witches broom	*Crinipellis perniciosa* (Stahel)Singer
Carrot bacterial blight	*Xantomonas campestris* pv. *carotae* (Kendrick) Dye

Carrot thin leaf	Virus
Cassava bacterial blight	*Xanthomonas campestris* pv. *cassavae* (Weihe & Dowson) Maraite & Weyns
Cedar-apple rust	*Gymnosporangium juniperi-virginianae* Schwein.
Celery bacterial leaf spot	*Pseudomonas syringae* pv. *apii* (Jagger) Young et al.
Celery late blight	*Septoria apiicola* Speg.
Celery leaf spot	See celery late blight
Cereal cyst nematode	*Heterodera avenae* Woll.
Charcoal rot	*Macrophomina phaseolina* (Tassi) Goidanich
Cherry leaf spot	*Blumeriella jaapii* (Rehm) Arx
Chestnut blight	*Cryphonectria parasitica* (Murr.) Barr
Chickpea Ascochyta blight	*Mycosphaerella rabrei* Kovachevski
Chickpea root rot	See pea root rot
Chickpea stunt	Virus
Chickpea wilt	*Fusarium oxysporum* Schlechtech. f. sp *ciceris* (Padwick) Matuo & k. Sato.
Choke disease	*Epichloe typhina* (Pers.) Tul. in Tul. & C. Tul.
Chrysanthemum leaf spot	*Diplodia* sp. Fr. in Mont.
Chrysanthemum rust	*Puccinia horiana* Henn.
Citrus anthracnose	*Glomerella cingulata* (Stoneman) Spauld. & H. Schrenk
Citrus blue mold	*Penicillium italicum* Wehmer
Citrus canker	*Xanthomonas campestris* pv. *citri* (Hasse) Dye
Citrus green mold	*Penicillium digitatum* (Pers.)Sacc.
Citrus melanose	*Diaporthe citri* F. A. Wolf
Citrus nematode	*Tylenchulus semipenetrans* Cobb
Citrus scab	*Elsinoe fawcettii* Bitamount & Jenk.
Citrus tristeza	Virus
Coconut lethal yellowing	Mycoplasma
Coconut wilt (red ring)	*Rhadinaphelenchus cocophilus* (Cobb) Goodey
Coffee berry disease	*Colletotrichum coffeanum* (Wallr.) S. J. Hughes
Coffee leaf spot	*Cercospora coffeicola* Berk. & Cooke
Coffee rust	*Hemileia vastatrix* Berk. & Br.
Collar rot	*Phytophthora cactorum* (Lebert & Cohn) J. Shröt.

Columbia root-knot nematode	*Meloidogyne chitwoodi* Golden et al.
Cork oak ink disease	*Phytophthora cimmamomi* Rands
Corn anthracnose	*Colletotrichum graminicola* (Ces.) G. W. Wils.
Corn bacterial stalk rot	*Erwinia dissolvens* (Rosen) Burkh. and *E. chrysanthemi* pv. *zeae* (Sabet) Victoria et al.
Corn brown spot	*Physoderma maydis* (Miyabe) Miyabe
Corn downy mildew	*Sclerospora graminicola* (Sacc.) J. Schröt.
Corn head smut	*Sporisorium holci-sorghi* (Rivolta) K. Vánky
Corn leaf blight	*Helminthosporium* sp. Link
Corn northern leaf blight	*Exserohilum turcicum* (Pass.) K. J. Leonard & E. G. Suggs
Corn smut	*Ustilago zeae* (Beckm.) Unger
Corn southern leaf blight	*Cochliobolus heterostrophus* (Drechs.) Drechs.
Corn stalk rot	*Gibberella* spp. Sacc.
Coryneum blight	*Stigmina carpophila* (Lév.) M. B. Ellis
Cotton angular leaf spot	*Xanthomonas campestris* pv. *malvacearum* (Smith) Dye
Cotton anthracnose	*Colletotrichum gossypii* Southworth
Cotton bacterial blight	See angular leaf spot
Cotton boll rots	Several fungi including species of *Aspergillus* P. Mich. ex Link:Fr., *Botryosphaeria* Ces. & De Not., *Diplodia* Fr. in Mont., *Fusarium* Link:Fr.
Cotton rust	*Puccinia cacabata* Arth. & Holw. in Arth.
Cotton wilt	*Fusarium oxysporum* Schlechtend. f.sp. *vasinfectum* (Atk.) W. C. Snyder & H. N. Hans.
Cowpea wilt	*Fusarium oxysporum* Schlechtend. f.sp. *tracheiphilum* (E.F. Sm.) W. C. Snyder & H. N. Hans.
Cranberry fairy ring	*Psilocybe agrariella* Atk. var. *vaccinii* V. Charles
Cranberry red leaf	*Exobasidium vaccinii* (Fuckel) Woronin
Cranberry rose bloom	*Exobasidium oxycocci* Rostr.
Crown gall	*Agrobacterium tumefaciens* (Smith & Townsend) Conn
Crucifer bacterial leaf spot	*Pseudomonas syringae* pv. *maculicola* McCulloch) Young et al.
Crucifer black leg	*Phoma lingum* (Tode) Desmaz.

Crucifer black rot	*Xanthomonas campestris* pv. *campestris* (Pammel) Dowson
Crucifer club root	*Plasmodiophora brassicae* Woronin
Cucumber angular leaf spot	*Pseudomonas syringae* pv. *lachrymans* (Smith & Bryan) Young et al.
Cucumber gummy stem blight	*Didymella bryoniae* (Auersw.) Rehm
Cucumber scab	*Cladosporium cucumerinum* Ellis & Arth.
Cucumber wilt	*Erwinia tracheiphila* (Smith) Bergey et al.
Cucurbit downy mildew	*Pseudoperonospora cubensis* (Berk. & M. A. Curtis) Rostovzev
Curly top	Virus
Daffodil basal rot	*Fusarium oxysporum* Schlechtend. f. sp. *narcissi* W. C. Snyder & H. N. Hans.
Damping-off	Several fungi including species of *Fusarium* Link:Fr., and *Pythium* Pringsh.
Dodder	*Cuscuta* spp. L.
Dogwood anthracnose	*Discula destructiva* Redlin
Dothistroma needle blight	*Dothistroma septospora* (Doroguine) Morelet
Douglas fir needle cast	*Rhabdocline pseudotsugae* Syd.
Dutch elm disease	*Ophiostoma ulmi* (Buisman) Nannf.
Dwarf mistletoe	*Arceuthobium* spp. M. Bieb.
Dwarf smut	*Tilletia controversa* Kühn in Rabenh.
Elm mottle	Virus
Ergot	*Claviceps purpurea* (Fr.) Tul.
Fire blight	*Erwinia amylovora* (Burrill) Winslow et al.
Flag smut	*Urocystis agropyri* (G. Preuss) J. Schröt.
Flavescence doree	Mycoplasma
Flax browning	*Aureobasidium lini* (Lafferty) Hermanides-Nijhof
Flax rust	*Melampsora lini* (Ehrenb.) Dezmaz.
Flax wilt	*Fusarium oxysporum* Schlechtend. f. sp. *lini* (Bolley) W. C. Snyder & H. N. Hans.
Fusiform rust	*Cronartium quercuum* (Berk.) Miyabe ex Shirai f. sp. *fusiforme* (Hedgc. & N. Hunt) Burdsell & G. Snow
Giberrella seedling blights	*Gibberella fujikuroi* (Sawada) Ito in Ito & K. Kimura and *G. zeae* (Schwein.) Petch

Gladiolus corm rot	*Fusarium oxysporum* Schlechtend. f. sp. *gladioli* (L. Massey) W. C. Snyder & H. N. Hans.
Golden nematode	*Globodera rostochiensis* (Woll.) Behrens
Gooseberry leaf spot	*Mycosphaerella ribis* Feltgen
Goss's bacterial wilt	*Clavibacter michiganensis* subsp. *nebraskensis* (Vidaver & Mandel) Davis et al.
Grape anthracnose	*Elsinoe ampelina* Shear
Grape bacterial canker	*Xanthomonas campestris* pv *viticola* (Nayudu) Dye
Grape black rot	*Guignardia bidwellii* (Ellis) Viala & Ravaz
Grape bunch rot	See gray mold
Grape corky bark	Virus
Grape downy mildew	*Plasmopara viticola* (Berk. & M.A. Curtis) Berl. & De Toni in Sacc.
Grape dying arm	*Eutypa armeniacae* Hansf.& M.V. Carter
Grape fanleaf	Virus
Grape leafroll	Virus
Grape powdery mildew	*Uncinula necator* (Schwein.) Burrill
Grass blind seed	*Gloeotinia grangigena* (Quél.) T. Schumacher
Grass seed gall nematode	*Anguina agrostis* (Steinbach) Filipjev
Gray mold	*Botrytis cinerea* Pers.
Hawthorn leaf spot	*Diplocarpon mespili* (Sorauer) Sutton
Hawthorn rust	*Gymnosporangium globosum* (Farl.)Farl. and *G. bethelii* F. Kern
Hollyhock rust	*Puccinia malvacearum* Bertero ex Mont. in Gay
Hop downy mildew	*Pseudoperonospora humuli* (Miyabe & Takah.) G. W. Wils.
Indian paint fungus	*Echinodontium tinctorium* Ellis & Everh.
Iris ink disease	*Bipolaris iridis* (Oudem.) C. H. Dickinson
Jarrah disease	*Phytophthora cinnamomi* Rands
Karnal smut	*Tilletia indica* Mit.
Lambert mottle	Virus
Laminated root rot	*Phellinus weirii* (Merrill) R. L. Gilbertson
Lentil Ascochyta blight	*Ascochyta lentis* Vassiljevsky
Lettuce big vein	Virus
Lettuce corky root	Unidentified bacterium

Lettuce downy mildew	*Bremia lactucae* Regel
Lettuce mosaic	Virus
Lima bean downy mildew	*Phytophthora phaseoli* Thaxt.
Little cherry	Virus
Melon stem rot	See white mold
Mimosa wilt	*Fusarium oxysporum* Schlechtend. f. sp. *perniciosum* (Hepting) W. C. Snyder & H. N. Hans.
Mint rust	*Puccinia menthae* Pers.
Mummyberry	*Monilia vaccinii-corymbosi* (Reade) Honey
Northern root-knot nematode	*Meloidogyne hapla* Chitwood
Oat crown rust	*Puccinia coronata* Corda
Oat loose smut	*Ustilago avenae* (Pers.) Rostr.
Oat stem rust	See wheat stem rust
Olive knot	*Pseudomonas syringae* pv. *savastanoi* (Smith) Young et al.
Onion bulb nematode	*Ditylenchus dipsaci* (Kühn) Filipjev
Onion downy mildew	*Peronospora destructor* (Berk.) Casp. in Berk.
Onion leaf blight	See gray mold
Onion pink root	*Phoma terrestris* E. M. Hans.
Onion smudge	*Colletotrichum circinans* (Berk.) Voglino
Onion smut	*Urocystis magica* Pass. in Thuem.
Onion white rot	*Sclerotium cepivorum* Berk.
Palm lethal yellowing	Mycoplasma
Papaya ringspot	Virus
Pea Aphanomyces root rot	*Aphanomyces euteiches* Drechs. f. sp. *pisi* W. F. Pfender & D. J. Hagedorn
Pea Ascochyta blight	*Mycosphaerella pinodes* (Berk. & Bloxam) Vestergr.
Pea enation mosaic	Virus
Pea powdery mildew	*Erysiphe polygoni* DC.
Pea root rot	*Fusarium solani* (Mart.) Sacc. f. sp. *pisi* (F. R. Jones) W. C. Snyder & H. N. Hans.
Pea streak	Virus
Pea wilt	*Fusarium oxysporum* Schlechtend. f. sp. *pisi* (J. C. Hall) W. C. Snyder & H. N. Hans.
Peach leaf curl	*Taphrina deformans* (Berk.) Tul.

Peach yellow leaf roll	Virus
Peach yellows	Virus
Peanut black rot	*Calonectria crotalariae* (C. A. Loos) D. K. Bell & Sobers
Peanut pod rot	*Fusarium* sp. Link
Peanut stem rots	Several fungi including *Rhizoctonia solani* Kühn, *Sclerotium rolfsii* Sacc., and *Sclerotinia minor* Jagger
Peanut Verticillium wilt	*Verticillium albo-atrum* Reinke & Berthier
Peanut web blotch	*Phoma arachidicola* Marasas, G. D. Pauer, & Boerema
Pear decline	Mycoplasma
Pear scab	*Venturia pyrina* Aderhold
Pear trellis rust	*Gymnosporangium fuscum* R. Hedw. in DC.
Phymatotrichum root rot	*Phymatotrichiopsis omnivora* (Dugger) Hennebert
Pine blue stain	*Ceratocystis* spp. Ellis & Everh.
Pine brown needle spot	*Mycosphaerella dearnessii* Barr
Pinewood nematode	*Bursaphelenchus xylophilus* (Steiner & Buhrer) Nickle
Pink snow mold	*Microdochium nivale* (Fr.) Samuels & I. C. Hallett
Poplar canker	Several fungi including *Ceratocystis fimbriata* Ellis & Halst., *Cryptosphaeria lignyota* (Fr.) Auersw., and *Hypoxylon mammatum* (Wahlenberg) J. H. Miller
Poplar leaf blight	*Marssonina populi* (Lib.) Magnus
Potato blackleg	*Erwinia carotovora* subsp. *atroseptica* (van Hall) Dye
Potato black scurf	*Rhizoctonia solani* Kühn
Potato Clostridium rot	*Clostridium* spp. Prazmowski
Potato common scab	*Streptomyces scabies* (Thaxter) Waksman & Henrici
Potato early blight	*Alternaria solani* Sorauer
Potato late blight	*Phytophthora infestans* (Mont.) deBary
Potato leaf roll	Virus
Potato powdery scab	*Spongospora subterranea* (Wallr.) Lagerh.
Potato ring rot	*Clavibacter michiganensis* subsp. *sepedonicus* (Spieckermann & Kotthoff) Davis et al.
Potato seed piece decay	*Fusarium* spp. Link:Fr.

Potato silver scurf	*Helminthosporium solani* Durieu & Mont.
Potato skin spot	*Polyscytalum pustulans* (M. N. Owens & Wakef.) M. B. Ellis
Potato virus Y	Virus
Potato wart	*Synchytrium endobioticum* (Schilberszky) Percival
Pseudomonas blight	See stonefruit bacterial canker
Pythium root rot	*Pythium* spp. Pringsh.
Quince rust	*Gymnosporangium clavipes* (Cooke & Peck) Cooke & Peck in Peck
Raspberry cane blight	*Diapleela coniothyrium* (Fuckel) Barr
Raspberry mosaic	Virus
Raspberry ringspot	Virus
Red clover anther mold	*Botrytis anthophila* Bondartsev
Rhizina root rot	*Rhizina undulata* Fr.:Fr.
Rhizoctonia stem rot	*Rhizoctonia solani* Kühn
Rhododendron web blight	*Rhizoctonia solani* Kühn
Rice bacterial grain rot	*Pseudomonas glumae* Kurita & Tobei
Rice bacterial leaf blight	*Xanthomonas campestris* pv. *oryzae* (Ishiyama) Dye
Rice Bakanae disease	*Fusarium* sp. Link
Rice blast	*Pyricularia grisea* (Cooke) Sacc.
Rice brown spot	See rice leaf spot
Rice leaf spot	*Bipolaris oryzae* (Breda de Haan) Shoemaker
Rice root nematode	*Hirschmanniella oryzae* (Lac & Goodey) Sher
Rice sheath blight	See Rhizoctonia stem blight
Rice smut	*Tilletia barclayana* (Bref.) Sacc. & Syd. in Sacc.
Rice stem rot	*Sclerotium oryzae* Cattaneo
Rice white tip nematode	*Aphelenchoides besseyi* Christie
Root-lesion nematodes	*Pratylenchus* spp. Filipjev
Rose black mold	*Chalara thielavioides* (Peyronel) Nag Raj & Kendrick
Rose black spot	*Diplocarpon rosae* F. A. Wolf
Rose cankers	Several fungi including *Diaporthe eres* Nitschke and *Coniothyrium fuckelii* Sacc.
Rose powdery mildew	*Sphaerotheca* spp. Lev.
Rose rust	*Phragmidium* spp. Link

Rosellinia root rot	*Rosellinia necatrix* Prill.
Sclerotinia stem rot	See white mold
Sesame bacterial leaf spot	*Xanthomonas campestris* pv. *sesami* (Sabet & Dowson) Dye
Slime flux	*Erwinia nimmipressuralis* Carter
Sorghum downy mildew	*Peronosclerospora sorghi* (W. Weston & Uppal) C. G. Shaw
Sorghum head smut	See corn head smut
South American leaf blight	*Microcyclus ulei* (P. Henn.) Arx in Müller & Arx
Southern blight	*Sclerotium rolfsii* Sacc.
Southern cone rust	*Cronartium strobilinum* Hedgc. & Hahn
Soybean bacterial blight	*Pseudomonas syringae* pv. *glycinea* (Coerper) Young et al.
Soybean mosaic	Virus
Soybean root rot	*Phytophthora megasperma* Drechs. f. sp. *glycinea* T. Kuan & D. C. Erwin
Speckled snow mold	*Typhula idahoensis* Remsberg
Spiral nematode	*Helicotylenchus multicinctus* (Cobb) Golden
Sprinkler rot	See collar rot
Stewart's wilt	*Erwinia stewartii* (Smith) Dye
Stone fruit bacterial canker	*Pseudomonas syringae* pv. *syringae* van Hall
Stone fruit brown rot	*Monilia fructicola* (G. Wint.) Honey and *M. laxa* (Aderhold & Ruhland) Honey
Stone fruit Cytospora canker	*Cytospora* spp. Ehrenb.
Stone fruit X-disease	Mycoplasma
Strawberry anthracnose	*Colletotrichum gloeosporioides* (Penz.) Penz. & Sacc. in Penz.
Strawberry red stele	*Phytophthora fragariae* C. J. Hickman
Strawbreaker foot rot	*Pseudocercosporella herpotrichoides* (Fron) Deighton
Sugarcane anthracnose	*Colletotrichum graminicola* (Ces.) G. W. Wils.
Sugarcane mosaic	Virus
Sugarcane pineapple disease	*Ceratocystis paradoxa* (Dade) C. Moreau
Sugarcane red rot	*Colletatrichum falcatum* Went
Sugarcane rust	*Puccinia polysora* Underw.
Sugarcane smut	*Ustilago scitaminea* Syd. & P. Syd.

Sunflower downy mildew	*Plasmopara halstedii* (Farl.) Berl. & De Toni in Sacc.
Sweetpotato black rot	*Ceratocystis fimbriata* Ellis & Everh.
Sweetpotato pox	*Streptomyces ipomoeae* (Person & Martin) Waksman & Henrici
Sweetpotato soft rot	*Rhizopus nigricans* Ehrenb. now *R. stolonifera* (Ehrenb.:Fr.) Vuill.
Sweetpotato wilt	*Fusarium oxysporum* Schlechtend. f. sp. *batatas* (Wollenweb.) W. C. Snyder & H. N. Hans.
Swiss needle cast	*Phaeocryptopus gaeumannii* (T. Rohde) Petr.
Sycamore blight	*Ceratocystis fimbriata* Ellis & Halst. f. sp. *platani* J. M. Walter
Take-all	*Gaeumannomyces graminis* (Sacc.) Arx & D. Oliver var. *tritici* J. Walker
Tea root rot	Several fungi including *Rosellinia arcuata* Petch (now *R. bothrina* (Berk. & Br.) Sacc.) and *Ustilina deasta* (Hoffm. ex Fr.) Lind
Texas root rot	See Phymatotrichum root rot
Tobacco black root rot	*Thielaviopsis basicola* (Berk. & Broome) Ferraris
Tobacco black shank	*Phytophthora nicotianae* Breda de Haan
Tobacco blue mold	*Peronospora tabacina* D. B. Adam
Tobacco brown rot	See black shank
Tobacco wildfire	*Pseudomonas syringae* pv. *tabaci* (Wolf & Foster) Young et al.
Tomato bacterial canker	*Clavibacter michiganensis* pv. *michiganensis* (Smith) Davis et al.
Tomato bacterial spot	*Xanthomonas campestris* pv. *vesicatoria* (Doidge) Dye
Tomato black ring	Virus
Tomato early blight	See potato early blight
Tomato leaf mold	*Fulvia fulva* (Cooke) Cif.
Tomato leaf spot	Several fungi including *Cercospora* spp. Fresen., and *Stemphylium* spp. Wallr.
Tomato mosaic	Virus
Tomato spotted wilt	Virus
Tomato wilt	*Fusarium oxysporum* Schlechtend. f. sp. *lycopersici* (Sacc.) W. C. Snyder & H. N. Hans.
Turfgrass fairy ring	*Marasmius oreades* (Bolt ex Fr.) Fr.

Turfgrass melting out	*Drechslera poae* (Baudys) Shoemaker
Turfgrass necrotic ringspot	*Leptosphaeria korrae* J. C. Walker & A. M. Sm.
Verticillium wilt	*Verticillium albo-atrum* Reinke & Berthier and *V. dahliae* Kleb.
Victoria blight	*Bipolaris victoriae* (F. Mechan & Murphy) Shoemaker
Walnut branch wilt	*Hendersonula toruloidea* Nattras
Watermelon wilt	*Fusarium oxysporum* Schlechtend. f. sp. *niveum* (E. F. Sm.) W. C. Snyder & H. N. Hans.
Western pine gall rust	*Endocronartium harknessii* (J. P. Moore) Y. Hiratsuka
Wheat bare patch	*Rhizoctonia solani* Kühn
Wheat black chaff	*Xanthomonas campestris* pv. *undulosa* (Smith, Jones & Reddy) Dye
Wheat black point	Several fungi including *Alternaria* Nees and *Bipolaris* Shoemaker
Wheat Cephalosporium stripe	*Hymenula cerealis* Ellis & Everh.
Wheat common smut	*Tilletia caries* (DC.) Tul. & C. Tul. and *T. laevis* Kühn in Rabenh.
Wheat dryland foot rot	*Fusarium culmorum* (Wm. G. Sm.) Sacc.
Wheat gall nematode	*Anguina tritici* (Steinbach) Chitwood
Wheat head blight	*Fusarium graminearum* Schwabe
Wheat leaf blight	*Septoria tritici* Roberge in Desmaz.
Wheat loose smut	See barley loose smut
Wheat powdery mildew	*Erysiphe graminis* DC. f. sp. *tritici* Ém. Marchal
Wheat scab	See wheat head blight
Wheat stem rust	*Puccinia graminis* Pers.
Wheat streak mosaic	Virus
Wheat leaf rust	*Puccinia recondita* Roberge ex Desmaz.
Wheat stripe rust	*Puccinia striiformis* Westend.
Wheat tan spot	*Pyrenophora trichostoma* (Fr.) Fuckel
Wheat yellow slime	*Clavibacter tritici* (Carlson & Vidaver) Davis et al.
White mold	*Sclerotinia sclerotiorum* (Lib.) de Bary
White pine blister rust	*Cronartium ribicola* J. C. Fisch.
Witchweed	*Striga* spp. Lour.

INDEX